8 GEOMORPHOLOGY TEXTS

Limestone geomorphology

General Editor: K. M. Clayton, University of East Anglia

Other titles in this series

---

1. Geomorphology F. Machatschek (out of print)
2. Weathering (Second Edition) Cliff Ollier
3. Slopes A. Young
4. Geographical variation in coastal development (Second Edition) J. L. Davies
5. Glacial and fluvioglacial landforms R. J. Price (out of print)
6. Tectonics and landforms Cliff Ollier
7. Rivers Marie Morisawa

8 GEOMORPHOLOGY TEXTS

Limestone geomorphology

Stephen Trudgill
Department of Geography
University of Sheffield

Longman
London and New York

Edited by
K. M. Clayton

To the memory of E. K. Tratman

Longman Group Limited
Longman House, Burnt Mill, Harlow
Essex CM20 2JE, England
Associated companies throughout the world

Published in the United States of America
by Longman Inc., New York

First published 1985

British Library Cataloguing in Publication Data
Trudgill, S.T.
Limestone geomorphology. — (Geomorphology texts; 8)
1. Landforms 2. Limestone
I. Title II. Series
552′.5 GB401.5

ISBN 0-582-30011-8

Library of Congress Cataloging in Publication Data
Trudgill, Stephen T. (Stephen Thomas), 1947–
Limestone geomorphology.

(Geomorphology texts; 8)
Bibliography: p.
Includes index.
1. Karst. 2. Limestone. I. Title. II. Series.
GB600.T78 1985 551.4′47 84-7943
ISBN 0-582-30011-8

Produced by Longman Group (F.E.) Ltd
Printed in Hong Kong

Contents

Acknowledgements

The appeal of limestone geomorphology is the compelling combination of research and adventure and I am happy that my introduction to caves and water sampling in the rain came from the late E. K. Tratman, a remarkable person; from Dingle Smith, with his enthusiasm for field work, and from my fellow Bristol research students: the encouragement and companionship of all these are gratefully acknowledged. Other people to be thanked are Pete Smart for early discussion of the book and for eating my cooking in Jamaica; Hans Friederich for so much sampling in Ireland; Joe Jennings, Julia James and Paul Williams for hospitality and tours of Australasian karst; fellow Aldabra Expedition members; Field Studies Council Staff for constant interest and encouragement; colleagues at Sheffield, together with the secretaries, photographers and cartographers (notably Rosemary Duncan who did most of the figures, and Paul Coles) and my Sheffield research students; Marjorie Sweeting for constant interest and encouragement; my mother and father for being there; Tony Waltham for not drowning me in Jamaica and the whole host of the rest of the caving and karst fraternity who are always friendly and willing to share their interest, especially Dick, Anni, Phil, Duncan, John Gunn, the UBSS, the Canberra Speleological Society and the Sydney Speleological Society. Finally, also thanks to my friends who keep me sane, especially Liz Cole, Nigel Coles, Dave Briggs, David Job, Tim Burt, Tim Mitcham, Adrian Pickles, Bob Crabtree, Frank Courtney, Tony Thomas and many others.

Stephen Trudgill
January 1984

We are grateful to the following for permission to reproduce copyright material:

Academic Press Inc (London) Ltd for figs 3.25 (modified), 3.29 from figs 5.9, 5.8b pp 166–7 (Atkinson & Smith 1976), 11.2 from fig 6.3 p 183 (Smith et al 1976); the Editor, American Zoologist for fig 9.8 from fig 5 p 898 (Craig et al 1969); Australian Institute of Marine Science & the Editor, D. J. Barnes for Table 10.1 from Table 3 ch 18, p 259 (Trudgill 1983b); University of Bristol Spelaeological Society for fig 5.11 from fig 9 p 69 (Tratman 1969); British Cave Research Association for figs 3.13–3.16 from figs 2–4, 10 pp 31–2 (Trudgill 1979b), 4.26 from (Trudgill et al 1979); Butterworths for fig 5.15 modified from fig 3.19 p 101 (Briggs 1977); Constable & Co for extracts from p 144 *Journey Through Britain* (1971) by John Hillerby; Elsevier Scientific Publishing Co for figs 9.21, 9.34 from figs 4b, 16 pp 175, 189 (Bromley 1978); Faber & Faber Ltd for an extract from the poem 'In Praise of Limestone' by W. H. Auden from *Collected Poems*; Gebruder Borntraeger for figs 4.13 from fig 3 p 71 (Trudgill 1979a), 8.2 modified from fig 3 p 101 (McDonald 1976), 8.5 modified from fig 2, p 27 (B. J. Smith 1978), 9.4, 9.7, 9.29, 10.1–10.3, 10.11–10.13 from figs 2, 4, 8, 14–19 pp 164, 174, 182, 189–93 (Trudgill 1976a); Longman Group Ltd for figs 3.25 (Modified), 4.7, 4.25 from figs 6.11, 6.15, 6.1 pp 97, 100, 89 (Curtis, Courtney & Trudgill 1976), 10.15, 10.16a from figs 27, 33 pp 42, 51 (Davies 1980); the author, Malcolm Newsom for fig 6.3 (Hanwell & Newsom 1970); Presses Universitaires de France for fig 10.14 from figs E1–4 p 63 (Guilcher 1958); Springer-Verlag, Heidelberg & the author, T. Rutzler for fig 9.24 from fig 2b, p 209 (Rutzler 1975); John Wiley & Sons Ltd for figs 3.33–4 from figs 13.2–4 pp 375, 384 (Smith & Atkinson 1976), 4.8 modified from fig 3.21 p 93 (Trudgill 1976c), 4.19, 4.20 from figs 4b, 6 p 31–3 (Glew & Ford 1980).

'Fortunately for poets and those who like to walk about in the open air, the beauty of the landscape is not something that can be reduced easily to basic geology or a few ready-wrapped phrases about what places are used for. Preference and prejudice creep in. Mine are apparent in a love of limestone. . . .'. 'Stand on the gritstone . . . and you are looking into the limestone country. The acid squelch is replaced by wholesome springy turf, the cotton grass by fescues that sheep enjoy. Flowers abound in the hedgerows: water avens, wild garlic, Solomon's seal and lilies of the valley; the bird song is clamorous and the local inhabitants talk about something more interesting than the probability of rain.'

John Hillerby. *Journey Through Britain*. Paladin 1970.

'When I try to imagine a faultless love
Or the life to come, what I hear is the murmur
Of underground streams, what I see is a limestone landscape'

W. H. Auden. *In Praise of Limestone*. Penguin 1958.

'Waves that from far off time
oozing with the sea weed's slime
always absorb the lime
Far in Fanore'

Michael F. Roberts; poem on Fanore Bay displayed in O'Donahue's Bar, Fanore, Co. Clare, Eire.

Preface

The aims of *Limestone Geomorphology* are to provide the reader with some basic concepts in the geomorphological study of limestone regions and to present a selected review of some recent research taken from the wealth of published research papers. The focus of the book is on recent work on geomorphological processes. Earlier work concentrated on the description and analysis of form and on theories concerning the stages of evolution of landforms. In the last 30 years or so more attention has been paid to processes, especially on the field observation of erosion rates and on laboratory experimentation on individual processes of erosion. More recently, there has been an indication of a trend to the application of this process work to the earlier descriptive and evolutionary work, aided notably by the increase in research on dating of sequences of cave deposits. This trend has yet to be fully developed and the immediate future should see the growth of studies on dated sequences of landforms. The book thus covers much of the recent work on processes, rates and experimentation rather than the earlier work on form, and is able to look ahead in terms of more tentative work on process–form relationships and on dated sequences of landform evolution. The book contains a discussion of published work together with the author's own research, often with a British bias in the examples given (many selected from the *Bibliography of British Karst*; Trudgill and Brack 1977) and with illustrative material based on the author's experience. Many of the principles, however, may be applied to other areas – without being tempted to propose universal theories from any one set of studies. The growth of world limestone geomorphology studies has meant that some assertions made in one area may be checked for another area and generalised statements proposed, but as more studies are undertaken, there also emerges a realisation of the complexity of combinations of different factors and contrasting evolutionary histories in different areas. The book, as any book, is an interim statement and other books will undoubtedly cover other aspects of the literature neglected here, as well as future studies, and be able to make further statements concerning the evolution of limestone landforms. In this book, one particular focus of attention is on soil processes because many studies have indicated that the soil–upper bedrock zone is often the most important in terms of the distribution of erosional activity. A further focus is on marine limestone geomorphology, a topic much neglected in previous limestone geomorphology texts. Chemical processes are treated in a general way for a geographical readership and reference may be made to the more specialised works cited in the text for a more advanced discussion. Similarly, caves are discussed in broad, outline terms and other texts exist which deal with caves alone. Applications of geomorphological knowledge are also briefly discussed. The book is thus primarily a selective assessment of recent work on limestone erosion processes, indicating, where possible, how process and form may be related.

1 Introduction

1.1 Process and form in geomorphology

The task of the geomorphologist is essentially one of providing explanations of landforms. Thus, the question being asked is 'why is it like it is?' and this question obviously has to be preceded by the question 'what is it like?'. Description of landform is a prerequisite of explanation. But what constitutes an explanation of a landform? Landforms can be seen as having a geometry which varies across their surfaces and explanations for these variations in geometric form can most usefully involve the spatially varied action of processes. Thus, differential erosion can be invoked to account for the evolution of upstanding and low-lying portions of landforms. The uneven action of processes on a homogeneous material or the even action of processes over heterogeneous material may be involved in explaining differential erosion (Fig. 1.1.).

This type of explanation involves a *process-response* relationship where the form is seen as a response to the processes operating on the material involved. The fundamental dilemma of testing this type of explanation lies in the following points.

Process measurements can be taken at the present day and the advantage of this is that empirical work can be undertaken where the relative importance of the operation of different current processes and

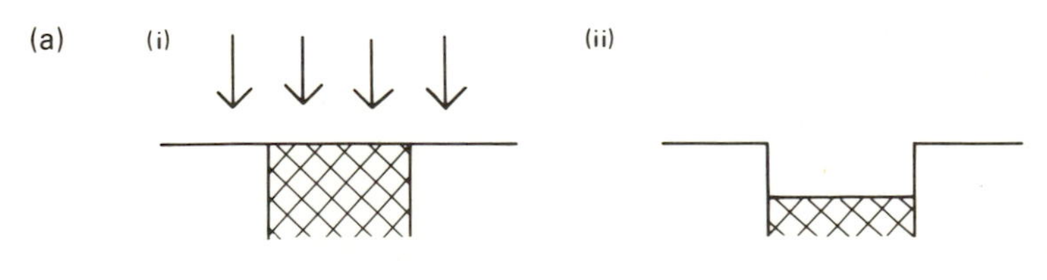

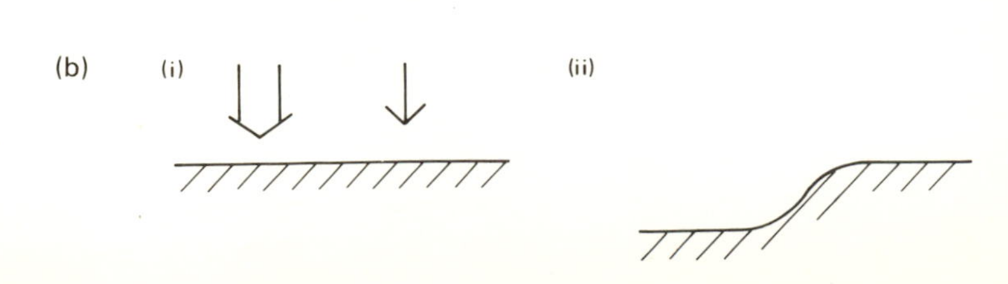

Fig. 1.1 Production of landforms by differential erosion.
(a) Uniform erosion and differential susceptibility of rock.
(b) Uniform lithology and differential forces of erosion
(i) initial conditions
(ii) landform production.

factors can be measured. However, landforms have usually been evolving over long periods of time – much longer than those represented by the time span available for present-day process measurements. If present-day process measurements do not successfully help in the explanation of observed landforms it is possible that important processes currently operating have not been measured. Alternatively, the landform may be essentially a fossil one. In this latter case the current processes are of no or only minor relevance to the development of the landform observed at the present day. Thus, the nub of the problem is that present-day process measurements may not explain observed landforms because of the historical legacy which the landform has. In this case attempts at explaining form from current processes is akin to trying to explain the condition of a second-hand car without reference to the activities of the previous owners.

The resolution of this dilemma stems from two alternative approaches. Firstly, accumulated historical evidence, such as a sequence of deposited sediments, may be available and interpretations from this may be made about past evolution. This is especially true if the sequence can be dated. Secondly, the questions of scale and materials are important. It may be possible to test process-response models in a rigorous fashion at a small scale or where very responsive material is involved. In these cases historical legacy tends to be minimised. A large landform, however, may be essentially an historical artefact. Current processes may only be modifying the smaller features but these small features will then provide a focus for study; micro-features may be forming at the present day and be adjusted to measurable present-day processes (Fig. 1.2). This will be especially true of responsive materials in dynamic situations such as easily moved loose sand on the bed of a river or loose earth exposed to intense rainfall. On limestones, features of the order of the millimetre scale can be produced in the time scales of 1–10 years (as described in Ch. 4, for example). Processes and rates can thus be measured at the micro-form scale of response. However, it may be difficult to extrapolate from such models back through time in order to provide explanations for larger scale historical landforms. Such an extrapolation may be invalid as conditions may have changed radically in

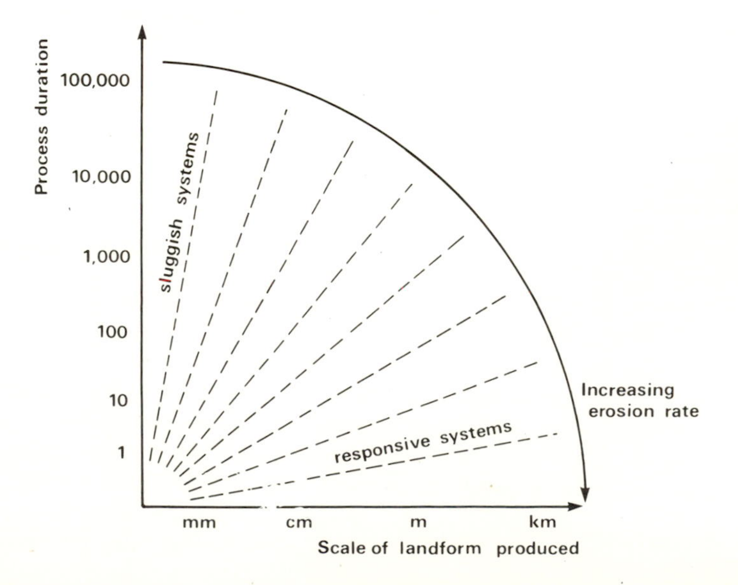

Fig. 1.2 A model of erosion scale and process duration. As process duration increases, features of larger scale are produced, increasingly so in responsive materials (modified from Trudgill, 1976c).

the recent past (Fig. 1.3) and in any case there is often no independent test of such extrapolations unless dated sequences are available. In the case of a dated sequence it may be possible to establish the relationships between the occurrence of erosional and depositional events, rates of erosion and landform production on a large scale (as, for example, discussed in Ch. 5).

Geomorphology can thus progress through the use of present-day measurement and experimental technique where hypothesis testing can involve the evaluation of process-form response, often at a small scale, or it can progress by inference about landform evolution from observation of sedimentary sequences laid down historically.

1.2 The study of process-form relationship in limestone geomorphology

In limestone geomorphology, dissolution processes and current erosion rates have been well researched but, as outlined above, such studies have tended to focus on the explanation of smaller landforms, such as those forming on bare rock surfaces, and to current modification of historically derived landforms. Extrapolations have been made over longer time scales but, as also discussed above, these may not always have been valid. However, recent developments in uranium series dating techniques for cave stalagmite deposits (Harmon, *et al.*, 1977; Atkinson, *et al.*, 1978) have meant that the formation of cave passages, and larger scale landforms, can be placed in time-based historical sequences of development.

Fig. 1.3 A conceptual model of climatic change and landform response; macroscale features exhibit greater lag and inheritance than microscale features (modified from Trudgill, 1976c).
(a) Fluctuating climate.
(b) Major change (e.g. glacial-pluvial, pluvial-semi arid transitions).

Horizontal axis is time, vertical axis is climate, e.g. rainfall amount.

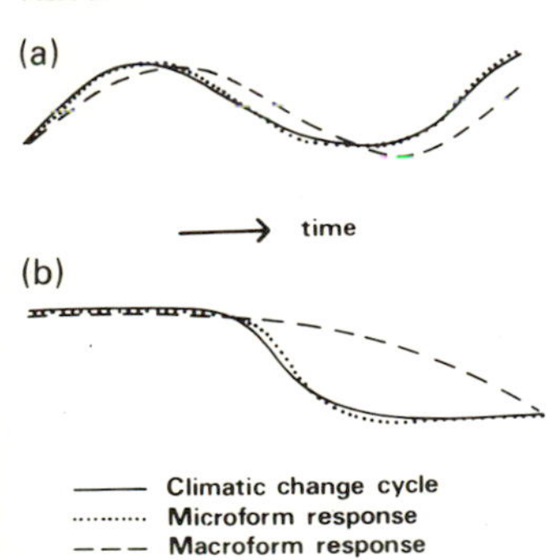

In limestone geomorphology, the processes of erosion are better understood than is the historical development of landforms (Williams, 1978a). The derivation and variations of the solute load of run-off waters in space and time has been extensively researched but the precise relationship of these to landform evolution is less well known. The challenge of providing detailed and substantive process-response models for limestone landforms thus still remains in many cases. Progress has been made, however, in the understanding of differential erosion within the landscape in relation to lithology and considerable knowledge exists concerning the process factors involved in determining the distribution of erosion rates in the present landscape. The advances in knowledge of limestone hydrology and chemistry have often elucidated the geomorphological relationships involved: limestone hydrology and solution chemistry are still better understood than limestone landforms are, however.

Limestone areas often form distinctive landscape features where caves, disappearing and reappearing streams (Fig. 1.4) and bare rock surfaces (Fig. 1.5) are common. However, while these features are common on limestones, the processes operating on limestones are not unique to these rocks. Limestone landforms are produced by combinations of processes, the emphasis of which is distinct from that of other rocks. The key points are as follows.

Limestones are highly soluble in acid waters; however, they do not dissolve equally over their surfaces and the action of dissolution is usually concentrated down joints and fissures. This solubility is the distinct emphasis of limestone land-forming processes but all the other weathering and erosion processes common to less soluble rocks are also operational and should not be neglected. These processes include hydration, frost action and abrasion (as discussed in Ch. 2). In addition, since biological processes involving root surfaces and the decay of organic matter are capable of producing acids, biochemical action may also

Fig. 1.4 The dry valley at Malham, Yorkshire, UK. Here, the surface stream once ran along the line now occupied by a wall. Pleistocene scree occupies much of the valley, with thin stony soils. Flow is now by underground route, though flow has been known in the dry valley on very wet occasions. (Photo: S. T. Trudgill.)

Fig. 1.5 Part of the Burren limestone district, County Clare, Eire, where glaciation has left a wide expanse of soil-free bare rock surfaces. (Photo: S. T. Trudgill.)

be important in limestone land-forming processes (as discussed in Ch. 3).

The concentration of dissolution processes down joints and fissures leads to the focusing of water into such joints and fissures (Fig. 1.6). Caves thus readily develop in the bedrock by the action of running water on bedrock openings. This development tends to leave limestone blocks standing up in the landscape while the processes of erosion are focused at some depth in the bedrock mass, as discussed in Chapter 6.

Most limestones have a high purity, consisting primarily of calcium carbonate, and thus, once dissolved, only thin residual soils are left. These may be easily eroded and limestone landscapes are often distinctive for their abundance of bare rock surfaces, unless a mantling of thicker drift occurs derived from some other source.

Fig. 1.6 Well-jointed Carboniferous Limestone exposed on a glaciated limestone pavement in County Clare, Eire. Water is focused down the joints both on the surface and underground. (Photo: S. T. Trudgill.)

Hence, although running water is vital to the formation of limestone landforms, all that is often visible at the surface may be a dry, fissured upstanding rock mass, leaving, for example, dry river valleys on the surface while the streams have been re-routed in fissures and caves underground.

There are many variations on this theme. Lithological variations in limestone type are common. Joint frequency, bedding frequency, chemical content, insoluble residue, grain size, crystal pattern and fossil content, as outlined in Chapter 2, all vary according to the environment of deposition and subsequent alteration of the rock. In addition, limestones exist in many parts of the world having contrasting climates and which have endured different climatic histories. Within climatic zones, limestones are exposed in a great number of terrestrial and marine environments all of which have characteristic assemblages of erosional processes.

It can be seen that in order to explain the existence of any one set of limestone landforms it is necessary to study the relationships between rock type, environment, the operations of present-day erosional processes and the legacy which past processes have left in limestone areas. The intention of this book is to outline research of these factors and their inter-relationships.

1.3 Limestone landforms, karst and book layout

The book deals with limestone landforms: such landforms have become known collectively as karst landforms from the Yugoslavian area, 'Kras' (translated as 'karst') where limestone landforms are well developed. Karst can be defined as a distinctive assemblage of landforms developed on soluble rocks (cf. Jennings, 1971, p. 1) and as such include landforms produced by solution processes on soluble rocks such as basalt. This book deals with limestone landforms, however.

Limestone is a carbonate rock, as is dolomite, which is also discussed in the book. The definition, classification and description of limestones and other carbonate rocks are discussed in Chapter 2, as is an outline of the solution chemistry of limestones. From this chapter, which describes fundamental principles, the controls on the spatial and temporal variations are discussed (Ch. 3). The smaller surface features are then considered in Chapter 4 – these features can often be most closely related to the controls and processes measurable at the present day and which are discussed in Chapter 3. The large-scale features are discussed in subsequent chapters, where time-scale of evolution and inheritance become important. Caves are discussed in Chapter 5 and the features diagnostic of karst and developed as a consequence of the routing of hydrological networks underground are discussed in Chapter 6 in the context of a fluvial landscape. Description, classification and, especially, morphometry form the focus for Chapter 7, leading on to a review of process-form relationships in Chapter 8. Coastal landforms are treated subsequently, with a discussion of processes in Chapter 9 and of process-form relationships in Chapter 10. Chapter 11 concludes by discussing limestone geomorphology in general terms and briefly outlines some applications of the subject.

2 Carbonate rocks and erosion processes

2.1 Introduction

If a rock contains more than 75% carbonate it can be classified as a carbonate rock (Folk, 1959; Bissel and Chilingar, 1967). Limestone is the commonest sedimentary rock example and is composed dominantly of calcium carbonate. The rock is often one of very high purity, with a $CaCO_3$ content of above 95%, and not uncommonly 99%. The other major carbonate rock is composed of the mineral dolomite, a calcium magnesium carbonate. Both the mineral, $CaMg(CO_3)_2$, and the rock may be termed dolomite, but the term dolostone is often also used to indicate a carbonate rock composed dominantly of the mineral dolomite. Magnesium content, crystal form, trace element composition, impurity and fossil content may vary widely in carbonate rocks according to their mode of origin and subsequent alteration.

The chemical and physical characteristics of the rock act to influence the effectiveness of weathering and erosion processes. Carbonate rocks are more soluble than many other rocks but they, like any other rock, are subject to a number of geomorphological processes, including mechanical processes such as frost action and abrasion. The processes involved in the weathering and erosion of carbonate rocks may thus be many and diverse but a common factor controlling their effectiveness is the ease with which the rock can be penetrated: the effectiveness of all the processes involved tends to increase with increasing internal and external surface area. Of especial importance is the frequency and number of penetrable joints and bedding planes as these provide lines of access to percolating water. Increased contact of internal and external surfaces with the environment encourages dissolution and hydration, as well as facilitating freeze–thaw action. Penetration of a rock by plant roots can also take place along joints and bedding planes. These lines of weakness may also form focal points for the action of abrasion. On a more detailed level, the surface area available for weathering reactions also increases with increasing porosity and permeability of the rock. At this level, the size of grains, their degree of interlocking and their cementation will be important factors influencing weathering and erosion. Thus, massive, well-cemented rocks with few joints tend to be more resistant to erosion while well-jointed, more porous rocks may break down more easily.

The relationship between joint frequency and porosity is a crucial one. High porosity and permeability will lead to a more uniform flow of water through the rock, making the presence of joints less important as water-flow routes; low permeability acts to focus water flow along joints and other fissures, making their frequency a more important influence upon erodibility.

The chemical and physical characteristics of the rock which act to influence the effectiveness of weathering and erosion processes are inherited from the original nature of the carbonate deposit and from the processes of lithification and diagenesis. During lithification a carbonate deposit becomes a harder rock mass. The term diagenesis embraces the chemical and physical alterations from the original state, both pre- and post-lithification. During these alterations, many processes can occur, including dewatering, compression, recrystallisation, cementation and chemical alteration. The influence of the chemical composition of solutions percolating through the rock mass is often a major factor in chemical and mineralogical alteration. In addition, tension or compression cracks may appear in relation to regional tectonic stresses.

It can be deduced that carbonate rocks exposed to weathering and erosion will show great diversity, according to their original nature and subsequent history. The interactions between rock type and erosional environment which lead to the production of landforms should therefore be studied not only in terms of the solubility of the carbonate minerals which are the primary constituents of the rock but also in terms of the detailed chemical, mineralogical and structural variations that are produced during the history of deposition and alteration.

2.2 Chemistry and mineralogy of carbonate rocks

The minerals commonly present in carbonate rocks are calcite, aragonite and dolomite. Calcite can be divided into high and low magnesium calcite; the term 'magnesium calcite' may also be used to indicate a high magnesium calcite.

Calcite is calcium carbonate ($CaCO_3$) in the rhombohedral class of the hexagonal crystal system (Read, 1962). Aragonite is $CaCO_3$ in the orthorhombic crystal form, a form often, but by no means exclusively, associated with carbonate deposits in shallow tropical seas. It may also be found in cave deposits where the precipitating solution is highly saturated with $CaCO_3$ (White, 1976).

High magnesium calcite is defined by Plummer and Mackenzie (1974) and Chilingar *et al.* (1967a) as calcite with a magnesium carbonate content above 4% (as 4 mole percent $MgCO_3$; see Appendix for explanation of units). It is a form often laid down biogenically.

Dolomite is extremely rare as a primary mineral, but may be deposited in hypersaline lagoons. Most dolomite rocks are formed by the introduction of magnesium into calcite from percolating groundwater or seawater (Randazzo and Hickey, 1978; Folk and Land, 1975; Chilingar *et al.* 1967b). The replacement of calcite by dolomite is often incomplete, leaving interstitial calcite between the dolomite crystals. The term magnesian limestone may be used to describe a dolomite-rich limestone but the rocks intermediate between limestone and dolomite have been more precisely defined by Bissel and Chilingar (1967) as shown in Table 2.1. Dolomite rock tends to be more porous than the original calcite-formed rock because there is a 12% reduction in molecular volume when calcite is replaced by dolomite rhombs.

Trace element concentration varies with environment of deposition.

Table 2.1 Classification of carbonate rocks by percentages of calcite and dolomite (modified from Chilingar *et al.*, 1967a)

Rock	*% Calcite*	*% Dolomite*
Limestone	95	5
Magnesium limestone	90–95	5–10
Dolomite limestone	50–90	10–50
Calcitic dolomite	10–50	50–90
Dolomite (dolostone)	10	90

Some organisms concentrate specific trace elements and carbonate sediments rich in organic matter may contain traces of lead, zinc, nickel, copper, phosphate and other elements which may affect carbonate rock solubility, as discussed in section 2.6. (p. 22). In addition, terrigenous components, such as clays and silicates, may be present, decreasing the purity of the carbonate rock. These, together with aluminium and iron oxides constitute the insoluble residue of a carbonate rock.

Mixtures of the various carbonate minerals may be found in a variety of environments of deposition. For example, in recent tropical carbonate sediments, aragonite coral fragments and high magnesium calcite shell fragments may be found in an aragonite cement. In older rocks, aragonite is rare because it is unstable, converting to calcite.

Aragonite is more soluble than calcite and small traces of magnesium may increase the solubility of calcite; dolomite is less soluble than calcite. These factors are discussed further in section 2.6, but in order to evaluate the lithological factor in limestone geomorphology it should be borne in mind that solubility may be a factor of minor importance in rock erodability: petrological characteristics can be of equal or greater significance. (See section 2.6.2, p. 21).

2.3 Petrological classification of carbonate rocks

Sedimentary carbonate rocks are formed from carbonate deposits laid down in coral reefs, shallow banks and deeper basins and by the reworking and secondary deposition of these deposits (Fig. 2.1). The sediments are laid down by organisms, mainly corals, molluscs, foraminiferae and algae. They may also be deposited by chemical precipitation from saturated solutions but here, again, the involvements of

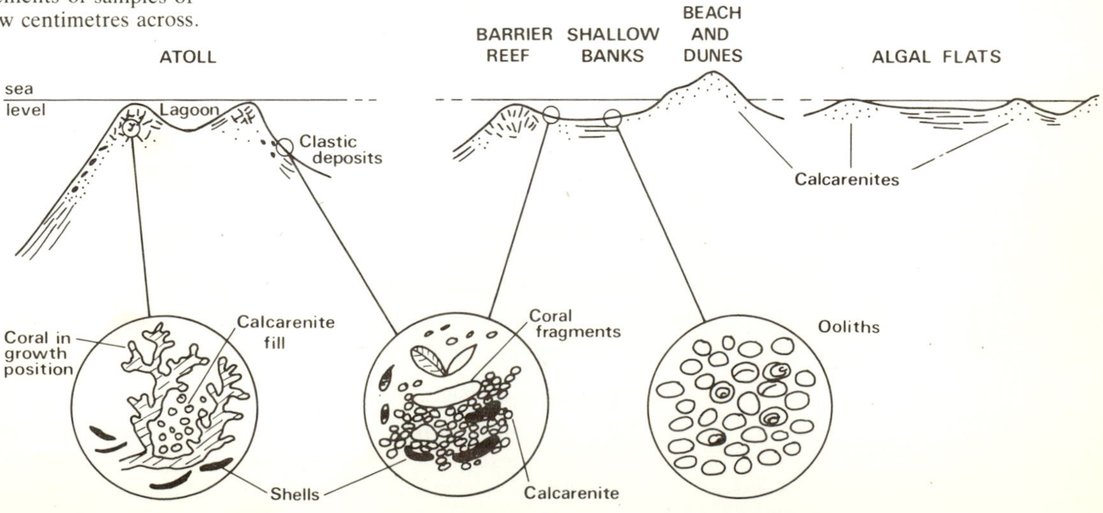

Fig. 2.1 Schematic diagram of environments of deposition of marine carbonate sediments. Insets show enlargements of samples of deposits a few centimetres across.

organisms has considerable influence. Biological processes may substantially alter aqueous carbonate equilibrium (as discussed in Ch. 9), increasing or retarding the tendency of waters to precipitate carbonates.

The particle sizes of carbonate sediments can range from the larger boulder, cobble or pebble sizes through the sand size and down to the finer silt and clay-sized particles. The larger particles are mostly derived from coral or they may be of separated portions of cemented sediments. The intermediate sizes are comprised of shells and broken fragments (termed clasts) of corals and shells or other carbonates such as algal limestones. The smaller sizes are derived by abrasion of larger particles during transport, giving sand- and silt-sized particles, or they may be the finer algal deposits, oozes, muds and precipitated cements. Carbonate rocks can thus be composed of corals, shells and other biogenic fragments mixed in varying proportions with what were originally carbonate muds and sands. The rock characteristics which influence the erodability of carbonate rocks are closely related to the relative proportions of these constituents and therefore limestone geomorphologists have found it useful to adopt the petrographic classification system of Folk (1959, 1962). This classifies limestones according to the dominance and particle size of each constituent. The classification of carbonate rocks is also discussed by Bissel and Chilingar (1967), who caution that the Folk classification applies to original textures and that its application may become difficult because of diagenetic alterations.

In the Folk classification system, terrigenous constituents (that is, those having a source outside the basin of deposition, such as quartz sand, feldspars, clays and resistant minerals such as zircon) are distinguished from allochthonous constituents. The latter covers the carbonate materials derived inside the basin of deposition and are divided into **allochems** and **orthochems**. Allochems are transported grains and orthochems are not transported but are formed *in situ*. Allochems are subdivided as follows:

1. **intraclasts:** reworked fragments of previous limestone and carbonate sediments
2. **oolites**: rounded, concentric formations
3. **pellets:** faecal material
4. **fossils:** shells and corals, for example.

Orthochems are subdivided as follows:

1. **micrite**: a micro-crystalline ooze of opaque, clay-sized particles
2. **sparite**: crystals of calcite spar (well-formed rhombic crystals) of

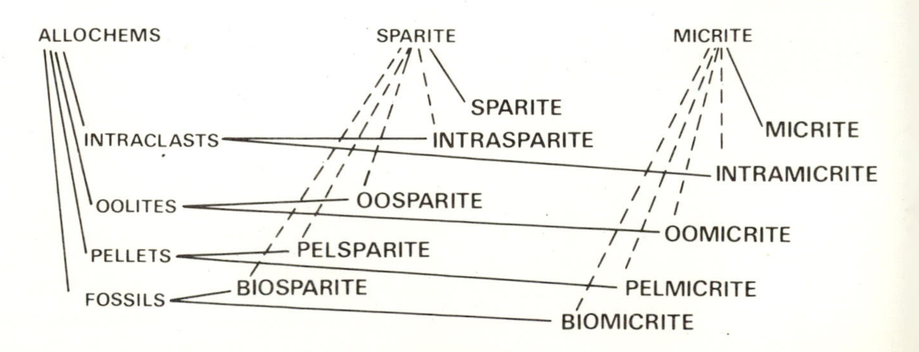

Fig. 2.2 Subdivisions of limestone constituents and rock types according to the scheme of Folk (1959).

Key: ——— primary constituents; - - - - - - - secondary constituents (matrix or cement).

10 μm or more in diameter and of greater clarity than micrite. The three major constituents, allochems, sparite and micrite, are used as a primary classification and the terms may be combined, as shown in Fig. 2.2, to describe rock types. A high fossil content is indicated by the prefix 'bio'. Thus, a biosparite is a fossiliferous limestone with a calcite spar matrix and an intrasparite is one of small, reworked fragments (clasts) in a spar matrix. A **biolithite** is a massive agglomeration of biogenic material, with coral colonies and algal mats in growth positions.

Limestones may also be subdivided on the basis of dominant grain size:

1. $>$ 2 mm – calcrudite (e.g. large shell or coral fragments)
2. 2–0.02 mm – calcarenite (sand sized grains)
3. $<$ 0.02 mm – calcilutite (fine grained material)

2.4 Joints, faults and bedding planes

Lines of weakness which may be exploited by erosional agencies result, locally, during deposition and lithification and, regionally, from tectonic stresses. Bedding planes often develop in relation to changes in sedimentation rate and type. The general principles of fault and joint development are discussed by Price (1966) and a useful case study on the orientation and origins of joints and faults is provided by Moseley (1973), working in northwest England. In this study, many principles are illustrated, but, naturally, for any other area the combinations of influences will vary.

In the northwest England study area, volcanic rocks form the basement to the limestone in the west and granite rocks form the basement to the east; volcanic rocks also outcrop to the west. The basement rocks, regional blocks, faults and joint sets are shown in Figure 2.3. The orientations in the basement volcanic rocks provide a major influence upon joint and fault orientation in the limestone above. Fractures in the form of NW and N orientated faults and NE to ENE cleavage occur in the basement and are transmitted upwards to the Carboniferous Limestone where the dominant fault trends are also NW, N and NE.

The area was subject to E–W stresses during Armorican (Hercynian) earth movements, giving rise to N–S folds in the limestone in the western part of the area, where it rests over the volcanic rocks. These earth movements were at their height towards the end of the Carboniferous period, subsequent to the period of major limestone deposition. In the east of the area, the more stable limestone blocks (the Alston and the Askrigg blocks) lie over granite and were unfolded. In all areas, joint trends can be related to the regional tectonic stresses. Joint trends parallel the fault trends, with NW, N and NE orientations predominating. These relate to the orientations in the underlying rock and the NW and the NE orientations are at angles of maximum horizontal stress, bisecting the E–W stress and the N–S folding. In addition, tectonic relaxation of stress induced N–S and E–W fractures. Interpretations of alignments are thus not simple in detail, but they may be made in terms of basement rock orientations and the compound effects of regional stresses.

Orientations of joint sets often dominate the development of caves (as discussed in Ch. 5) and surface landforms (as shown in Fig. 1.6 for

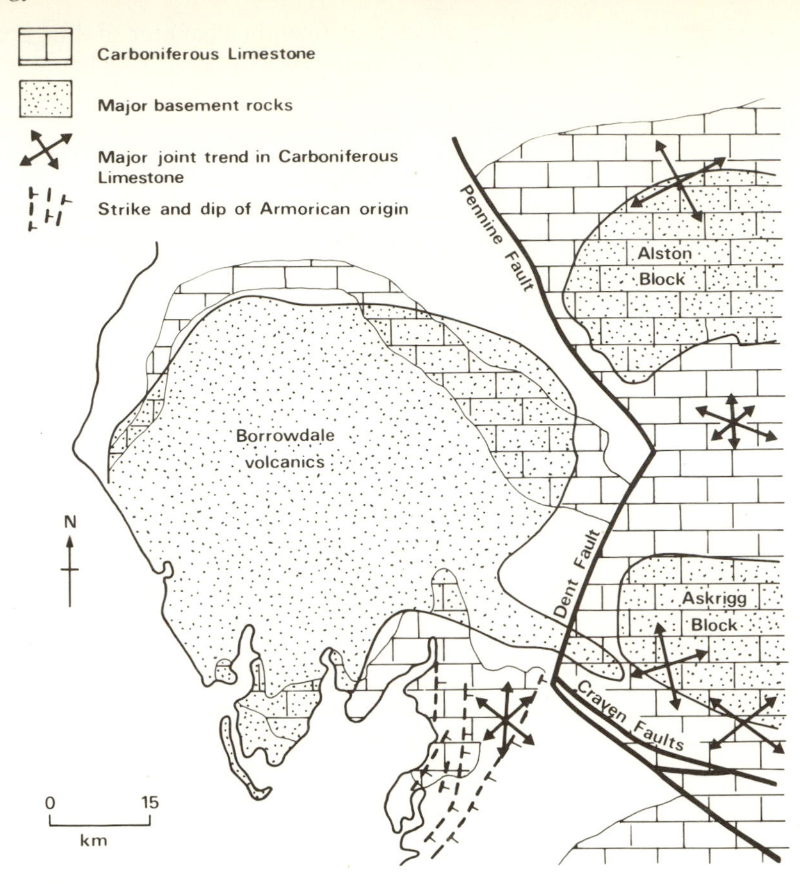

Fig. 2.3 Joints and faults in Carboniferous Limestone, northwest England in relation to basement rocks and regional stresses (modified from Moseley, 1973).

the Burren district of Co. Clare, Eire, where N–S (196°) and E–W (270°) sets exist, with some other oblique sets). Folding has a similar influence in terms of the orientation of flow within the bedrock down inclined bedding planes.

Joints, faults and bedding planes may all provide lines of weakness for the penetration of erosion agencies. However, this is an oversimplistic view since throughout geological time they also provide flow lines for the penetration of mineralising hydrothermal fluids from below. Deposition of lead, copper, tin, fluorspar and silicate minerals is common in joints and faults. In addition, joints may be sealed by deposition from solutions percolating from above. Additionally, many joints are sealed by secondary calcite deposition. Thus, the common generalisation that joints and other fissures provide lines of weakness is by no means a universal one; in some areas they may provide lines of greater resistance, especially if silicate deposits are present in the fissures. Calcite infilling may be preferentially eroded or form an upstanding vein; both situations are observed in the caves of northwest County Clare, Eire (Tratman, 1969). Here, the vein thickness and mechanical strength in relation to the degree of crystal interlocking are important factors. A protruding resistant vein may lead to increased turbulence in cave waters and locally higher dissolution rates around it. Protruding veins may persist until they are removed by the abrasive action of transported bedload material such as cobbles or sand.

Joints and faults may thus be seen as tension or stress fractures producing lines of weakness for preferential erosion to exploit; if infilled they may provide lines of resistance. Generally speaking, however, they provide the major routeways exploited by solutional and other erosion processes.

2.5 The solution chemistry of calcium carbonate

2.5.1 Modelling procedures

It can already be seen that the erosion of limestones may involve a large number of lithological factors, including carbonate rock chemistry, mineralogical variations, petrological composition and joint frequency. However, like all complex situations, it is often best approached by an investigation of simple, model situations which may be used to assist in the basic understanding of the multivariate natural situation. Thus, because limestone is dominantly comprised of calcium carbonate the solution chemistry of calcium carbonate has been extensively investigated as a first step in the understanding of more complex limestone erosion processes. It should be emphasised, however, that essential and fundamental as this first step is, it is only a first step. The important further step is to establish how far the solution behaviour of pure calcite can be used to predict the behaviour of impure limestones.

There is a tendency to regard limestone as if it were pure calcium carbonate, but the important step from calcite to limestone is still not fully investigated and it is true that the solutional behaviour of calcite is better understood than the solutional behaviours of different limestones. Further studies on the departures of limestone from ideal calcite behaviour are still needed, with calibrations of principles for different situations.

In the study of calcite dissolution, laboratory experiments have been crucial in the fundamental understanding of control variables and the interactions of processes. The testing of the universal applicability of this understanding is still far from complete and, as Miller (1952) cautions, '. . . as for so many laboratory experiments, direct application of data obtained to natural processes is complicated by numerous factors of unknown magnitude'. Watson (1972) considers that this is especially true in terms of time scales, when extrapolation is made from theoretical and short-term laboratory work to the long-term field situation. Thus study of calcite solubility provides an essential, sound, theoretical basis for the understanding of limestone solubility. However, there may be a difference between calcite solubility and limestone solubility, because of petrological, mineralogical or trace element variations.

2.5.2 Definition of terms

Texts covering calcite solubility are many, but reference can usefully be made to the summary of Stumm and Morgan (1970, pp. 174–83) and the more detailed papers of Picknett (1964) and Roques (1969). Within the subject of limestone geomorphology, limestone solution processes have been treated with varying degrees of simplicity and complexity. In the present volume, some terms will be defined and a basic outline given; The Appendix also deals with some aspects of carbonate chemistry. Here, the word 'solution' is used as a noun referring to the aqueous medium surrounding the solid phase, thus a '**solution process**' is one taking place in a solution; the phrase 'limestone solution' is often used to indicate the process of **'dissolving'**: here the term **'limestone dissolution'** will be used to indicate that limestone is dissolving in a solution.

Dissociation is the separation of the constituent solid phase ions of a salt in solution, yielding **cations** (positively charged) and **anions** (negatively charged). An **acid** is a substance which dissociates in water to yield hydrogen ions (H^+). Hydrogen ions, also termed **protons**, actually exist in water as **hydroxonium** ions (H_3O^+) but are normally discussed as the separate H^+ ion. An **alkali** is a substance which dissociates in water to yield **hydroxyl** ions (OH^-) either directly or by reaction with water. A **metal** is an element which yields cations in water. A broad, simple definition of an acid is a substance that can donate a proton to any other substance; a **base** is then defined as any substance which accepts a proton.

2.5.3 Calcium carbonate in pure water

In pure water calcium carbonate dissociates into a metal cation and an anion during the process of dissolution (Fig. 2.4):

$$\underset{\text{Solid phase calcium carbonate}}{CaCO_3} \xrightarrow{\text{dissociation}} \underset{\text{calcium cation}}{Ca^{2+}} + \underset{\text{carbonate anion}}{CO_3^{2-}} \qquad [2.1]$$

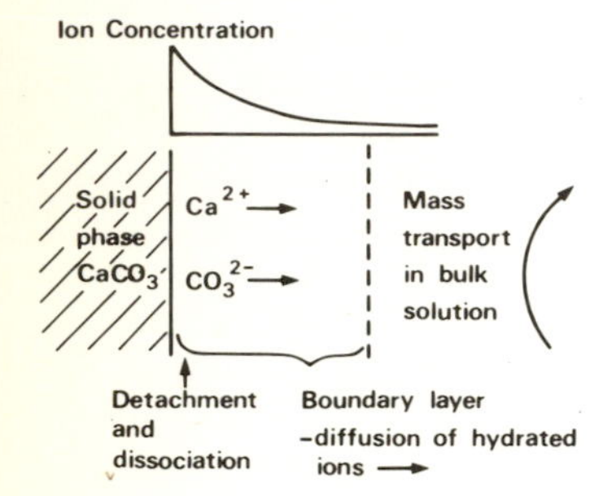

Fig. 2.4 Simple dissolution of calcium carbonate in pure water; dissociation and diffusion down the concentration gradient are the main processes involved; mass transport in the bulk solution may act to maintain a steep concentration gradient.

The carbonate ion can react with water, accepting a proton, thereby classifying the carbonate as a base; the overall reaction (equations [2.1] and [2.2]) yields OH^- ions, thereby classifying calcium carbonate as an alkaline substance:

$$\underset{\text{carbonate ion}}{CO_3^{2-}} + \underset{\text{water}}{H_2O} \xrightarrow{\text{hydrolysis}} \underset{\text{hydrogen carbonate ion}}{HCO_3^-} + \underset{\text{hydroxyl ion}}{OH^-} \qquad [2.2]$$

Reactions of cations or anions with water is termed **hydrolysis**.

For equation [2.1], over time, the movement of ions into solution is initially rapid when solid phase calcium carbonate is introduced into pure water. The movement is rapid as there is a steep diffusion gradient from the solid to the liquid. The solute concentration (the concentration of the dissolved solid in the liquid) gradually increases over time and the gradient decreases until the movement is minimal. In a closed system (that is, with no transport of solutes or water into or out of the system, such as would be the case for a solid dissolving in a closed beaker of water) solute concentration over time will take the form shown in Fig. 2.5 with a rapid and then gradual increase to given level. This level is termed the equilibrium level or concentration at saturation. The overall reaction is defined by the equation (Roques, 1969):

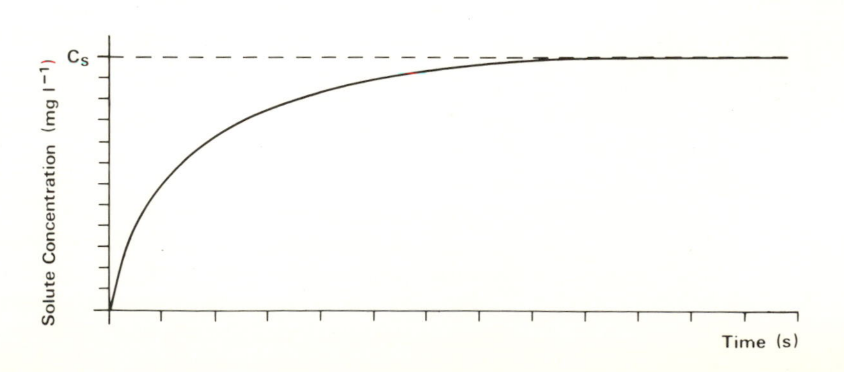

Fig. 2.5 Increase of solute concentration over time in a closed system. The concentration C_S is the saturation concentration. The units will be defined for any one solute under given conditions.

$$\frac{dm}{dt} = SE(C_s - C) \qquad [2.3]$$

where

$\frac{dm}{dt}$ = rate of transfer of matter over time (i.e. the amount of transfer from the solid phase to the liquid phase defined by points on the curve on Fig. 2.5).

C = solute concentration in the liquid at any one time (i.e. at any point on the curve in Fig. 2.5).

C_s = solute concentration at saturation (this level is given by the value on the vertical axis opposite the upper horizontal part of the curve on Fig. 2.5).

S = surface area of contact

E = molecular permeability coefficient, i.e. the rate of diffusion of molecules; molecules with a smaller hydrated radius tend to move faster than those with a larger one.

This may also be rewritten:

$$\frac{d[Ca^{2+}]}{dt} = \frac{SE([Ca^{2+}]s - [Ca^{2+}])}{V} \qquad [2.4]$$

where V = volume of solution, (Roques, 1969).

The use of [] in the equation indicates concentration.

At equilibrium, solute ions are moving from the solid to the liquid at a rate equal to the reverse movement. Thus dissolution is equal to precipitation, as shown in Fig. 2.6.

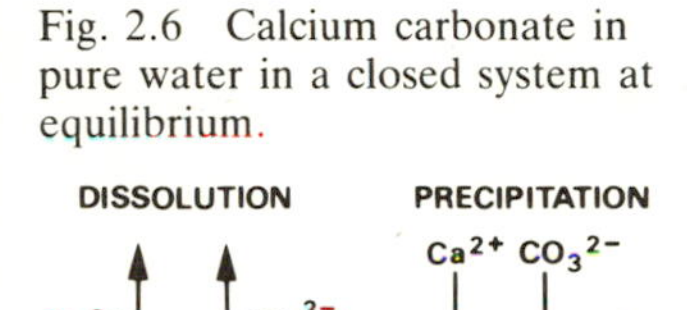

Fig. 2.6 Calcium carbonate in pure water in a closed system at equilibrium.

2.5.4 Calcium carbonate in acid waters

The simple dissolution process described above for pure water accounts for a very small proportion of the amounts of calcium and carbonate present in solutions occurring in the natural environment. A value for calcium carbonate dissolved in pure water is quoted as 14 mg l^{-1} (Stringfield & Legrand, 1969, p. 375); actual values dissolved in natural water systems are often in the region of 200–250 mg l^{-1}. The dissolution of calcium carbonate is greatly increased by the presence of protons; these allow equations [2.1] and [2.2] to proceed to the right. The hydrolysis of the carbonate ion by hydoxonium ions is a key step, with the excess of H^+ ions over H_2O in H_3O^+ as a controlling factor. The reaction can thus be written either as:

$$CaCO_3 + H_3O^+ \longrightarrow Ca^{2+} + 2HCO_3^- + OH^- \qquad [2.5]$$

or

$$CaCO_3 + 2H^+ \longrightarrow Ca^{2+} + 2HCO_3^- \qquad [2.6]$$

The dissolution of calcium carbonate thus proceeds in relation to the acidity of water, with calcium ions in solution being balanced by hydrogen carbonate (bicarbonate) ions in solution, the amount of hydrogen carbonate being limited by the supply of hydrogen ions present to combine with carbonate ions.

There are several sources of acidity in natural waters; for example, some bacteria produce sulphuric acid (see section 3.2.2, p. 34), but there are two main sources of acidity: first, organic acids; second, the dissolution of carbon dioxide in water.

Organic acids can be proton donators and dissociate in water to yield hydrogen ions, as discussed in section 3.2.2.

The gas carbon dioxide exists in the atmosphere (approximately 0.03%) and can be present in solution as the carbon dioxide molecule. This is normally referred to as $CO_{2(aq)}$ to signify its presence in the aqueous medium as distinct from the gaseous phase $CO_{2(g)}$. A proportion of the $CO_{2(aq)}$ combines with water to form carbonic acid, H_2CO_3:

$CO_{2(g)}$	$+ H_2O$	$\longrightarrow$	$CO_{2(aq)}$	$+ H_2O$	[2.7]
carbon dioxide in the free atmosphere	water		carbon dioxide dissolved in water		

$CO_{2(aq)}$	$+ H_2O$	$\longrightarrow$	H_2CO_3	[2.8]
carbon dioxide in solution	water		carbonic acid	

Carbonic acid can rapidly dissociate to form H^+ and HCO_3^- ions, the former actually present as H_3O. This yields a source of acidity in solution:

H_2CO_3	$\longrightarrow$	H^+	$+$	HCO_3	[2.9]
carbonic acid		hydrogen ion		hydrogen carbonate ion	

The amount of dissociation varies with acidity, the hydrogen carbonate ion being the dominant form in circumneutral solutions (Fig. 2.7). The hydrogen ion from equation [2.9] becomes the source of acidity capable of reacting with calcium carbonate as shown in equations [2.5] and [2.6]. The overall process is illustrated in Fig. 2.8, with each divalent cation of calcium being balanced by two anions of hydrogen carbonate in the final solution.

2.5.5 Reactants, products and equilibrium

Calcium carbonate dissolution is governed by the supply of protons derived from carbon dioxide and other acids. Thus, it is possible to define the amounts of calcium carbonate which will dissolve for given levels of acidity or given levels of carbon dioxide. The amount of calcium carbonate in solution at equilibrium will be defined in terms of the reactants and the products. The **reactants** diffuse towards the solid phase surface down a concentration gradient as they are used up during the reactions listed above; **products** diffuse away from the solid surface as they are

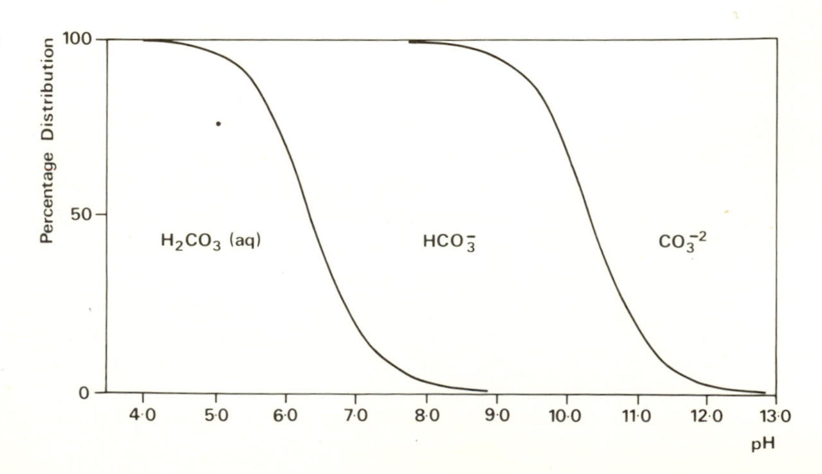

Fig. 2.7 Carbonate ion speciation as a function of pH (equations 2.7–2.9 and 2.10) (25 °C, 1 atm pressure) (from Hem, 1970).

Fig. 2.8 Reactants and products during calcium carbonate dissolution in the presence of carbon dioxide: $CaCO_3 + CO_2 + H_2O \longrightarrow$ $\boxed{Ca^{2+}}$ + $\boxed{2HCO_3^-}$. $\boxed{}$ = products in solution.

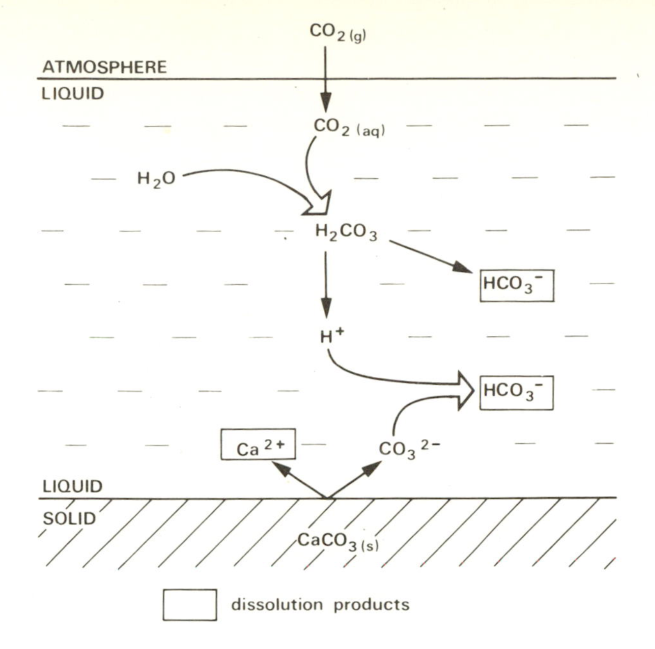

Fig. 2.9 Diffusion of reactants to and products from the solid–liquid interface.

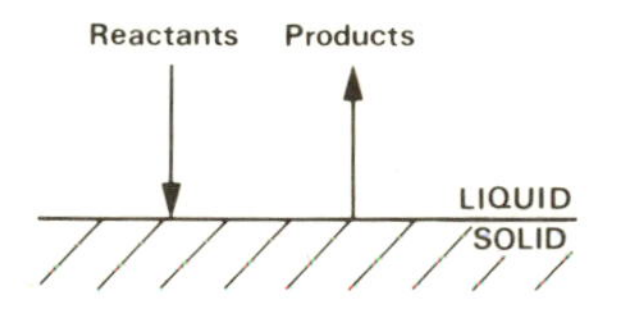

Fig. 2.10 At equilibrium the forward reactant, R_f, is equal to the back reaction, R_b.

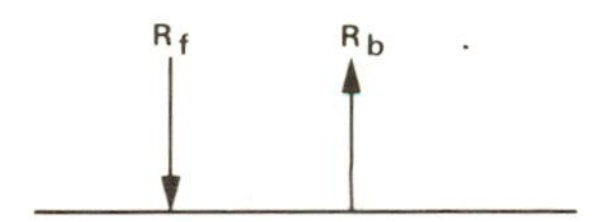

produced by the reactions. Thus, H^+ and CO_2 will diffuse towards, and Ca^{2+} and HCO_3^- will diffuse away, from the solid surface, in both cases diffusing from areas of higher to lower concentration (Fig. 2.9). In a closed system, that is where carbon dioxide is initially available for reaction with no renewable supply, the amount of calcium in solution will reach an equilibrium with the amount of available carbon dioxide. Here, an equilibrium condition is defined when the arrival of reactants – the forward reaction (R_f) – is equal to the movement of the products away from the solid – the backward reaction (R_b) (Fig. 2.10). The concentration at saturation is defined as the concentration obtaining under these conditions and can be defined relative to the amount of acidity in solution. Thus, calcium carbonate solubility values at equilibrium can be expressed in terms of **pH**, where pH is a logarithmic expression of the hydrogen ion concentration (concentration being denoted by square brackets):

$$pH = -\log_{10}[H^+] \qquad [2.10]$$

The relation between pH and saturation concentration values is shown in Table 2.2, for conditions where pH is controlled by carbon dioxide.

True equilibrium states are, in fact, rarely found in nature (Mercado and Billings, 1975) as it may take many years to achieve this state. However, geomorphologists have often adopted an empirical approach to define whether natural waters are far from equilibrium or near to equilibrium, the former indicating that there is a potential for dissolution to proceed further. A far from equilibrium condition will be indicated by excess H^+ in the system described by equations [2.5] and [2.6]. This condition can be identified by adding powdered calcium carbonate to a water sample and measuring any uptake of calcium into solution (Stenner, 1969; Trudgill, 1983a and see Appendix). An uptake is often referred to as **'aggressiveness'**. In practice, such uptake may not simply involve an assessment of the amount of calcium carbonate needed to reach equilibrium with free carbon dioxide ($CO_{2(aq)}$ and H_2CO_3) but also any calcium taken up because of the presence of organic acids. In ad-

Table 2.2 Values of C_s for calcium carbonate at given pH values where pH is controlled using carbon dioxide, (modified from Picknett, 1964) (mg l^{-1})

pH	$CaCO_3$	Ca^{2+}
6.48	577.3	231.2
6.58	528.2	211.5
6.72	422.1	177.0
6.71	410.1	164.2
6.80	406.4	162.7
6.82	370.8	148.5
6.87	342.8	137.3
6.92	316.2	126.6
7.18	241.5	96.7
7.27	212.9	85.2
7.83	92.9	37.2
8.95	76.9	30.8
8.27	52.2	20.9

dition, rocks do not always behave in the same way as pure calcium carbonate, and it is therefore also useful to measure departures from equilibrium not only with a standard, pure calcium carbonate but also with crushed samples of any country rock under investigation. The method is described by Trudgill (1983).

2.5.6 Solution processes and water flow

In closed systems with static water, reactants move to and products move from the solid surface. The reactants become adsorbed on to the surface, chemical reaction takes place and the products become desorbed (Fig. 2.11). The overall rate of the reaction may be controlled by the transport rates of the reactants and products to and from the surface or by the rates of chemical reactions and sorption processes. The former is termed transport control, the latter chemical control. However, if the water is moving and if the system is open, while the same processes still operate, the overall rate of reaction becomes influenced by the rate of water flow and the chemical composition of that water.

Fig. 2.11 Surface processes; ① + ⑤ = transport processes, ②, ③ + ④ = chemical processes (modified from Mercado and Billings, 1975, and Trudgill, 1977c).

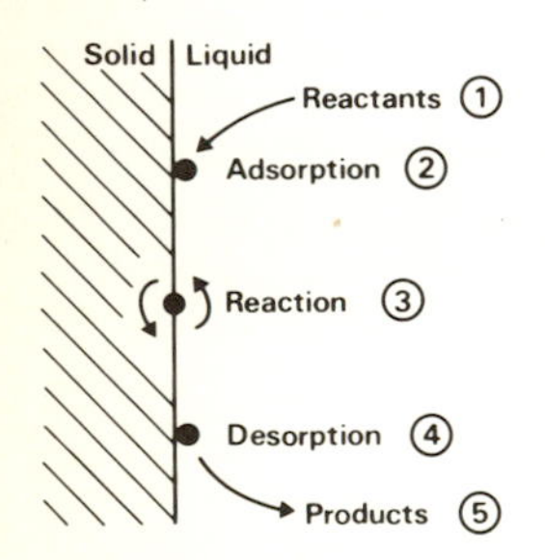

If it is assumed initially that water arriving at any one reaction site is low in solutes, flow has the effect of not permitting the reaction products to build up at that site. Since the rate of mass transfer from the solid to the liquid is in part controlled by diffusion gradients, the removal of products keeps the diffusion gradient steep and the rate of mass transfer high.

If the water arriving at a site is already high in solutes, the rate of mass transfer from the solid to the liquid will tend to decrease because of a decreased diffusion gradient; however, because the water is flowing, the removal of products is still encouraged and the rate of mass transfer will be greater in a flowing high solute concentration water body than would be the case in a static water body of similar concentration.

In describing the effect of solvent motion on limestone solution processes, Kaye (1957) illustrates how the diffusion layer adjacent to the solid decreases in thickness as the rate of velocity increases (Fig. 2.12). This has the effect of decreasing the transport distances for the reactants and products and the overall rate of the reaction will become controlled by the chemical processes of adsorption, reaction and desorption. Dissolution rates may thus be controlled by either the rates of transport of ions or by the rates of surface reactions. In rapidly flowing water any dissolution products can become rapidly transported and so the rate-limiting processes are the surface reactions (Berner, 1978). The important factor is the rate of flow relative to the reaction rate. Thus, moving from

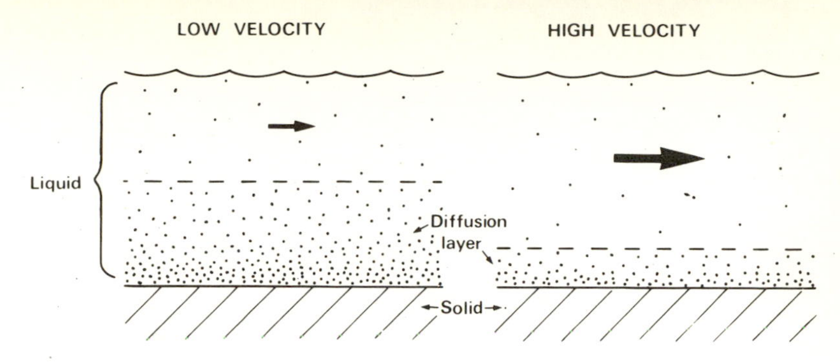

Fig. 2.12 Thickness of diffusion boundary layer and solution velocity. With a thinner boundary layer (right) dissolution rate is increased (modified from Kaye, (1957).

a situation of static water to one of water flow (which may be continuous or intermittent), movement increases the rate of dissolution if the reaction rates are faster than the transport rates (this is analogous to people arriving at a bus queue at a rate faster than that at which buses arrive; increasing the number of buses increases the rate of movement away from the bus stop). However, if water-flow rate is fast relative to the reaction rate, the overall rate becomes independent of flow and becomes reaction rate limited (this is analogous to a situation where buses are more frequent than the arrival of people at the bus stop, many buses will be empty and the movement of people from the bus stop will be equal to the rate of the arrival of people). The concept of flushing frequency is used by Berner to describe the frequency with which water in contact with a solid is changed:

$$\frac{dc}{dt} = R - \frac{F}{V}c \qquad [2.11]$$

where:

c = concentration of dissolved solute
t = time
F = flushing rate (= inflow−outflow), volume per unit time
V = fixed volume of water at concentration 'c'
R = rate of dissolution (mass per unit volume of solution per unit time).

Dissolution rate is accelerated by flushing because of a drop in c unless the rate of flushing is large relative to the mineral reaction rate which means that maximum reaction rate is reached and acts as a control on the overall process (Fig. 2.13). These concepts are discussed further in Chapter 3 (Section 3.3, p. 40).

Equilibrium considerations can be viewed in a downstream context, for example for water flowing over a subaerial rock surface, or water flowing down a cave (White, 1977). Equilibrium is approached as a water body progressively picks up dissolution products as it sequentially contacts rock surfaces. Thus, Fig. 2.5 could be redrawn with contact time with a cave wall, or length of cave passage as the horizontal axis. Thus, when water is flowing, it is important to consider whether the system is being studied from the point of view of a particular location on a rock surface or from the point view of the chemical evolution of water as it flows past a whole sequence of locations on rock surfaces.

In terms of the progressive evolution of cave waters, the decrease in aggressiveness and increase in calcium concentrations can be plotted (Fig. 2.14), though much of the increase in calcium down a cave passage

Fig. 2.13 Dissolution rate and flow: ① Mass transfer is related to flow and dissolution rate increases with flow rate. ② Mass transfer is related to reaction rate and reaction rate is limiting factor irrespective of flow. S = optimum conditions for saturation where flow rate is in balance with reaction rate. From S ⟶ 2 concentrations in solution decrease because flow rate is faster than reaction rate; from S ⟶ 1 concentrations decrease as flow rate decreases.

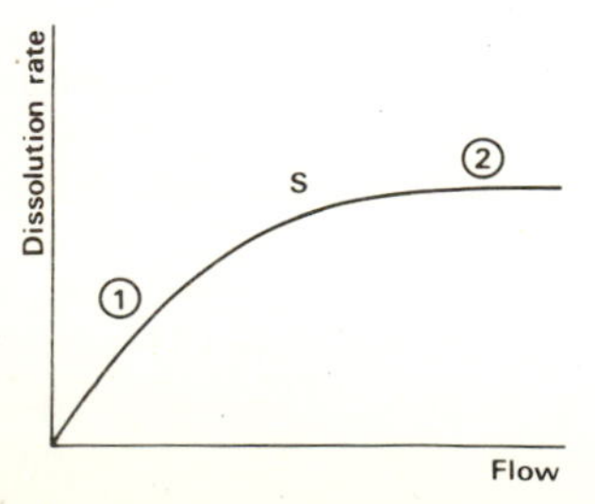

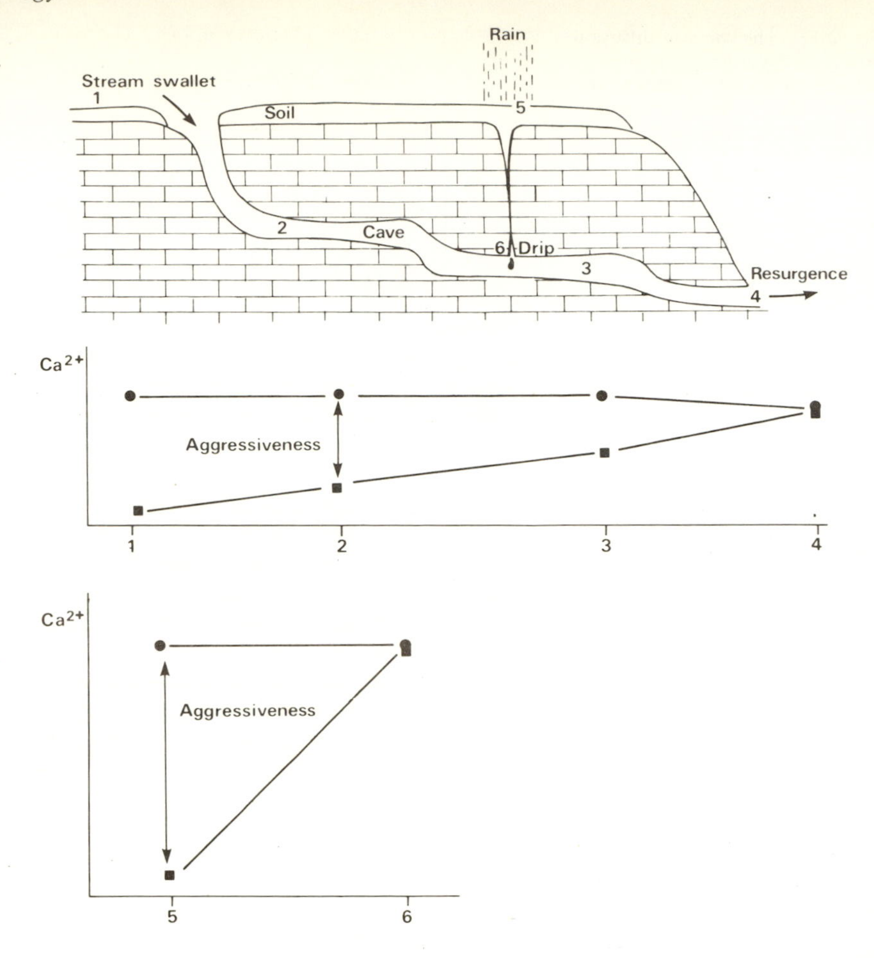

Fig. 2.14 Changes in aggressiveness in limestone system. ● – potential calcium concentration. ■ – Actual calcium concentrations.

is solely due to calcium derived from drips in the roof and not from dissolution of the cave wall.

2.6 The lithological factor in limestone erosion

Limestone landforms are produced by interactions between the rock and the erosional environment; lithological variation is therefore a major factor influencing landform evolution at a number of scales. Micro-scale variations in solubility and porosity influence micro-topography, while joints, beds, faults and folds have considerable influence upon the more major landforms.

2.6.1 Applications of carbonate solution chemistry to limestone dissolution

In the history of limestone geomorphology, emphasis has always been placed on solution processes and on the assumption that limestone solubility can be substantially predicted by calcite equilibrium solubility. A great deal of effort has gone into the measurement of calcium in solution in limestone run-off waters. However, a full understanding of the departures of limestone solubility from calcite solubility has not been fully achieved.

In terms of equilibrium carbonate solution chemistry, Morgan (1967) stresses the open nature of natural water systems and questions the usefulness of equilibrium models for the understanding of such systems; processes operating in open systems are illustrated in Fig. 2.15. Morgan

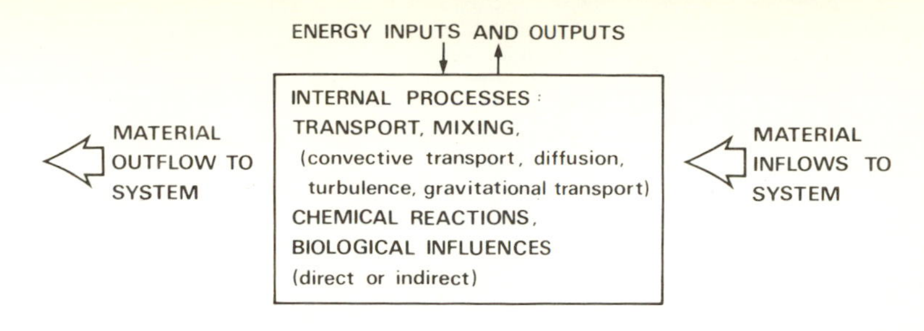

Fig. 2.15 Generalised model for chemical dynamics in an open natural water system (modified from Morgan, 1967).

suggests that gradients exist in many systems and that equilibrium in carbonate systems may be approached only in very long residence time systems, such as may be found in micro-pores in groundwater. It is stressed by Curl (1977) that diffusion-controlled processes limit the attainment of equilibrium, with the consequent penetration of non-equilibrium water far into karst aquifers. In addition, White (1977) challenges a simplistic approach to carbonate dissolution, indicating that the kinetic regime may differ at different levels of undersaturation, with a rapid increase in dissolution rate just below saturation. In addition, the mixing of carbonate waters (Bögli, 1964; Wigley and Plummer, 1976) not only produces an increased dissolution potential because of the non-linear relationship between CO_2 and Ca^{2+} in solution (Fig. 2.16); mixing may also lead to a change in kinetic regime, with a greatly increased dissolution rate (White, 1977).

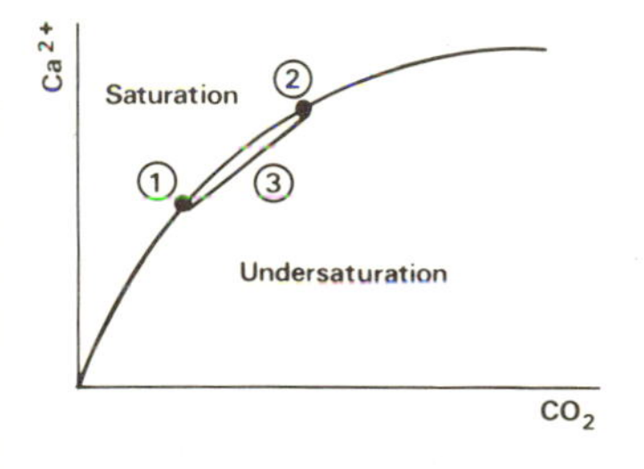

Fig. 2.16 Schematic diagram of mix-corrosion. The curve represents the equilibrium concentration of Ca^{2+} in solution at given CO_2 levels; mixing of waters at ① and ② occurs along the line ③ in the area of undersaturation (modified from Bögli, 1964).

In a study of saturated calcite solutions, Picknett (1973) showed that many equilibrium calculations were in error since the ion species present were not only Ca^{2+} and HCO_3^- but also 10% of the calcium present may be in the form $CaHCO_3^+$. Picknett (1973) not only discusses the qualifications applicable to pure calcite solutions he also stresses (Picknett, 1964; 1973, p. 77) the role of impurities present in natural limestones which can substantially limit the application of calcite equilibrium solubility calculations. The metals found to have an inhibiting effect on calcite solubility are, in decreasing order of effectiveness: lead, lanthanum, yttrium, scandium, cadmium, copper, gold, zinc, germanium and manganese. Many of these occur in limestone and a range of trace element compositions is illustrated in Table 2.3. The precise effects of these elements is as yet not fully established and Picknett (1964) indicates that for dilute waters in contact with solid calcite the effects may be slight; however, micro-scale variations in trace element composition of limestones may be crucial in controlling the evolution of micro-topographic features during limestone dissolution.

Calcite equilibrium solubility calculations may therefore be of limited application to the topic of limestone solution processes for two reasons. Firstly, kinetic considerations may be more appropriate and, secondly, trace element composition may have a significant influence on dissolution processes.

2.6.2 Lithology and erosion

Differential erosion – a term implying that some portion of a rock mass is eroded more than another – is a subject of major importance in the understanding of the evolution of relief, as outlined in Chapter 1. In younger, less well-lithified rocks many of the original constituents are still well preserved and thus shells, corals and calcarenite may be closely juxtaposed, giving rise to differential erosion because of differences in solubility. Many recent carbonates are comprised of aragonite and high magnesium calcite which are more soluble than calcite (Berner, 1967;

Table 2.3
(a) Example of trace element composition of clastic components of recent carbonate rocks (modified from Trudgill, 1979a)

Clasts	*Zn (ppm)*	*Pb (ppm)*	*P (%)*	*Si (ppm)*
Mixed corals	16.7	8.3	0.5	0.0
Platygyra coral	82.2	–	1.8	1442
Mixed shells	343.2	–	1.8	0.0
Goniastrea coral	68.0	–	0.44	0.0
Phosphates	440.8	8.9	44.0	4869

= no data available; 0.0 = reading of zero

(b) Data modified from Wolf *et al.* (1967) in ppm.

	Limestone	*Dolomite*	*'Carbonates'**
Co	34	17	300
Co	88	260	500
Mn	3200	4100	2800
Na	150	240	–
Ni	70	14	100
Pb	100	8	200
Sr	1600	3000	6000
Y	13	1	80
Zn	700	200	500

* mostly recent, partially lithified deposits.

Trudgill, 1976b; Picknett, 1972). Aragonite is more soluble than calcite in the ratio of 1.59 : 1, while traces of magnesium carbonate in solution can increase the solubility of calcium carbonate. Thus mixtures of calcite, aragonite and high magnesium calcite in a rock can readily give rise to differential solutional erosion.

In older rocks, aragonite has usually been converted to calcite and if the rock has become recrystallised many of the original fossil structures will have been lost. Thus the rock itself is more homogeneous but, here, differential erosion is related more to joint frequency and bedding.

The effects of lithology on dissolution rate have been investigated by Rauch and White (1977) in a kinetic context of relevance to dissolution in open, natural water systems. They found that dissolution rate decreases as the percentages of dolomite and dispersed insoluble material increased. Maximum dissolution rates were found in carbonate rocks with 1.0–2.5% MgO content (Fig. 2.17). Picknett (1972) suggests that calcium carbonate solubility is at a maximum at 4–7% $MgCO_3$ in solution. Dolomite is much less soluble than calcite (Busenberg and Plummer, 1982; Thrailkill, 1977), (Fig. 2.17).

There is no necessary relationship between laboratory solubility and erodability in nature. Rauch and White (1977) note that impurities in the form of silty streaks may actually encourage dissolution in natural systems by acting as pathways of weakness which may be exploited by flowing water. In addition, silty streaks in limestone may also increase surface roughness, giving rise to increased solution turbulence, thus enhancing dissolution. Also, sparite is a highly soluble carbonate rock, but tends to be resistant in the landscape because of its greater mechanical strength (Rauch and White, 1977; Sweeting, 1966, and see Fig. 2.18).

Rocks which may be composed of the more soluble carbonates may be present as more upstanding features in a landscape simply because of their structure. Thus a coherent, coral structure, composed of more soluble aragonite may be more upstanding than a less soluble loosely ce-

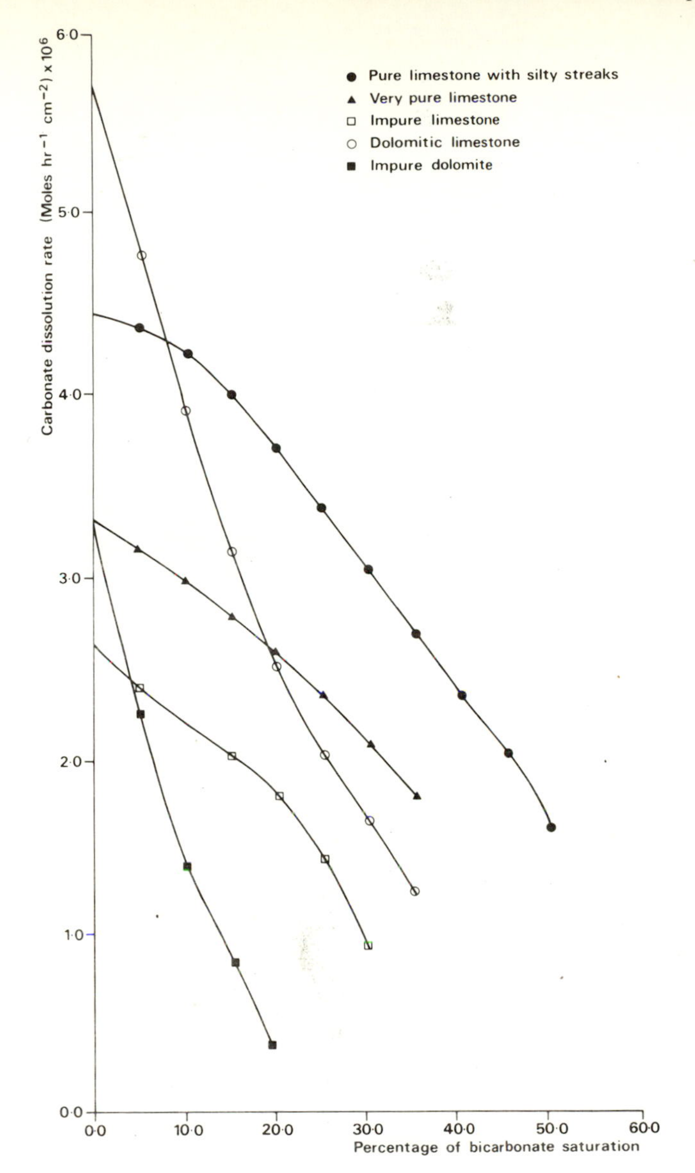

Fig. 2.17 Examples of dissolution rates for a range of carbonate rocks (modified from Rauch and White, 1977).

mented calcite calcarenite (Table 2.4). Thus, in many ways, the focus of geomorphologists upon solution chemistry and solubility has possibly detracted from the main geomorphological task of specification of relative erodability of rocks under given erosion regimes. Very little work has been carried out on this topic. Jakucs (1977), Sweeting (1972) and Jennings (1971) all focus upon solution processes and the relative solubility of different limestones, for example. Abou-El-Enin (1973) comments that the more thinly-bedded limestones form less resistant rock masses. Sweeting (1974) notes that the porcellanous band, a micrite, in north Yorkshire is harder and less soluble than the sparite, but because it is traversed by numerous fractures is less resistant than the sparite rocks. Both Trudgill (1979a, 1976b) and Spencer (1981) use the term 'solutional disintegration' to record the fact that the erosion process may be a combination of solutional and mechanical processes. Further stud-

Fig. 2.18 An exposure of the Carboniferous Limestone, Malham, Yorkshire, UK. The rock face has been eroded glacially and modified by subsequent solutional and frost action. The more massive and resistant upper layers are composed of sparite and the lower, more well jointed layers are micritic. Sparites tend to be more massive and resistant to erosion (Sweeting, 1966, 1972). (Photo: S. T. Trudgill.)

Table 2.4 Lithology and erosion (modified from Trudgill, 1976b)
(a) Micro-erosion rates and lithology of measurement point

Lithology	*Erosion rate (mean value)*
Coral fragments	0.09 mm a^{-1}
Calcarenites	0.39 mm a^{-1}
Algal limestones	0.10 mm a^{-1}

(b) Micro-erosion rates, mineralogy and cementation

Rock type	*Erosion rate (mean value)*
* Dominantly low calcite calcarenites	0.35 mm a^{-1}
* Dominantly high magnesium calcite and aragonite calcarenites	0.51 mm a^{-1}
Porcellanous, well cemented algal limestone	0.09 mm a^{-1}
Loose, poorly cemented algal limestone	0.11 mm a^{-1}

* Mineralogy by carbonate staining (Wolf *el al.*, 1967) and X-ray diffraction

ies, such as those by Day (1981) on rock hardness and topography, may help to establish the nature of the relationship between lithology and erosion on a more systematic basis. Such generalisations as are possible at the present time are that solubility departures of limestone from calcite are common; kinetic considerations are often more important than equilibrium solubility considerations; limestones with small traces or magnesium carbonate present appear to be the most soluble, assisted by surface roughness associated with impurity; dolomites are less soluble than calcitic limestones; and that *there is no necessary relationship between solubility and erodability* and the most upstanding, resistant rocks in the landscape may simply be those which are less frequently jointed and better cemented, irrespective of solubility. Of all the lithological factors, joint frequency, porosity and permeability are probably the most important factors influencing the shape, height and progressive evolution of limestone landforms as they control the access of the erosion agencies to the rock mass. Where solution processes dominate the erosion regime, factors affecting the variations in dissolution rate relative to waterflow rate and saturation status are probably the greatest determinants of differential erosion.

3 Controls on solution processes in limestone areas

3.1 Introduction

In the weathering and erosion of limestones the rock interacts with its environment and any landforms produced as a result of this interaction can thus be expected to be interpretable in terms of lithological and environmental characteristics, given the qualifications that the stage of evolution and the degree of adjustment of the landform to environmental changes will also be important. Lithological factors have already been discussed in Chapter 2 and it is clear that the important factors include solubility, heterogeneity, mechanical strength and joint frequency. In terms of environmental factors it will also be clear from the discussion in Chapter 2 that the availability of carbon dioxide is a primary factor to be considered and that other sources of acidity should also be taken into account. In addition, from the study of solution kinetics in Chapter 2, it will be clear that contact time between solvent and solid is a significant control. These factors find their expression in terms of environmental controls through the existence of soil covers with varying slope angles and in temperature and rainfall regime. Thus, limestone solution processes can be expected to vary with climate and soils as well as lithology and topography.

Rainfall regime acts as a control on soil moisture levels and combines with slope to influence run-off rates which, in turn, govern the supply of water and solution reactants over rock surfaces and the removal of solution products. Soil cover acts as a producer of and reservoir for carbon dioxide and as a store for percolating rainwater. Organic soils yield supplies of organic acids which may also be picked up by percolating water. Soil temperature has an important effect on the rate of respiration of soil organisms, on root respiration and on organic matter decomposition and, together with soil moisture, acts to control the supply of soil carbon dioxide. The important interrelationships between the key variables are shown in Fig. 3.1, and they can be viewed at local, regional and global scales.

In addition to the rock–environment interactions, other factors come into consideration when attempts are made at providing explanations of landforms, especially tectonic uplift, climatic change and stage of erosion. Thus, the factors of soil carbon dioxide, organic acids, water-flow rates, topography and climate discussed below cannot provide a full ex-

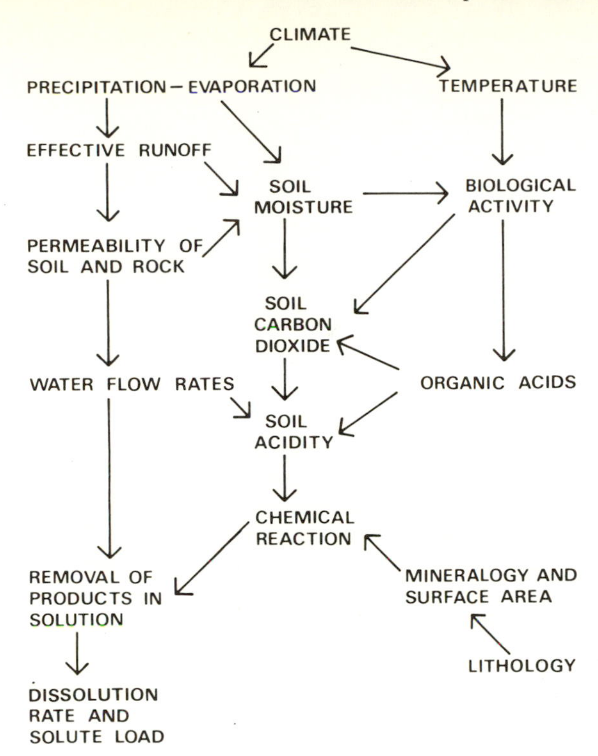

Fig. 3.1 Key variables in the control of limestone solution processes. ——→ = an influence on.

planation for limestone landforms but simply provide convenient – and well investigated – topics for initial discussion.

3.2 Solution chemistry

3.2.1 Soil carbon dioxide

A great deal of research effort has gone into the measurement of soil carbon dioxide in recent years (for example, Miotke, 1974) and this effort is based on the assumption that it is the principal source of dissolution potential for limestone. In fact, there are other important sources of acidity in the soil, especially organic acids, and high soil carbon dioxide levels do not necessarily mean that high rates of limestone weathering are taking place beneath a soil cover. This latter factor can be seen to be especially true if the factor of soil permeability is taken into account. Carbon dioxide may build up in a soil either because of high carbon dioxide productivity (such as in an organic-rich soil) or because it cannot escape due to low soil permeability. Anaerobic soils have a low oxygen content and high carbon dioxide content because gaseous exchange is limited in some way and they have a restricted circulation with the free atmosphere or with oxygenated waters. Thus, if a soil has a high carbon dioxide level it is likely that this is due to low soil permeability (Fig. 3.2). This also means that the free circulation of water from the soil to the bedrock is unlikely, with the water charged with carbon dioxide remaining in the soil under relatively stagnant conditions. Thus, measurements of high carbon dioxide levels in clay soils and waterlogged soils may well have little relevance to limestone solution processes and all soil carbon dioxide measurements are of limited value unless data for soil permeability or the movement of water through the soil are also available.

The methods used for carbon dioxide measurement in the soil all have to take into account the fact that it is extremely easy to alter the variable

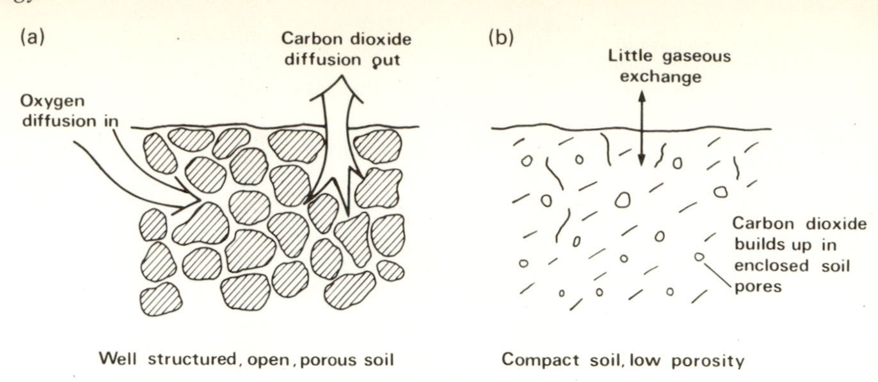

Fig. 3.2 The influence of soil permeability on gaseous exchange. (a) Rapid exchange is possible in open-structured soil and CO_2 concentrations will only be high if CO_2 productivity is high. (b) Carbon dioxide concentration is high because of low gaseous exchange; permeability to water will also be low and therefore transport of weathering products will also be limited.

being measured. Most methods involve some disturbance of the soil and inevitably some contact with the atmosphere; this can readily lead to gaseous exchange and a lowering of the measured soil carbon dioxide content unless great care is taken. The earliest methods involved the use of a pump and a gas analyser; here care was necessary in order to make sure that the pumping rate was not excessive so that air was drawn in from the surface. However, it is not always easy to establish what an excessive rate is and it is possible that all methods should be regarded as giving a minimum level. Sophisticated permanent underground installations have been made at Rothamsted Experimental Station, UK with sealed tanks and a circulating air system; the air is passed through soda lime which increases in weight as carbon dioxide is taken up. A similar apparatus is described by Edwards and Sollins (1973) (Fig. 3.3). Geomorphologists have been less interested in permanent installations for monitoring carbon dioxide production but more interested in portable measurement devices which can be used to compare one site with another. Nicholson and Nicholson (1969) used the permeability of polythene to carbon dioxide in their installation of tubes in the soil for 1–2 weeks during which the soil air equilibriates with the tube air (the time of equilibriation being established by the successive measurements of tubes over time until a near constant reading is reached, usually over 5–7 days). The tube is then analysed on a gas analyser such as Haldane apparatus involving soda lime uptake. More recently, portable carbon dioxide probes have become available, and the Draeger apparatus and the Gastec probe have become widely used. These involve a portable hand pump and the use of crystals which are colour sensitive to acid gases. Air is drawn in from the soil through a replaceable gas tube containing the crystals. The Draeger apparatus uses a small set of bellows which can be used to draw the air up from depth in the soil using a steel tube while the Gastec probe has a short tube directly on to the pump. The design of the probe tip is crucial to the prevention of contamination of soil air with atmospheric air. Miotke (1974) has a design which has a retractable sheath and a similar design for the Gastec probe is shown in Fig. 3.4. The closed tips are inserted into the soil and they are opened when in the soil. The pumping rate should always be as slow as possible and leakage of atmospheric air down the side of the tube can be minimised by sealing the tube to the surface soil with clay or other soil. Medical research commonly uses carbon dioxide analyses in both respired air and blood; the techniques available can be adapted for use in cave air and limestone drainage waters. Miserez (1970) describes the use of a medical carbon dioxide probe based on a semi-permeable mem-

Fig. 3.3 Design of apparatus for monitoring carbon dioxide production from a soil surface (modified from Edwards and Sollins, 1973).

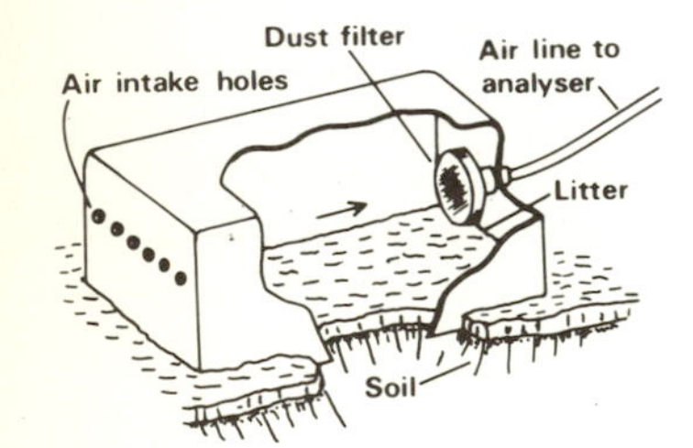

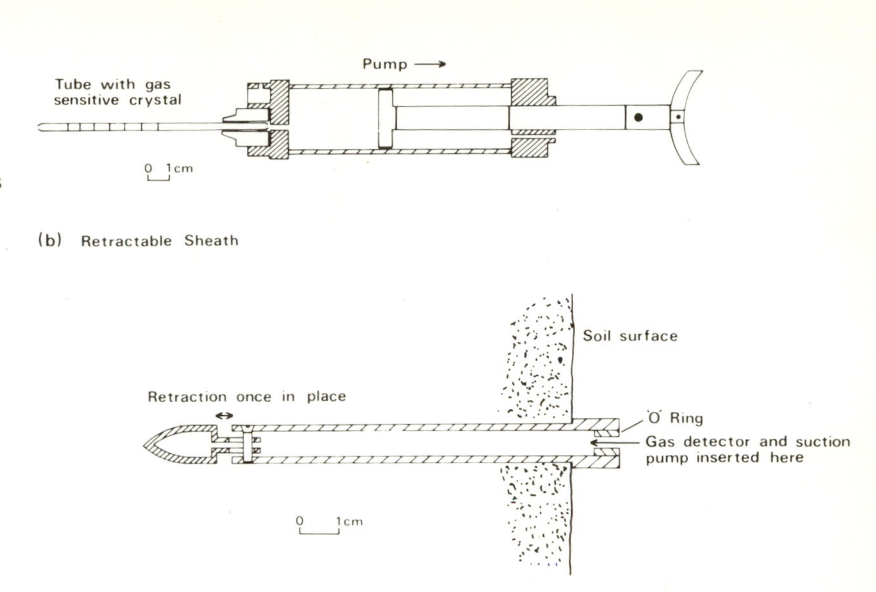

Fig. 3.4 (a) Gastec probe for the measurement of carbon dioxide content of air. Air is drawn through the replaceable gas sensitive tube by withdrawing the pump handle. (b) Sheath for use in a soil environment. The tube is pushed into the soil and the barrel retracted once in place. This opens the gap near the tip of the tube, allowing soil air to contact the gas sensitive tube. The pump is operated slowly and the gas detector is sealed from the atmosphere by a rubber 'O' ring inside the sheath.

brane, through which carbon dioxide may pass, and an internal solution of sodium bicarbonate; changes in gas concentration equilibriate with the solution and the electro-motive force generated is measured on a potentiometer. Miserez reports that in test solutions there was good agreement between calculated and measured pCO_2. The use of this probe has not been widely reported in soil environments.

Soil carbon dioxide levels normally increase with depth, again as a factor of soil permeability and exchange with the atmosphere (Fig. 3.5). Whether in cultivated or uncultivated soils, soil pore space is larger nearer the soil surface and the soil is more oxygenated. Thus, the depth of measurement is a very important consideration when one set of carbon dioxide measurements are compared with another. However, the pattern may not always be a simple one of an increase with depth. As root respiration is a source of carbon dioxide, high values are to be expected around roots in the soil, thus if roots are concentrated in a par-

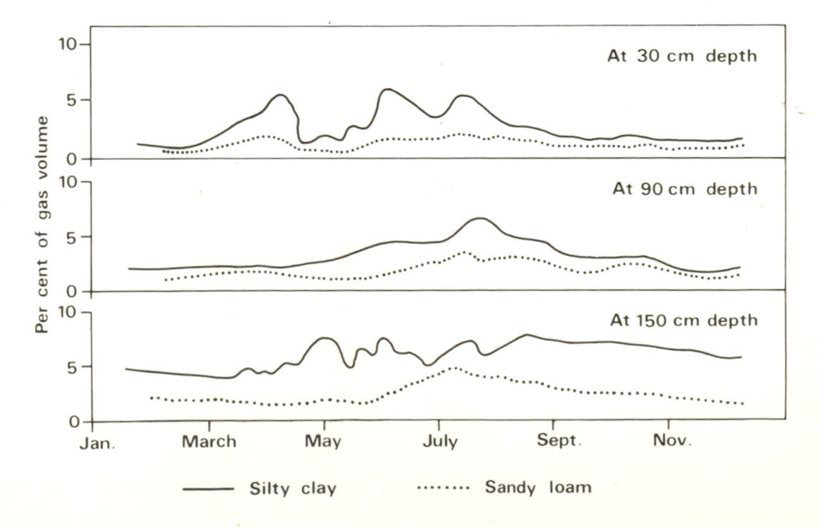

Fig. 3.5 Soil carbon dioxide content at three depths of two soil types under apple orchard usage. Note (1) that the sandy loam soil always has lower values because of its more open, porous structure (see Fig. 3.2), and carbon dioxide is building up more in the less porous silty clay soil; (2) the increase of concentration with decreasing permeability as depth increases and the high values in the surface, more organic soil; (3) seasonal variations, organic matter decay and respiration being greater in the moist and warmer months (modified from Russell, 1961).

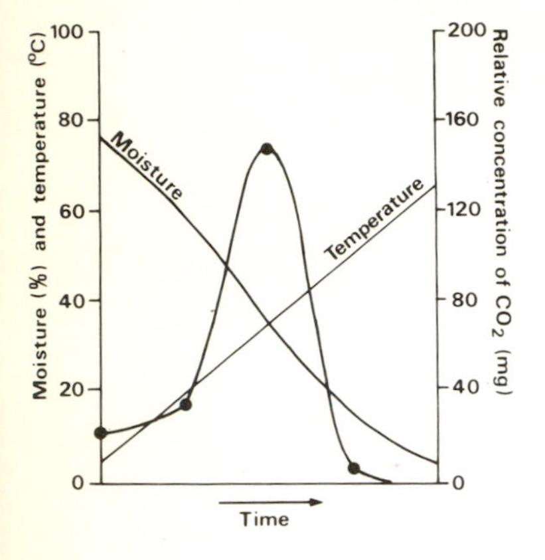

Fig. 3.6 Carbon dioxide evolution during gradual heating and drying of soil, showing optimum productivity at intermediate levels of both temperature and moisture (modified from Kononova, 1966).

ticular layer in the soil, carbon dioxide levels may increase down to this layer and decrease below it. This also applies to soils with organic surface layers, where the carbon dioxide may decrease with depth away from the organic substrate, the organic matter providing oxidisable carbon which the micro-organisms transform to carbon dioxide.

It is also important not to confuse carbon dioxide *levels* with carbon dioxide *productivity*. A highly productive soil rich in organic matter may be open textured and allow gaseous exchange, for example within loosely packed leaf litter; thus low levels may be measured but percolating soil waters during rain storms may pick up large amounts of carbon dioxide when the litter layer is wet. Conversely, in impermeable soils, low productivity may in time build up high levels. In many ways, measurements of carbon dioxide productivity would be more useful than the measurements of level normally taken.

Since carbon dioxide is produced by bacterial oxidation of carbon the rate of production is higher at higher temperatures and under moist conditions. Also, diffusion of carbon dioxide out of the soil is inhibited by higher moisture contents, which reduce the gas-filled pore space. Thus, according to the data of Kostychev (reported by Kononova, 1966) which are illustrated in Fig. 3.6 the rate of carbon dioxide production when the soil was gradually heated and dried in the laboratory was at a maximum at temperatures of about 25 to 35 °C and moisture contents of 30 to 40%.

The production of carbon dioxide in soils under natural conditions has been studied by Froment (1972) in mixed oak forests, Garrett and Cox (1973) for oak-hickory forests and Edwards and Sollins (1973) for poplar forests, all in northern temperate environments. They stress the importance of soil moisture and temperature. When soil temperatures are high in summer, soil carbon dioxide production increases with increasing soil moisture and when soils are wet productivity can increase with increasing temperature. Edwards and Sollins concluded that night-time production was greater than daytime production, with 48% of the carbon dioxide being produced from litter decomposition and 35% from root respiration. Maximal production occurred in the autumn (Fig. 3.7). Froment concluded that soil temperature exerted a more important influence upon soil carbon dioxide evolution than soil moisture did, the former having a correlation coefficient of $r = 0.54$ at 1 cm depth and 0.58 at 10 cm (Fig. 3.8) while for the latter it was $r = 0.39$. Temperature is clearly the most important factor, unless moisture becomes limiting – but this only appears to be the case when drought stress occurs during late summer or unusually prolonged dry periods of weather.

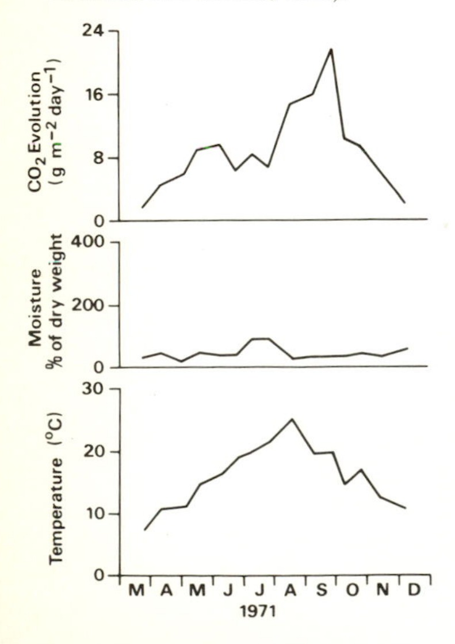

Fig. 3.7 Seasonal pattern of carbon dioxide production, as assessed by the apparatus shown in Fig. 3.3 (modified from Edwards and Sollins, 1973).

These kinds of relationships have been extrapolated by climatic geomorphologists who have sought to provide explanations for the occurrence of different landforms in different climates in terms of climatically determined dissolution potential. However, there is not a simple relationship between climate and soil carbon dioxide production. Measurements suggest that there may be as much variation within climatic zones as between them. The data available were summarised by Atkinson and Smith (1976) and Atkinson (1977) and are shown in Table 3.1. The data are for tropical and temperate situations, though not all are for limestone soils. It is clear that the ranges for temperate and tropical soils are overlapping. Miotke (1974) has reported on a survey of soil carbon dioxide levels and showed a pattern of variations with climatic zones in the USA (Table 3.2). It is difficult to translate these kinds of data into causal inferences concerning the evaluation of landforms in

Soil/Vegetation	*Soil depth* (cm)	*Usual %* CO_2	*Summer*	*Winter*	*Extreme values* (cm)
Tropical					
Evergreen forest	10	0.5 – 1.0	–	–	–
	200	3.4 – 6.3	–	–	–
Bamboo forest	10	0.2 – 3.5	–	–	–
	200	4.1 – 10.8	–	–	–
			Wet Season	Dry Season	
Cacao plantation, Trinidad	10	–	2.8– 6.5	0.2–0.8	–
	25	–	3.0– 8.5	0.8–1.7	–
	45	–	4.2– 9.7	1.4–3.8	–
	90	–	4.5–14.3	3.4–7.6	–
	120	–	3.6–17.5	3.7–6.8	–
Limestone soils, Jamaica	15	0.3 – 1.6	–	–	–
	30	0.4 – 3.0	–	–	–
	60	–	–	–	–
Temperate					
Arable		0.9	–	–	–
Pasture		0.5 – 1.5	–	–	0.5 –11.5
Sandy arable		0.16	–	–	0.05– 3.0
Arable loam		0.23	–	–	0.07– 0.55
Moorland		0.65	–	–	0.28– 1.4
Arable		0.1 – 0.2	–	–	0.01– 1.4
Manured arable		0.4	–	–	0.03– 3.2
Grassland		1.6	–	–	0.3 – 3.3
Dark Chestnut	7	0.1	–	–	–
	300	1.7	–	–	–
Steppe: tree coenoses		2.5 – 3.4	–	–	–
herbaceous		1.2 – 2.0	–	–	–
Sandy loam	30	–	2.5	0.3	0.2 – 3.6
	30	–	1.5	0.1	0.1 – 1.9
Loamy sand	50	–	0.8	0.2	0.2
					0.2 – 1.1
	20	–	0.9	0.1	0.05– 2.0
Brown earth on limestone		0.27 – 0.41		–	0.08– 0.7
Orchard/grass	30	–	1.5 – 2.5	0.1– 1.0	–
	90	–	2.5	1.3	–
	150	–	4.9	2.6	–
Valley bog	5	–	1.0 – 3.5		–
'Sandy Soil'		1.06			–
'Manured sandy soil'		9.74			–
'Black clay'		0.66			–
'Fertile moist soil'		1.79			–

Table 3.1 Carbon dioxide concentrations in soil air (modified from Atkinson and Smith, 1976)

different climates, since other factors such as lithology and water flow through time may also differ from region to region. It is more profitable to consider here the detailed links between soil carbon dioxide production and solution processes before tackling the wider issues of climatic geomorphology.

Chemical analyses of waters from limestone springs commonly show concentrations of dissolved calcium of between 40 and 120 mg l^{-1}. Moreover, it is usually found that the same waters contain approximately twice as many hydrogen carbonate ions (HCO_3^-) as they do calcium

Table 3.2 CO_2 – Concentrations in soil atmospheres of America (modified from Miotke, 1974) (Based on nearly 500 measurements from 1969–1972)

	Vol. % CO_2
Puerto Rico (February 1969)	
Mogotes slopes (relative dry)	0.08 – 1.2
Cockpit soil, clay, water saturated	1.5 – 7.0
Florida (April/May)	
Key Largo, stony soil	0.2 – 0.4
Everglades, grass	0.7 – 2.2
Cape Kennedy, wood	0.4 – 0.5
Waverly, wood, dry sandy soil	0.15 – 0.3
Waverly, meadow, dry sandy soil	0.16 – 0.2
The measured data represent the end of the dry season at the beginning of the rainy summer season.	
Rocky Mountains, Southeast Canada (July/August)	
Crowsnest Pass and Mt Castleguard,	
Banff National Park (1500–2600 m)	
Atmosphere	0.018
In perennial snow	0.01 – 0.01 – 0.014
Frost debris	0.04 – 0.1
Alpine tundra	(0.04) – 0.15 – (0.38)
Alpine meadow	0.1 – 0.4
Wood, conifers	0.1 – (0.5)
Seward Peninsula, Alaska (At the end of August)	
Tundra, mostly above permafrost	
(a) Marine terrace	0.18 – 0.4 – 1.2
(b) Frost debris, polygone soil	0.05 – 0.15
(c) Humid moss, pulvinate above moraines	dry 0.18 – 1.5 – 3.8 wet
In perennial snow patches	0.02

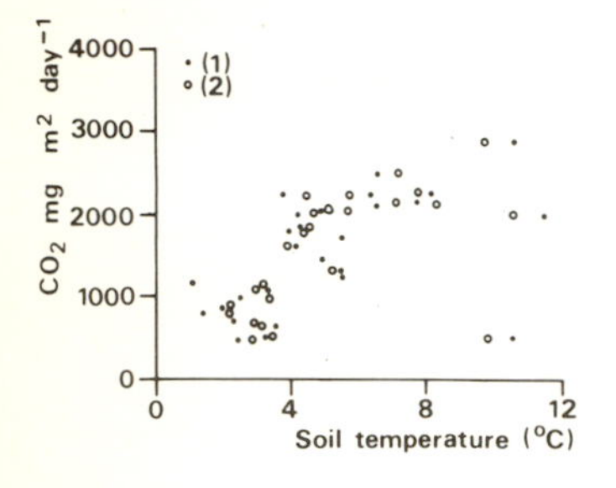

Fig. 3.8 Soil carbon dioxide production. (1) 1 cm depth. (2) 10 cm depth (modified from Froment, 1972).

ions. This indicates that the calcium, which is derived from the solution of limestone, is present almost entirely in the form of calcium balanced by hydrogen carbonate (p. 15). It has been widely inferred, therefore, that the principal mechanism by which limestone is dissolved involves the reactions (see also Ch. 2, p. 17):

$$CO_2 + H_2O \longrightarrow H_2CO_3 \longrightarrow H^+ + HCO_3^- \quad [3.1]$$

$$\text{and } CaCO_3 + H^+ + HCO_3^- \longrightarrow Ca^{2+} + 2HCO_3^- \quad [3.2]$$

The normal partial pressure of carbon dioxide in the atmosphere at sea-level is 0.0003 atmospheres or approximately 0.03% by volume. Using the equilibria established for the reactions occurring in the water–carbon dioxide–calcite system it can be shown that water in equilibrium with the normal atmosphere can dissolve up to 74 mg l^{-1} of limestone (expressed as mg l^{-1} $CaCO_3$, see Appendix). Limestone spring waters commonly hold a solute concentration of calcium which can be expressed as equivalent to 100 to 300 mg l^{-1} $CaCO_3$, which is far greater than the figure of 74 mg l^{-1} which can be achieved by water in equilibrium with the normal atmosphere. Adams and Swinnerton (1937) noted this fact and suggested that the source of the extra carbon dioxide required lay in the interstitial air of the soil, where the concentration of carbon dioxide was known to be much greater than 0.03%. This is now widely accepted, although there may be other sources of acidity involved and the carbon dioxide content of interstitial air in bedrock may play an important role as well as soil air (Atkinson, 1977).

That the soil profile is certainly the source of much of the chemical aggressiveness of karst waters was demonstrated by Groom and Ede (1972) who describe experiments in which water was passed through natural soil profiles in the laboratory, and analysed for its dissolved limestone content. Stream waters and tapwater with initial concentrations of about 30 mg l^{-1} $CaCO_3$ rose to concentrations of 75 mg l^{-1} when allowed to stand in contact with limestone chippings and in contact with the atmosphere. Water passed through a 50 cm column of brown earth soil with a low limestone content and with a natural growing vegetation increased to 69 mg l^{-1} $CaCO_3$. In contrast, water passed through 50 cm of identical soil overlying 75 cm of limestone chippings dissolved 326 mg l^{-1} $CaCO_3$, demonstrating that the sources of soil acidity dissolved 257 mg l^{-1} limestone from the chippings.

Carbon dioxide will dissolve in soil water in proportion to its concentration in the gas phase. H^+ and HCO_3^- will form in solution. When the solution dissolves calcite, the concentration of calcium and hydrogen carbonate ions increases, and the concentration of free carbon dioxide and H^+ decreases. The solution is thus in disequilibrium with the gas phase from which the carbon dioxide was originally dissolved, and further carbon dioxide is dissolved from the air. This fresh supply causes further dissolution of the calcite, which in turn reduces the amount of carbon dioxide and H^+ in solution, so that more carbon dioxide again dissolves from the gas phase. By repetition of this process the system will eventually achieve an equilibrium in which the amount of calcium carbonate in solution will be exactly balanced by the carbon dioxide in the gas phase. Thus it is possible to plot a graph relating the carbon dioxide content of the gas phase to the total amount of limestone that an equilibrium solution can dissolve. Several such graphs have been presented by Picknett (1964), Jacobson and Langmuir (1972), Roques (1969) and Smith and Mead (1962). Figure 3.9 is a plot of the potential calcium concentration of a solution against the carbon dioxide content of the associated gas phase (from Smith and Mead, 1962). It should be noted that in an open system carbon dioxide can be dissolved from the gas phase at the same time as calcite is dissolved from the solid phase. All three phases are in contact at equilibrium. The open system condition is termed 'aerobic' by Smith and Mead and is illustrated by the upper curve on Fig. 3.9. The lower curve shows the case in which the water is brought into equilibrium with the gas phase and then the solid phase separately. In this case, fresh carbon dioxide cannot dissolve continuously as limestone is dissolved, and the potential calcium content of the water at equilibrium is much less. This initially open and subsequently closed system is termed 'anaerobic' by Smith and Mead (1962). The solution of limestone within the soil profile and in the zone of bedrock immediately beneath it will normally conform to the open system case with fresh carbon dioxide available.

Fig. 3.9 Potential calcium carbonate concentrations in solution in equilibrium with carbon dioxide in open systems, where gaseous exchange is possible, and closed systems, initially open to CO_2 supply but then isolated from further supply (modified from Smith and Mead, 1962).

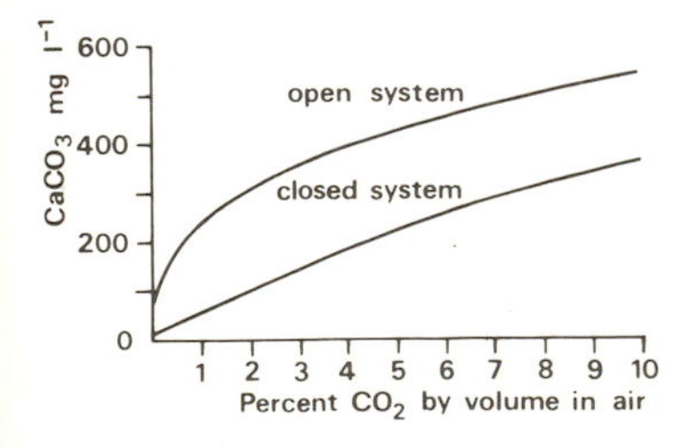

Table 3.3 shows the values of potential $CaCO_3$ concentrations which could be achieved in the open system condition by solutions in contact with selected values for soil air. While the values show a wide range, they compare well with the 100 to 300 mg l^{-1} $CaCO_3$ found in most limestone springs, suggesting that, in general, carbon dioxide from soil air is an important source of soil water aggressiveness.

It is worth noting that some, but by no means all, springs show a seasonal fluctuation in hardness which may be tentatively correlated with variations in the soil temperature and soil air carbon dioxide. For example, Pitty (1966, 1968) finds a range of 160 to 200 mg l^{-1} $CaCO_3$ be-

Soil/vegetation		*Usual % CO_2*	*Extreme values*	
Arable		0.9 (242)		
Pasture		0.5 – 1.5 (200 – 285)	0.5 – 11.5	(200 – 547)
Sandy arable		0.16 (139)	0.05 – 3.0	(96 – 355)
Arable loam		0.23 (156)	0.07 – 0.55	(107 – 206)
Moorland		0.65 (218)	0.28 – 1.4	(166 – 279)
Arable		0.1 – 0.2 (119 – 149)	0.01 – 1.4	(57 – 279)
Manured arable		0.4 (186)	0.03 – 3.2	(81 – 363)
Grassland		1.6 (291)	0.3 – 3.3	(170 – 367)
Brown earth on limestone		0.27 – 0.41 (164 – 188)	0.08 – 0.7	(111 – 223)
		Summer values	*Winter values*	
Orchard/grass	30 cm soil	1.5 – 2.5 (285 – 336)	0.1 – 1.0	(119 – 250)
	90 cm soil	2 – 5 (312 – 419)	1 – 3	(250 – 356)
	150 cm soil	4 – 9 (390 – 506)	2 – 6	(312 – 444)

Table 3.3 Carbon dioxide content of soil air in temperate regions (modified from Atkinson, 1977), with corresponding open system solubilities of calcite (the solubilities are taken from the data of Smith and Mead, 1962 and are shown in mg l^{-1} in brackets)

tween winter and summer in Goredale Beck, Yorkshire, and 220 to 240 mg l^{-1} in Crowdwell Rising, Derbyshire. From the Jura Mountains, Aubert (1969) reports a seasonal fluctuation in nine springs from 180 to 210 m l^{-1}. However, when such fluctuations are present they are always smaller, and often much smaller, than the likely variation in soil air carbon dioxide would lead one to expect and it is probable that passage through the bedrock leads to a dampening of temporal fluctuations present in the soil (Atkinson, 1977).

In summary, the theory of carbonate equilibria suggests that measurements of soil carbon dioxide levels should indicate the potential for limestone dissolution. Indeed, there appears to be an agreement between the levels of calcium in solution in springs fed solely by water percolating through limestone and levels of soil carbon dioxide. However, there are many intervening variables and the relationship is not a simple one. Measurements of high carbon dioxide levels in soils of low permeability may overestimate the potential for limestone dissolution since the soil waters are liable not to be mobile and therefore not to reach the limestone beneath. Conversely, measurements of low carbon dioxide levels in open structured organic soil layers such as leaf litter may tend to underestimate the potential for dissolution since organic acids may provide a substantial dissolution potential in this case. In addition, the effects of soil processes may be considerably modified by the slow passage of percolation through bedrock; temporal fluctuations may be smoothed out and some organic matter oxidation may occur and reactions will proceed under closed system conditions; for these reasons soil processes and measurements may not be a reliable indicator of groundwater and percolation spring behaviour. The measurement of soil carbon dioxide remains a useful indicator of limestone dissolution processes but the measurements should be made and interpreted in the light of the qualifications outlined above.

3.2.2 Organic acids

The importance of organic acids as weathering agents is not fully known, but their effectiveness under certain conditions is undoubted. Organic acids are produced during the decomposition of organic matter and may also be derived as a bark or leaf leachate from trees. They can dissociate to produce hydrogen ions, when they can weather carbonate minerals by hydrolysis, or they can incorporate calcium or magnesium cations into their organic structures, a process known as **chelation**. Organic weathering processes can include direct active attack upon minerals by bacteria, fungi and roots by hydrolysis or chelation or the less direct action in solution where percolating waters pick up organic acids which then weather limestone as the percolating waters move into contact with the rock. This would be the case when water percolates from an organic surface soil horizon to a lower mineral carbonate horizon or from a peaty deposit over a lower-lying limestone outcrop.

Soil minerals, including calcite, may be attacked by the activity of autotrophic bacteria (Alexander, 1961) or by the acidic byproducts of their physiology. The dissolution of calcite, aragonite, magnesite and dolomite by lactic-acid producing bacteria has been observed experimentally (Kononova, 1966). Calcite may also be decomposed by butyric-acid bacteria, nitrifying bacteria and anaerobic nitrogen-fixers. Duff *et al.* (1963) have reported the solubilisation of minerals by bacterial products, notably 2-ketogluconic acid. Kuznetzov (1962) cites the geological significance of the acid-producing bacterium *Thiobacillus oxidans* which will oxidise organic sulphur to sulphuric acid:

$$3O_2 + 2S + 2H_2O \xrightarrow{T.\ oxidans} 2H_2SO_4 \qquad [3.3]$$

The sulphuric acid will attack any carbonates in the soil. *Thiobacillus oxidans* occurs in peats, the run-off from which can be extremely aggressive towards limestone. An important effect, stressed by Russell (1961) and Waksman (1931), is that microbial transformations involving calcium carbonate usually release carbon dioxide, which may further increase the aggressiveness of the soil water. An example of such a transformation and the resultant production of carbon dioxide is the action of *Thiobacillus denitrificans* cited by Russell (1961):

$$6KNO_3 + 5S + 2CaCO_3 \xrightarrow{T.\ denitrificans} 3K_2SO_4 + 2CaSO_4 + 2CO_2 + 3N_2 \qquad [3.4]$$

The action of soil fungi is discussed by Waksman (1931). Fungi can synthesize organic acids from carbohydrates available in the soil, and may produce more carbon dioxide than do bacteria. Citric and oxalic acid-producing fungi are cited by Henderson and Duff (1963) as acting to decompose soil minerals.

Processes in the soil close to plant roots provide a further example of direct attack upon minerals. Calcium and other metals may be exchanged between root surfaces and mineral fragments in the soil. Richards (1974) describes how this is achieved by the movement of hydrogen ions from rootlets (possibly via negatively charged clay colloids) to the mineral grains, with a transfer of calcium ions in the reverse direction. Ions appear to migrate in response to concentration gradients set up by the roots.

Plants growing on limestone, but which usually grow on carbonate free substrate, may create pockets of acid soil around their roots, partly by root processes and partly by the acid nature of their litter. These pockets are particularly well displayed by common heather or Ling (*Calluna vulgaris*) (Grubb *et al.*, 1969; Grime and Hodgson, 1969) (Fig. 3.10). Such pockets of acid soil may have an effect in weathering the limestone beneath them.

The organic matter of soils may contain substances which weather minerals by chelation. In this process, calcium is incorporated directly into the molecular structure of an organic compound. The process is described by Chaberek and Martel (1959) and Manskaya and Drozdova (1969). Schnitzer and Skinner (1963) discuss the relationships between metallic ions and soil organic matter. Stable complexes are formed incorporating calcium and magnesium. The complexes tend to be most

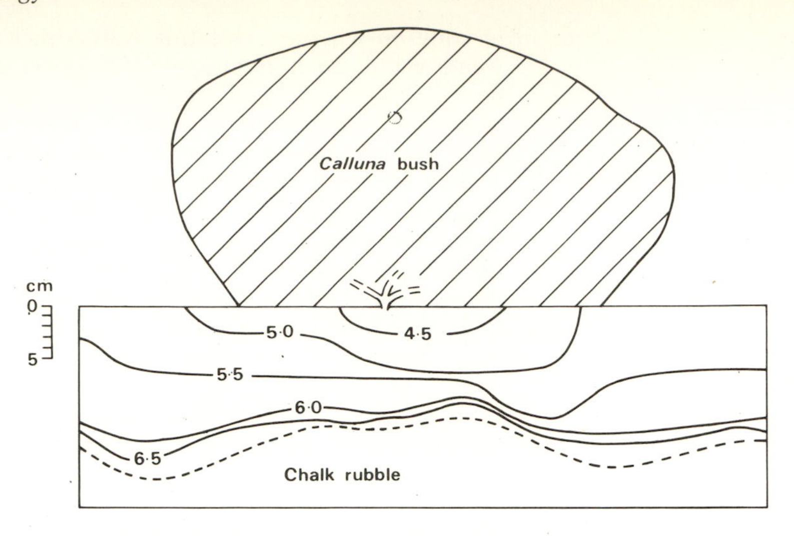

Fig. 3.10 Section through chalk soil under a specimen of *Calluna vulgaris*, demonstrating the greater acidity in the surface soil under *Calluna* (modified from Grubb *et al.*, 1969).

soluble at low pH, and thus acidity encourages the removal of ions by chelation and leaching (cheluviation).

Lichens, which are very common on bare limestone surfaces, are well known as chelating agents and their role in pedogenesis is discussed by Schatz (1963) and Syers (1964). Lichens may produce organic acids and the possibility that these lichenic acids are responsible for the simple dissolution of limestone is questioned by Hale (1967) on the grounds that they are too weak and produced in too small quantities to be effective, and Hale therefore suggests that chelation is the primary mechanism. A possible mechanism for the chelation of calcium by a lichen acid, lecanoric acid, is shown in Fig. 3.11.

Fig. 3.11 A possible mechanism for the chelation of calcium by lecanoric acid. (a) suggested structure of free lecanoric acid. (b) a possible calcium–lecanoric acid chelate (modified from Hale, 1967).

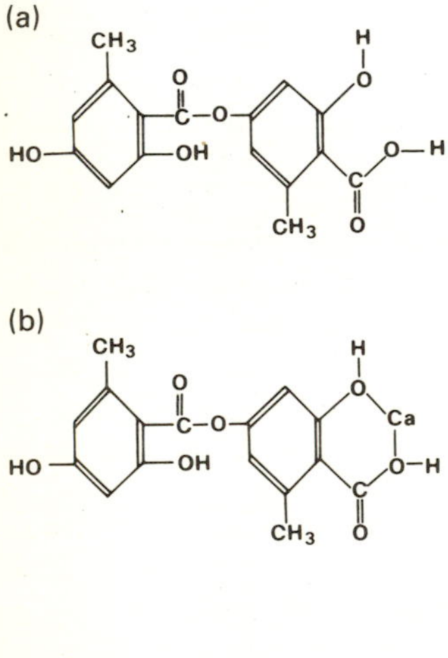

(a)

(b)

The processes of bacterial and fungal activity, root exchange and chelation by organic matter may each be important in determining the weathering potential in certain circumstances. In some cases calcium may be simply adsorbed on to the clay-humus molecule. Organic matter, containing humic acids and clay-humic acid colloids, is ubiquitous in soils, and forms an important source of hydrogen ions. Chemical weathering of limestone by humic acids is likely to be a general occurrence, although its effect may not always be large in comparison to that of carbonic acid.

The organic acids found in soil are divisible into humic, fulvic and hyetomelanic acids, which form a successive sequence produced at different stages of organic decay, with humic acid as a relatively stable product. In a study specifically concerned with limestone weathering Murray and Love (1929) described how the 'growth acids', which are products of plant metabolism, and the humic decomposition acids may attack the limestone to form humates. The stability of humates of very high molecular weight is questionable, and most may break down further to form simpler acids. The list of the simpler organic acids found in the soil includes butyric, lactic, acetic, propionic, gluconic, oxalic, fumaric, succinic, pyruvic, citric and tartaric acids (Murray and Love, 1929; Waksman, 1931; Handley, 1954; Kononova, 1966.) Each may attack calcite with the concomitant release of carbon dioxide. For example:

Oxalic acid

$$C_2O_4H_2 + CaCO_3 \longrightarrow Ca\,(CO_2)_2 + CO_2 + H_2O \qquad [3.5]$$

Acetic acid

$$2CH_3.COOH + CaCO_3 \longrightarrow Ca\,(COOH_3)_2 + CO_2 + H_2O \qquad [3.6]$$

or generally, using 'X' to denote the acid radical:

$$2HX + CaCO_3 \longrightarrow CaX_2 + CO_2 + H_2O. \qquad [3.7]$$

The direct adsorption of calcium on humus and clay-humus structures is implied by the high-base exchange capacity of humus acids. Kononova (1966) shows how this capacity increases as decomposition proceeds along the decay sequence to humic acid. Base exchange capacities for calcium per 100 g of organic matter range from 253 to 394 meq/100 g, which is up to one hundred times greater than the capacity of some clay soils.

Mention should be made here of the work of Bray (1969, 1971, 1972) on the chemistry of several streams fed from peat-covered areas in South Wales. He records a positive correlation between the aggressiveness of stream and cave waters and their chemical oxygen demand. The latter is a measure of the organic matter content of the water. In a number of samples which were analysed for pH, oxygen demand, aggressiveness, $CaCO_3$ content, alkalinity and non-alkaline salts, the aggressiveness is well in excess of that calculated from the pH of the solution. However, if it is assumed that all of the oxygen supplied in the tests for chemical oxygen demand is used in the production of carbon dioxide by oxidisation of organic matter, a new aggressiveness may be calculated. It is found that the actual aggressiveness of the samples is equal to the sum of the acid aggressiveness (calculated from pH) and that calculated from the oxygen demand. Bray suggests that bacterial oxidation may be responsible for the rapid conversion of organic matter to carbon dioxide observed in some streams.

In terms of field observations, deeply etched micro-runnels may be visible at the base of trees when bark drainage waters (stemflow) runs over limestone (Fig. 3.12). Stemflow pH values frequently lie in the

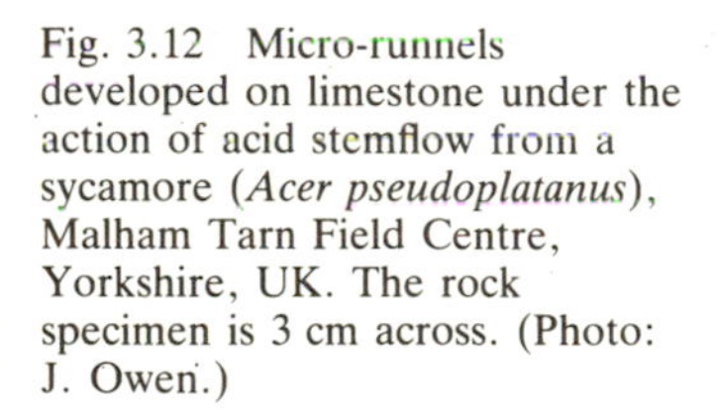

Fig. 3.12 Micro-runnels developed on limestone under the action of acid stemflow from a sycamore (*Acer pseudoplatanus*), Malham Tarn Field Centre, Yorkshire, UK. The rock specimen is 3 cm across. (Photo: J. Owen.)

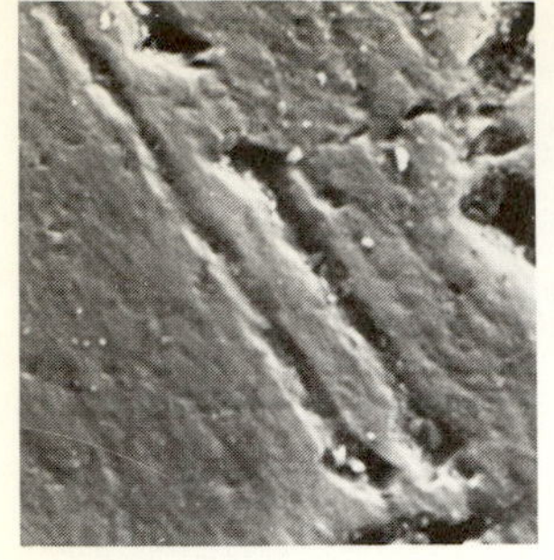

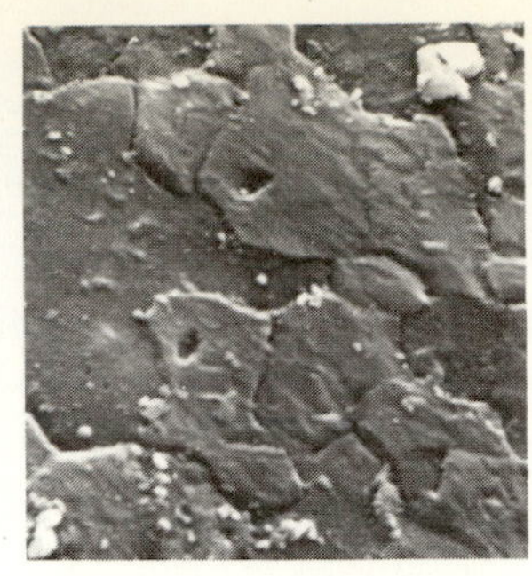

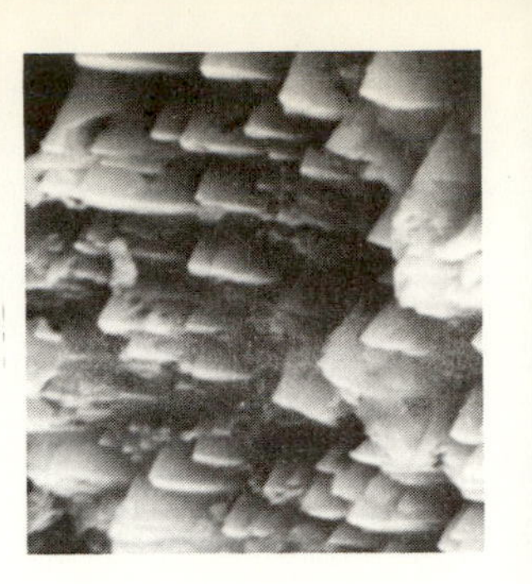

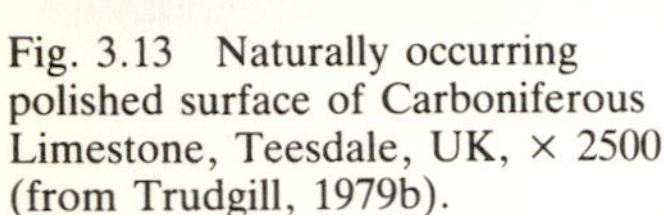

Fig. 3.13 Naturally occurring polished surface of Carboniferous Limestone, Teesdale, UK, × 2500 (from Trudgill, 1979b).

Fig. 3.14 'Polished', reflective, smooth surface produced by 0.2 M tartaric acid, × 2500 (from Trudgill, 1979b).

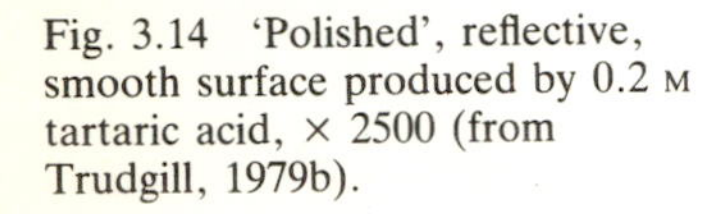

Fig. 3.15 Smooth surface produced by etching with 0.02 M hydrochloric acid, × 2500 (from Trudgill, 1979b).

Fig. 3.16 Etched surface produced by 0.02 M tartaric acid, × 2500 (from Trudgill, 1979b).

range of pH 3–5 and thus dissolution of limestone beneath woodland cover must be increased by stemflow acidity and leaf leachates.

Chemical polish can often be seen where extremely acid waters drain over limestone, especially from peat; this is a very smooth reflective surface produced by the even dissolution of limestone. It has been suggested by Trudgill (1979b) that the smoothness of the surface is related to high acidity (low pH), with natural smooth surfaces (Fig. 3.13) being reproduced by concentrated organic acids (Fig. 3.14) and mineral acids (Fig. 3.15) but not by weak organic acids, where etching occurred (Fig. 3.16). Organic acids are thus a powerful agent of, but not a necessary prerequisite of, limestone dissolution. Further work is necessary on the relative importance of organic acids and carbon dioxide in surface limestone landform production.

3.2.3 Soil pH

From a knowledge of the reactions series:

$$\begin{array}{l} CO_2 + H_2O \longrightarrow H_2CO_3 \longrightarrow H^+ + HCO_3^- \\ CaCO_3 \longrightarrow Ca^{2+} + CO_3^{2-} \\ \qquad\quad \downarrow \qquad\quad \downarrow \\ \qquad\quad Ca^{2+} + 2HCO_3^- \end{array} \quad [3.8]$$

it is clear that the H^+ ion plays a key role in the process of calcium carbonate hydrolysis. It is also clear that the dissociation of organic acids may produce hydrogen ions. It follows that measurements of soil water pH may arguably be a better predictor of the potential for limestone dissolution than soil carbon dioxide levels.

The relationships between pCO_2, pH and base status of a soil are discussed by Baver (1927), Cole (1957), Whitney and Gardner (1943), Shaw (1960) and Seatz and Peterson (1964). Russell (1961, p. 107) gives the relation:

$$2pH = \text{constant} + pCa + pCO_2 \quad [3.9]$$

where pCa and pCO_2 are negative logarithms analogous to pH. As the partial pressure of carbon dioxide is increased and pCO_2 decreases, the pH falls and the amount of calcium held upon charged clay particles decreases as calcium ions move into the soil solution. There is thus a mutual interaction between pH, base status and the ambient partial pressure of carbon dioxide. If one component of the system is altered the others will also change in order to reach a new equilibrium.

Similarly, Whitney and Gardner (1943) have shown that under laboratory conditions of constant soil moisture content and base status, the pH of soil water is a straight line function of the logarithm of the partial pressure of carbon dioxide. If the carbon dioxide pressure is held constant, variations in soil moisture have virtually no effect upon pH.

Under natural conditions, an increase in soil moisture due to rainfall can dilute the soil solution and cause a rise in pH. This is called the suspension effect by Seatz and Peterson (1964). If no further change in moisture content or carbon dioxide pressure occurs, carbon dioxide will dissolve in the diluted solution, and the pH will fall to its orginal value. The suspension effect is thus a disequilibrium produced by dilution.

Since there is a close relationship between soil pH, carbon dioxide pressure and base status, the pH should in theory be an accurate predictor of the amount of solution of limestone likely to occur at the base of the soil. In practice the pH varies, just as soil CO_2 does, with temperature and moisture content (because of the suspension effect), and may be buffered by the reserve of hydrogen ions on clay-humus particles or organic matter. As Whitney and Gardner (1943) point out, any single pH determination should be regarded as one of a range of values rather than as a constant property of the soil: this criticism applies equally to soil CO_2, but is often overlooked by geomorphologists. This variation could be assessed by studying the pH and CO_2 content of soils throughout the year, but this is often impractical.

The magnitude of the source of hydrogen ions can be indicated by the buffer capacity of the soil, which provides a longer term measure for comparing one soil with another (Seatz and Peterson, 1964). Buffer capacity can be measured by air drying and weighing out a specified amount of the soil, which is then stirred into suspension in a standard volume of distilled water, and its pH measured. Successive aliquots of 0.1 N sodium hydroxide or hydrochloric acid are then added and the pH noted after the addition of each aliquot. In this way, the available reserve of hydrogen ions (or hydroxyl ions) per unit weight of soil may be calculated for any given pH. The greater the acid reserve, the more sodium hydroxide solution must be added to bring the suspension to a constant pH. Figure 3.17 shows laboratory buffer capacity curves for distilled water, three humus soils and one mineral soil. As the figure makes clear, the acid buffer capacity of the humus soils is thirty to forty times greater than that of the mineral soil at pH 10. This buffer capacity is due not to carbonic acid, as this will have been lost during drying, but to weak organic acids and clay-humus colloids.

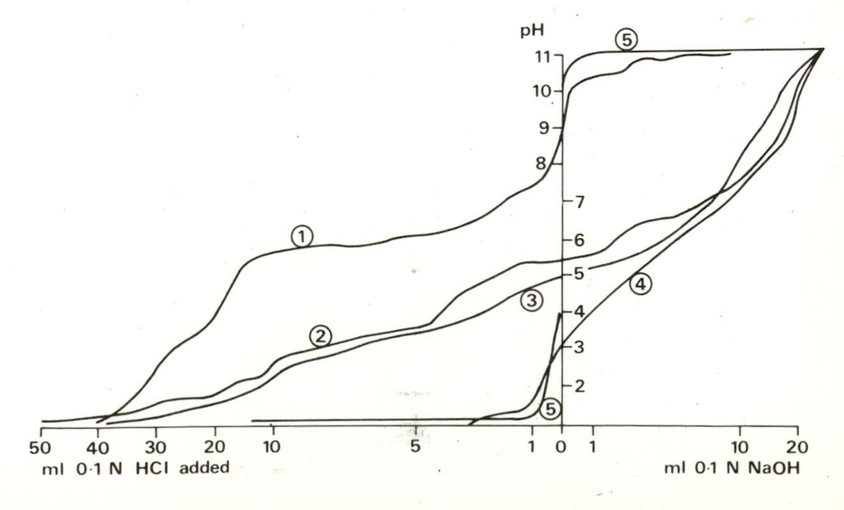

Fig. 3.17 Buffer capacity curves for soils, County Clare, Eire. ①, calcareous brown earth. ②, *Calluna* + peat ranker over limestone. ③, *Thymus* + humus rendzina over limestone. ④, *Calluna* + peat over sandstone. ⑤, standard buffer curve using distilled water.

The relationship between soil pH, soil carbon dioxide content and limestone erosion rate has not been widely researched but Trudgill (1977a) suggested that erosional weight loss of limestone tablets correlated very poorly with measured soil CO_2 levels. However, weight loss correlated with soil pH, the highest erosion rates being found under acid organic soils. This pattern is thought to be due to the low permeability of the soils with the high CO_2 level, and discussed, in section 3.2.1, and demonstrates the value of soil pH measurements as predictors of limestone dissolution rate (Fig. 3.18).

In summary, the evaluation of limestone dissolution potential has traditionally involved the investigation of soil carbon dioxide levels, based upon a knowledge of carbonate equilibria. In practice, soil carbon dioxide levels may vary over time and space and this variation can only be gauged by extensive work. In addition, dissolution by organic acids may be an important factor to consider. Furthermore, soil permeability and water flow rates may combine with levels of soil acidity to govern the actual dissolution of limestones. Accordingly, the general topic of limestone dissolution in relation to water-flow rate is discussed below.

3.3 Hydrology

3.3.1 Water-flow rates and dissolution processes

Water-flow rate is a key factor since it governs the supply of reactants to mineral surfaces and the removal of weathering products (Ch. 2). Solid-solvent contact time will also be governed by water-flow rate, rapid flow leading to short contact times and little opportuntiy for dissolution, except for the most rapidly dissolving mineral constituents; slow flow will enhance the possibility of chemical saturation occurring. In rapid flow, kinetic considerations will apply, and in slow flow, equilibrium considerations will be more important (Ch. 2). The theory of dissolution in relation to flow rate is discussed in detail by Berner (1978), Mercado and Billings (1975) and Mercado (1977). There will be an optimum flow rate where contact times are long enough to allow dissolution to proceed to a large extent but also where flow rates are rapid enough to allow weathering products to be removed. If weathering products accumulate, further dissolution will be inhibited.

On bare limestone surfaces flow rates are rapid and any factor which slows down flow, such as low slope angle, rough surface texture or vegetation, will enhance dissolution processes in the water which is retained on the surface. Under soils, and in any other porous medium, flow rates are slower and thus dissolution will be encouraged. In fact any increase in flow rate in porous media will tend to encourage dissolution as weathering-product removal will be encouraged. Thus, on bare surfaces, flow is rapid and decreasing it encourages dissolution; in porous media, flow rate is slow and increasing it generally encourages dissolution.

Porous media may, however, be fractured or fissured. Water flow may occur in large pores or other routeways which bypass the bulk of the medium. The pores between these preferential pathways may thus not be wholly operative in terms of dissolution processes. The details of the route of water flow through a system are thus important considerations.

In rapid flow, any dissolution product can be moved away quickly (as discussed in Chapter 2). Thus the provision of products from the mineral surface is the limiting factor for the overall dissolution process; this situation is termed reaction limited (Fig. 3.19(a)). Conversely, in slow flow, dissolution products may be formed but not carried away in solution; this situation is termed transport limited (Fig. 3.19(b)). The

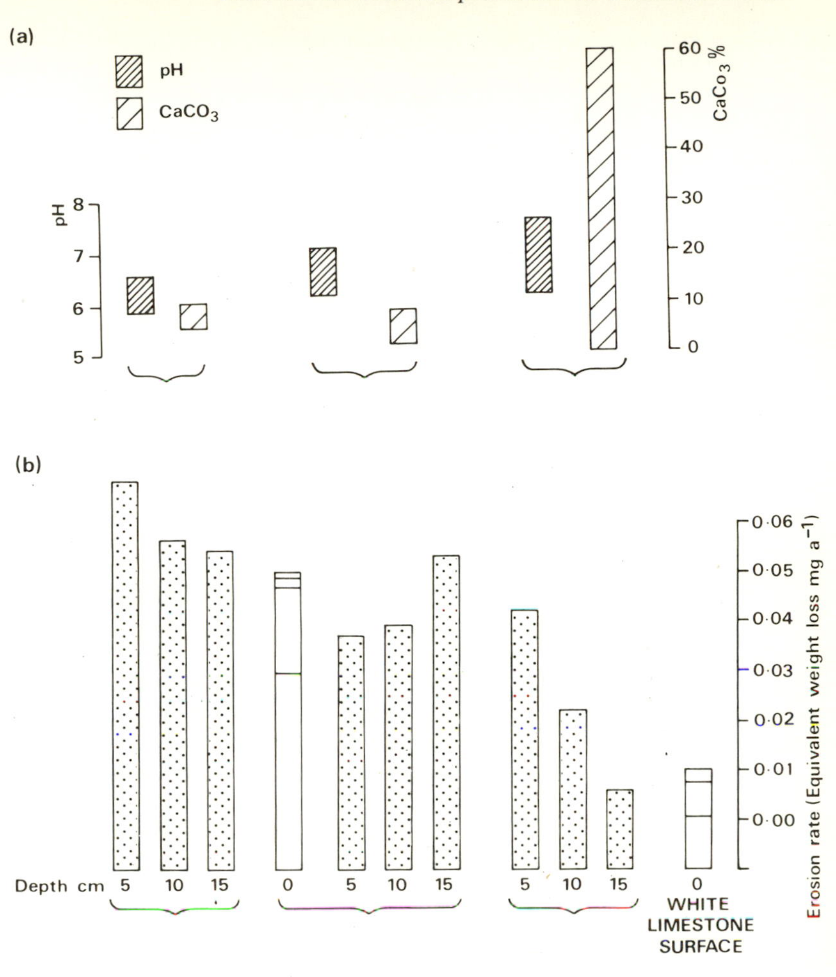

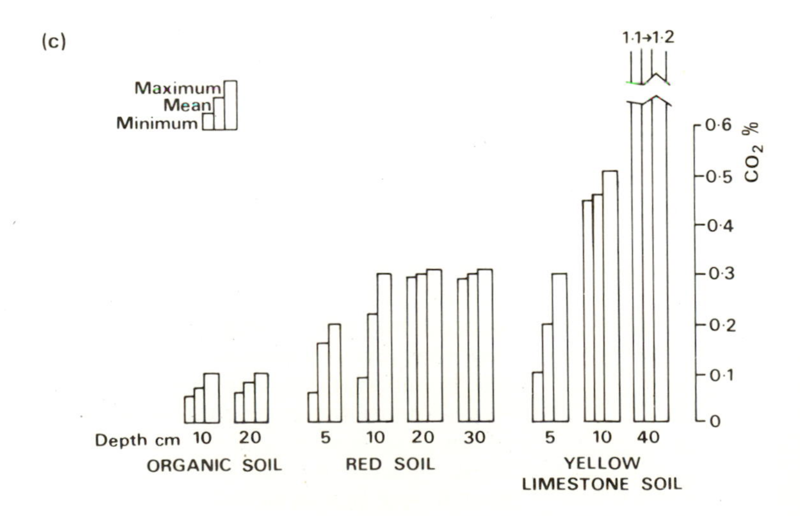

Fig. 3.18 Subsoil limestone erosion rates (weight loss tablets), Cockpit Country, Jamaica. (a) soil pH and $CaCO_3$ ranges of measured values from 10 samples. (b) erosion rates, *shaded*: tablets in soil, *unshaded*: tablets on surfaces. (c) soil dioxide (Gastec probe) (modified from Trudgill, 1977a).

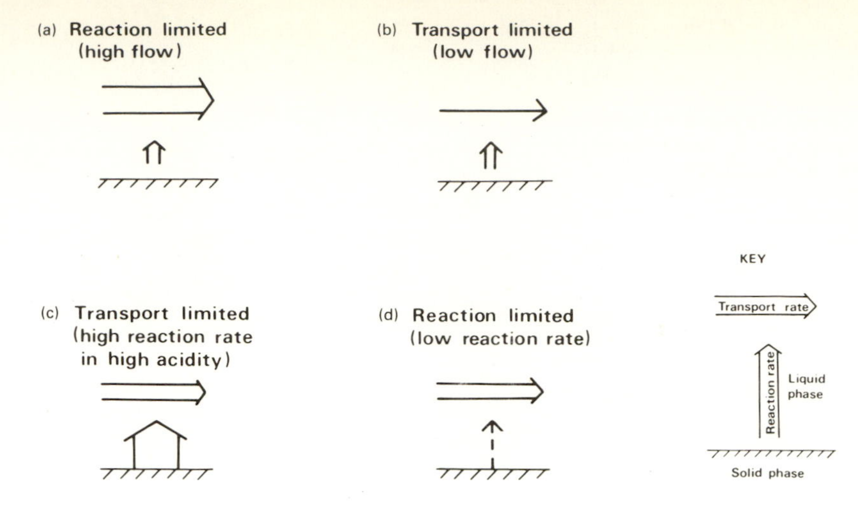

Fig. 3.19 Diagrammatic representations of reaction limited and transport limited situations during dissolution (see text for explanation).

transport-limited situation need not necessarily be produced by very low flow rates: reaction rates could simply be extremely high. Very high acidity may provide a supply of products at a high rate (Fig. 3.19(c)). Similarly, the rate-limited situation could be due to low acidity as much as to rapid flow (Fig. 3.19(d)).

Under transport-limited conditions (high acidity and/or low flow) all mineral constituents tend to dissolve since acidity is sufficient to dissolve to equilibrium and the low flow permits adequate solid-solvent contact time for equilibrium to tend to occur. This situation tends to produce smooth surfaces (chemical polish, see section 3.2.2 Fig. 3.13) with all mineral constituents dissolved equally. By contrast, differential erosion is encouraged in reaction rate-limited situations (low acidity and/or rapid transport) when only the most soluble (or rapidly soluble) mineral constituents dissolve, saturation having been reached or the movement of the solvent occurring before the lower or slower solubility constituents have time to dissolve. In most cases rapid flow is of less importance than low acidity.

This explains why some limestone surfaces are etched differentially (low acidity and/or short solvent–solid contact times) and some dissolved equally and smoothly (higher acidity and/or longer contact times). This topic is discussed further in Chapter 4.

3.3.2 Soil water percolation and leaching

A discussion of the soil leaching process is important to an understanding of the way in which soil water weathering potential may change in the short term. Clearly, solutional erosion of the underlying bedrock can proceed only when water drains from the soil under gravity. It is at this time that leaching also takes place and it is the downward removal of metals and metal-organic compounds in solution which determines the weathering potential of the soil in contact with the limestone.

The first factor which should be considered is the nature of the rainfall input. Anions, such as chloride and sulphate are present in rainwater, as well as atmospheric CO_2. Pollutants, such as sulphuric acid, may also acidify rain. The anions may combine with cations in the soil and bring them into solution. A second important factor is that rainfall often has 2 to 5 mg l^{-1} of calcium already present in it as it falls (Gorham, 1955,

1957, 1961; Stevenson, 1968). In limestone areas (where dust from lanes and quarries may be present in the air) or in areas near the sea (where spray of 800 to 900 mg l^{-1} $CaCO_3$ may be blown into the air) the rain may contain up to 40 mg l^{-1} $CaCO_3$ when it falls. Such concentrations will have to be subtracted from calculations of net calcium removal from the soil.

The effect of organic matter in enhancing leaching was demonstrated by the experiments of Turner *et al.* (1958) and Kerpen and Scharpenseel (1967). More calcium was lost from calcareous soils when organic matter was incorporated in the experimental leaching columns. Groom and Ede (1972) describe experiments in which over 300 mg l^{-1} $CaCO_3$ was lost by leaching of natural calcareous soils. Similarly, the litter layer at the surface of the soil has an important role in acidifying water entering the profile. Handley (1954) found that the pH of aqueous extracts of the litter of various plants was almost always acid. The presence of *Calluna* on limestone is again of interest for the marked acidity of its litter, the aqueous extract of which gave a pH of 3.4 when freshly fallen and 4.2 when decomposed. Thus, the effect of the litter and humified organic matter is to acidify water as it enters the soil.

Leaching has been studied by Trudgill (1976d) in thin soils over limestone in the Mendip Hills, England, and County Clare, Eire. Temporal variations in pH at various depths were used as an indicator of the leaching process. Figures 3.20 and 3.21 show results from two typical profiles in similar *Calluna* peat soils, one over siliceous Devonian sandstone the

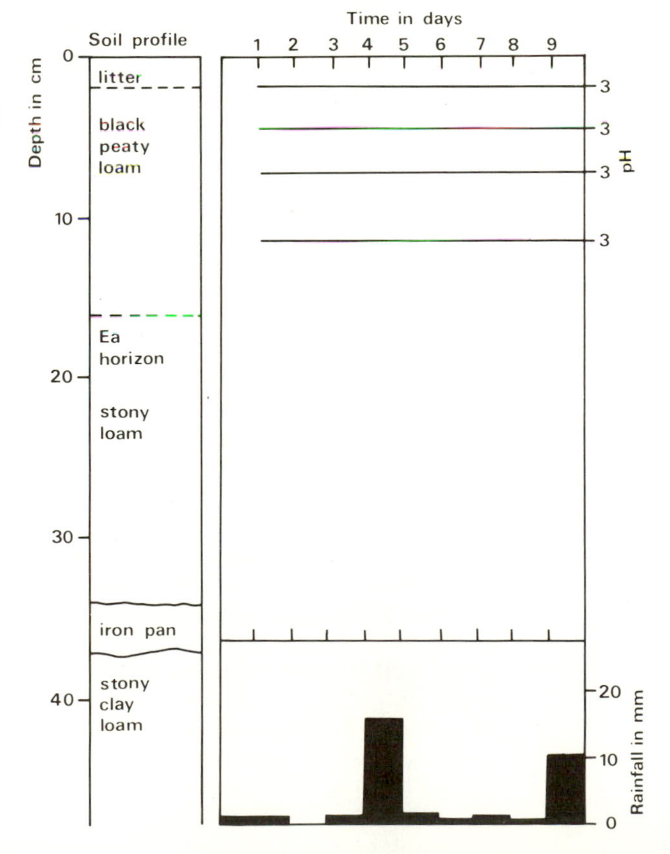

Fig. 3.20 Daily measurements of pH in a peaty soil with *Calluna* over sandstone during and after rainfall. pH values are constant throughout rainfall events (modified from Trudgill, 1976d).

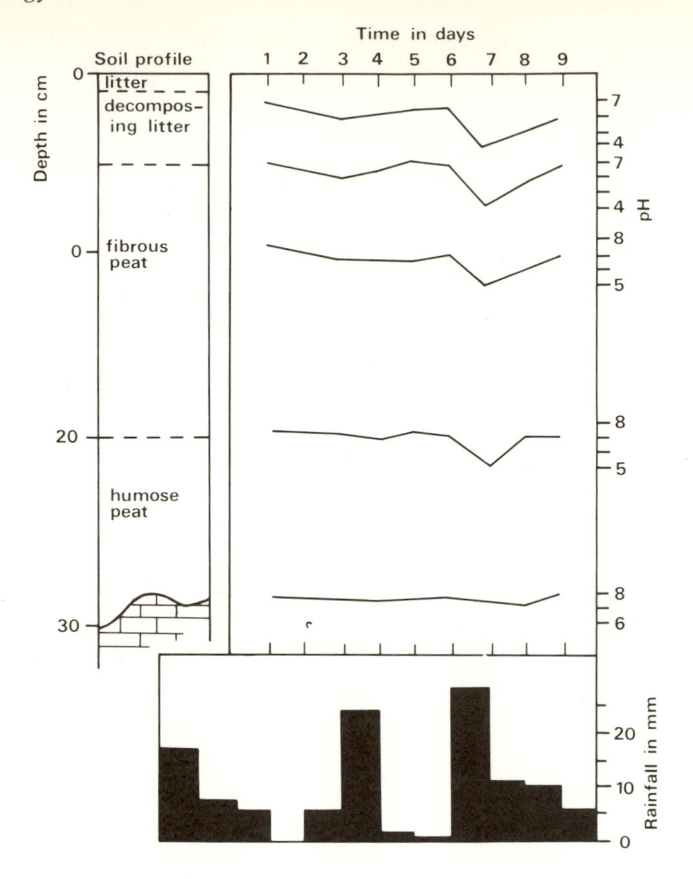

Fig. 3.21 Variations in the pH of a peaty soil over limestone during and after rainfall. Marked decreases are seen in the surface soil after the second rainfall event (modified from Trudgill, 1976d).

other over Carboniferous Limestone. The extent of variation in pH at a given depth depends upon the amount of rainfall in the preceding 24 hours. The fluctuation is greatest where *Calluna vulgaris* occurs over limestone and less for other plants. Over sandstone there are no fluctuations, indicating that the variations in pH reflect a limestone dissolution and leaching process, and not simply a wetting and drying process. The greatest daily variation at each of six sites is plotted against the preceding rainfall in Fig. 3.22 which shows a small rainfall can produce a very much more marked change in peat beneath *Calluna* than in humus or peat without *Calluna*. There is little change in pH at sites on sandstone.

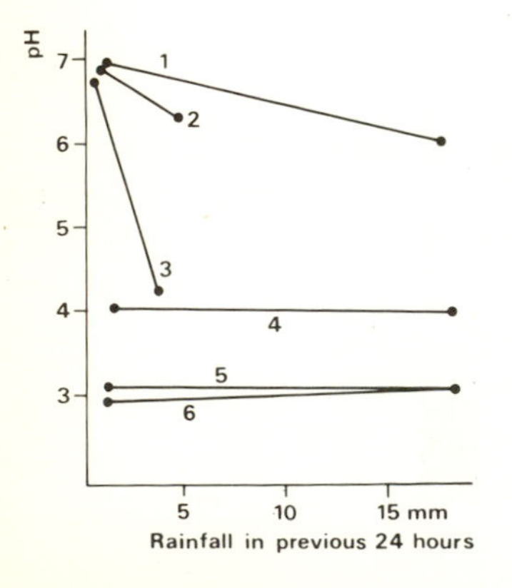

Fig. 3.22 Changes in soil pH following 20 mm rainfall for six sites in County Clare, Eire. 1, Humus rendzina over limestone. 2, peat ranker over limestone. 3, *Calluna* on peat ranker over limestone. 4, Thin humus over sandstone. 5, Peat over sandstone. 6, *Calluna* on peat over sandstone.

The leaching process represented by these results is illustrated diagrammatically in Fig. 3.23. Wetting at first causes a rise in pH because of the suspension effect. As the soil becomes saturated anaerobic conditions may occur, favouring the build-up of carbon dioxide. Gravitational drainage following high rainfall then removes calcium in solution until the humus is acidified down to the bedrock when further dissolution occurs. On drying out, capillary rise brings calcium to the surface, alkalising the humus. It is notable that the soil does not exhibit the usual high buffer capacity of organic matter, although the ion pumping mechanism of *Calluna* may be an important cause of the larger changes in pH, which are much smaller when *Calluna* is absent. These results are supported by those of Baver (1927) who describes a fall in pH of a base-rich soil after heavy rainfall in twelve cases out of eighteen.

Samples of gravitational water from soils in the Mendip Hills have

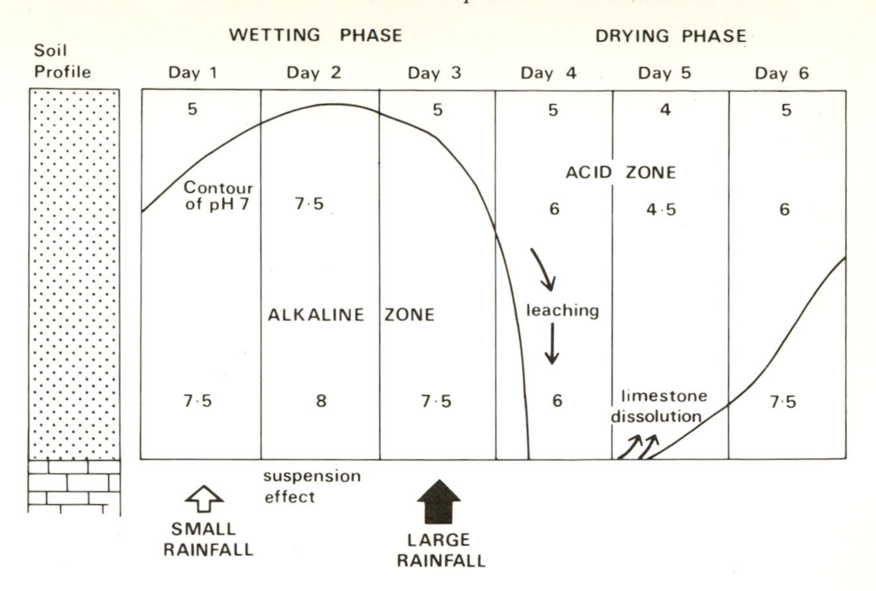

Fig. 3.23 A generalised model of the leaching process in high rainfall areas and in peaty soils (humus rendzinas and rankers) over limestone. The numbers on the diagram refer to the pH of the soil at the point shown and the contour of pH 7 has been sketched in.

been collected by Newson (1970) and Atkinson (1971).. The method of collection was to insert a polythene tube, perforated at the required depth and stoppered at its lower end, into a vertical auger hole in the soil. Gravitational water collects at the bottom of the tube and can be pumped out at intervals.

Table 3.4 summarises the results of some of this work. It can be seen that in general waters from soils with a solid calcium carbonate content of 10% or greater have high values of $CaCO_3$ in soil water and little or no weathering potential towards limestone. Only those soils containing little or no solid carbonates gave rise to waters with weathering potential.

In terms of calcium uptake in soils Bradfield (1941, p. 12) shows that half of the calcium in a clay soil studied was held in calcium-clay adsorption complexes. Kononova (1966) and Waksman (1931) state that up to half of the calcium in soil water may be chelated with organic matter or be present as organic acid salts. Moreover, many soil waters are yellowish and discoloured, indicating the presence of organic matter in solution or fine suspension.

From these results it can be suggested that rainwater percolating through soil has an acidifying effect on calcareous soils, especially in the upper horizons. Carbonate-rich waters are carried down the soil profile,

Table 3.4 The hardness of soil water samples, Mendip Hills, Somerset (modified from Newson, 1970; Atkinson, 1971)
Soil series and carbonate analyses from Findlay, 1965

Soil type	*Carbonate in soil at depths given*	*mg l^{-1}* $CaCO_3$ *in water*	*Saturated* mg l^{-1} $CaCO_3$*	Potential mg l^{-1} $CaCO_3$*
Ellick Series	11% 30 cm	150 – 200	150 – 200	None
(Brown Earth)	10% 60 cm	180 – 250	180 – 250	None
Nordrach Series (Brown Earth)	trace* 23, 60 and 75 cm	53	195	142
Lulsgate Series (Brown Earth)	trace* 23 cm	44	183	139
Nordrach Series (Stony phase)	80% –	265	270	5
Brown calcareous soil	25% –	60	65	5

* From saturation measurements using powdered calcite (see section 2.5.5)

much of the calcium being combined in organic complexes. In calcareous soils, solutes are derived from dissolution and leaching from within the soil profile; in acid soils the percolating soil waters remain acid till they contact the bedrock, where dissolution then occurs (see Ch. 4 p. 57).

3.3.3 Regional run-off regimes

Water flow rates can be seen to be important at small scale levels of considerations, on bare surfaces and in soils. It can also be proposed that flow rates and flow volumes are important at a regional scale. Erosion rates in $m^3\ km^2\ a^{-1}$ were plotted by Atkinson and Smith (1976) against mean annual run-off in mm a^{-1} (Fig. 3.24). Run-off data are used rather than rainfall because this allows for evapotranspiration losses, which can be considerable in tropical zones; despite higher rainfall, run-off may be lower in tropical zones than temperate ones. The erosion rates used were calculated using the formula proposed by Corbel (1959):

$$\text{Erosion rate } (m^3\ km^2\ a^{-1}) = \frac{(P - E).\bar{H}}{1000.s} \qquad [3.10]$$

where
P = mean annual precipitation in mm
E = mean annual evapotranspiration in mm
$\bar{H}$ = mean $CaCo_3$ content of water samples*, mg l^{-1}
s = specific gravity of the rock

(*This is stated by Atkinson and Smith to be of at least 50 samples, spread throughout the year, but is far better to be based on a solute rating curve as Gunn (1981a) has done (see also section 8.2, p. 104).

Clearly, there is a relationship between erosion rate and run-off, partly through autocorrelation since the erosion rate calculation involves run-off (P – E), but fundamentally because it appears that solute supplies are not readily exhausted and thus higher flow simply means that more solutes can be removed. Exhaustion does appear to occur in arctic or alpine regions, however, where there is little soil cover, dissolution potential is lacking and solvent-solid contact times may also be short.

In summary, water-flow rates at all levels influence dissolution processes; where solute supplies and dissolution potential are limited, slower flow may enhance dissolution; with high dissolution potential and ample solute supplies, increased flow rate and volume act to increase solutional removal.

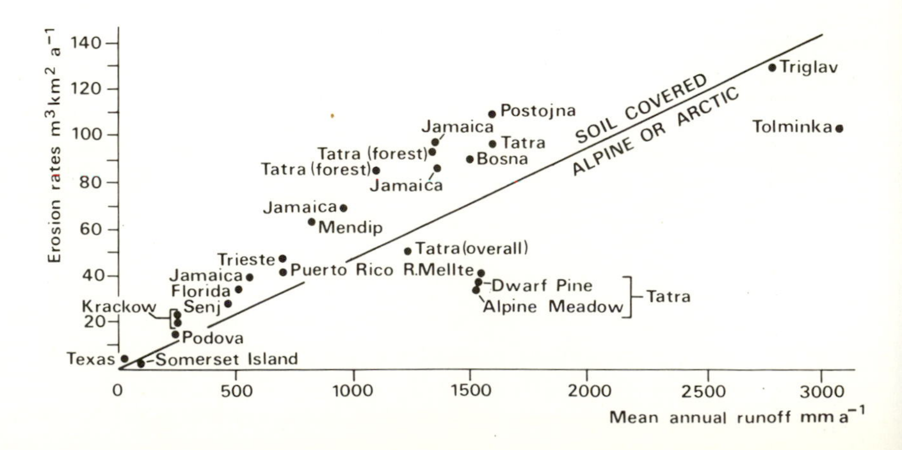

Fig. 3.24 Data for limestone erosion rates as a function of run-off (modified from Atkinson and Smith, 1976).

3.4 Topography

3.4.1 Downslope changes

Physical and chemical processes combine on hillslopes to make soils more alkaline at the base of slopes. Mechanical processes such as soil creep and slope wash move rock and soil material to the base of the slope, often leading to an accumulation of limestone rubble there. In addition, slope processes in past periglacial regimes have moved considerable amounts of calcareous material to the base of slopes in areas which are now temperate. These slope processes are illustrated in Fig. 3.25. While much of the water in limestone areas moves down through the bedrock (Fig. 3.26), any water that moves laterally through the soil will reinforce the mechanical processes, leading to the accumulation of chemical weathering products at the base of the slope. On the top of the slope vertical water movement predominates, leading to leaching of soils, while downslope mechanical movement helps to mix the soils and bring back to the surface carbonates which have been leached by infiltrating water (Fig. 3.27). This process is assisted by the larger soil organisms such as moles and earthworms which tend to bring soil material back up to the surface (Curtis, Courtney and Trudgill, 1976, Ch. 6).

These processes combine to produce slopes which tend to be acid upslope, but where water flow is diffuse, and alkaline downslope, but where water flow is more marked. Recent work (Crabtree, 1981) (Fig. 3.28) has suggested that the net effect of these processes is that erosion by solution processes is more marked upslope, soil acidity being the key factor. This implies that water flow at the top of slopes is sufficient to remove weathering products and that while water flow is liable to be much greater downslope, the fact that the water is alkaline means that in net effect weathering is limited here. The weathering products are derived in the soil profile at a level where leached acid soil contacts

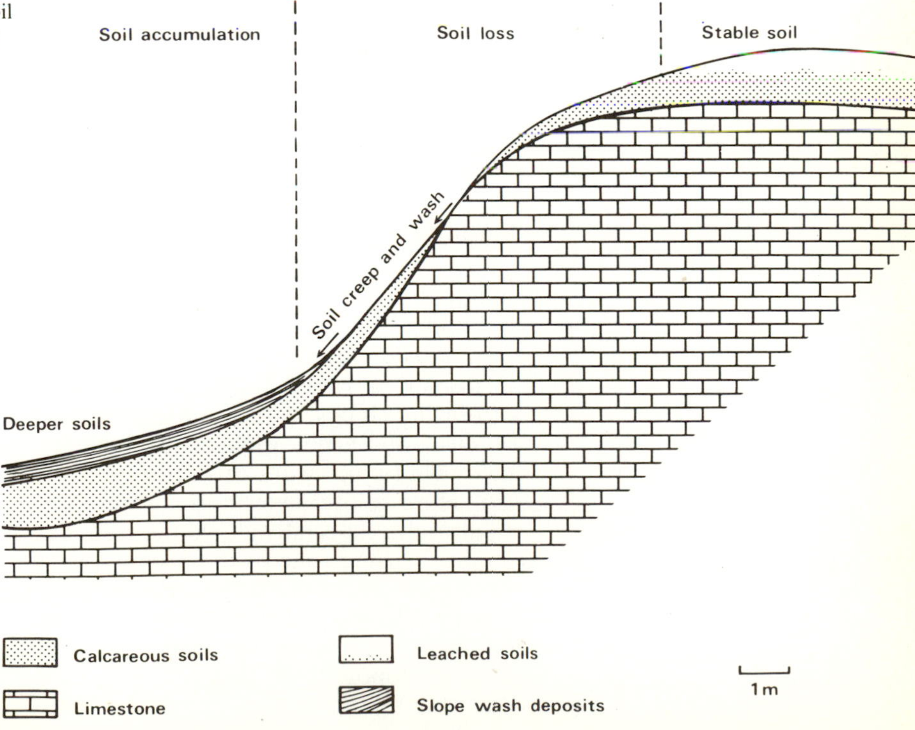

Fig. 3.25 Effect of slope on soil stability (from Curtis, Courtney and Trudgill, 1976).

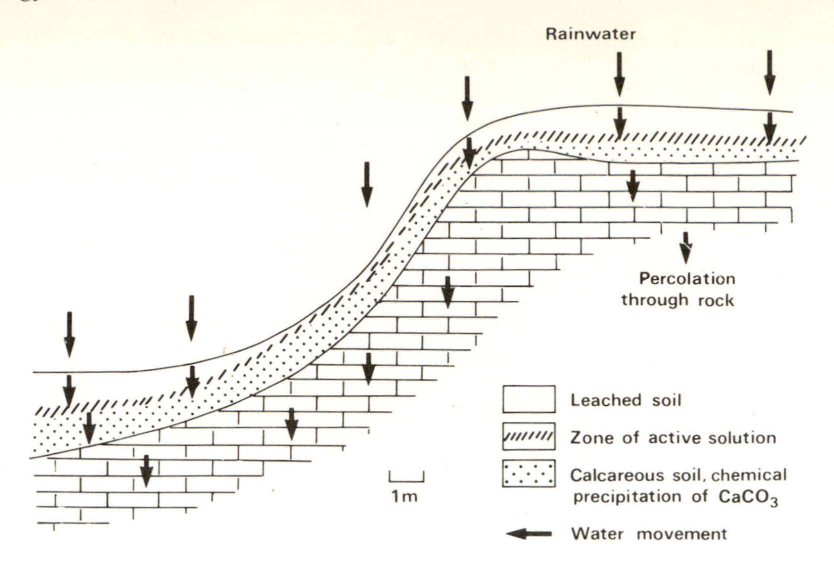

Fig. 3.26 Water movement on a limestone slope (modified from Curtis, Courtney and Trudgill, 1976).

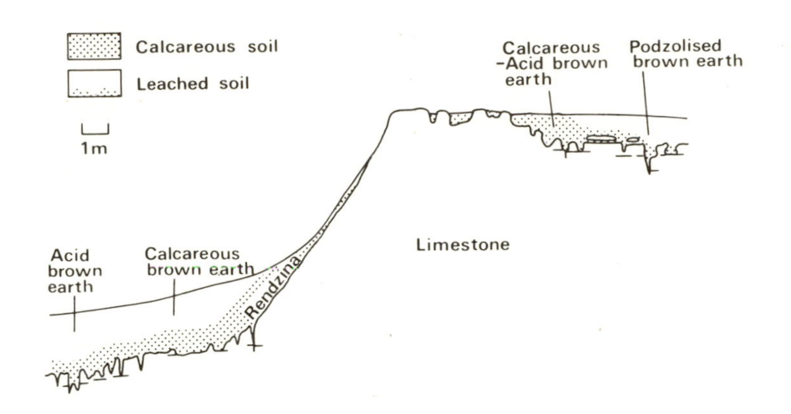

Fig. 3.27 Soils on a limestone slope (modified from Curtis, Courtney and Trudgill, 1976).

calcareous material. This position will vary from situation to situation, but the implication is that slope evolution proceeds by the preferential erosion of rock under the most acid soils.

3.4.2 Altitude and aspect

Topographic factors can also influence micro-climatic factors and therefore biological processes, notably organism respiration and therefore soil carbon dioxide levels. It can be expected that on colder slopes receiving low insolation (north-facing slopes in the northern hemisphere and south-facing slopes in the southern hemisphere) and at higher, colder, altitudes, soil carbon dioxide productivity levels will be lower. This is well shown by the work of Ford (1971a) in the Canadian Rockies (Fig. 3.29). Here solute levels and carbon dioxide levels decrease markedly with altitude, especially above the treeline. In addition, a similar pattern is shown by limestone in the Himalayas (Fig. 3.30).

Extensive micro-climatic work has been reported by Jakucs (1977) in Hungary who showed that soil temperatures varied markedly within dolines, hence soil moisture, and carbon dioxide levels will vary across such microtopographic features, influencing limestone dissolution processes (Fig. 3.31).

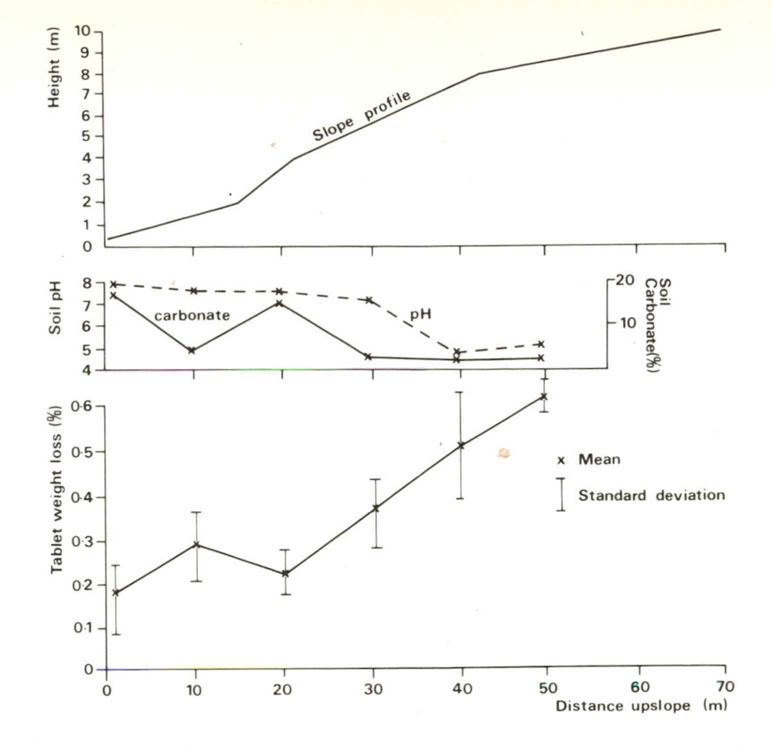

Fig. 3.28 Erosion patterns on a limestone hillslope derived using limestone tablet weight loss techniques. (Trudgill, 1975, 1983a). Note the increase in erosional weight loss upslope (modified from Crabtree, 1981).

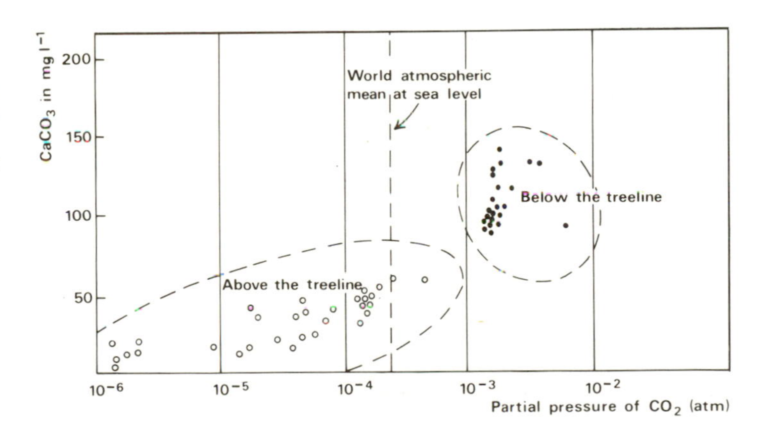

Fig. 3.29 Carbon dioxide and solute concentrations above and below the treeline in the Canadian Rockies. CO_2 and solute levels are lower at higher, colder altitudes above the tree line (from Atkinson and Smith, 1976).

3.4.3 The distribution of solution processes in the landscape

The factors influencing carbon dioxide production, soil characteristics and run-off combine to vary the distribution of limestone-solution processes in the landscape. Both vertical and spatial variations can be seen, with the distribution of the erosion varying within the soil profile and bedrock profile and also over topographic forms such as hollows and spurs. The distribution of limestone erosion processes in the landscape have been summarised by Atkinson and Smith (1976) and their data are shown in Table 3.5. Atkinson and Smith have produced a diagram of the distribution of limestone erosion based on data for run-off and solute loads (Fig. 3.32). All the data point to the importance of dissolution in the zone of the soil, including the lower soil profile and the upper bedrock. Beckinsale (1972) has emphasised this point, in that calculations of overall limestone erosion rates (section 3.3.3) do not reveal

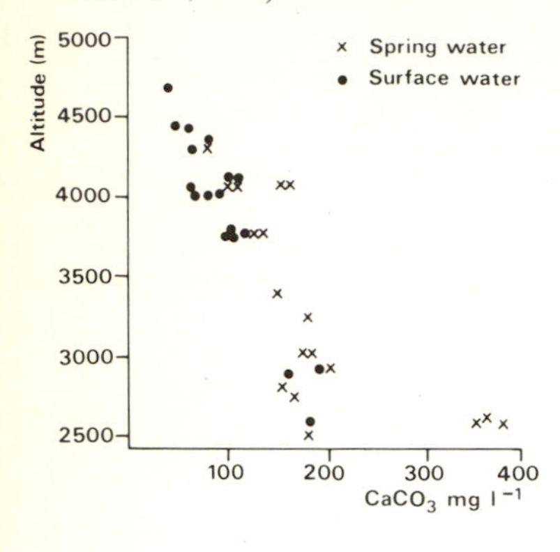

Fig. 3.30 Decreasing solute concentrations with higher altitude, Himalayas (modified from Waltham, 1971).

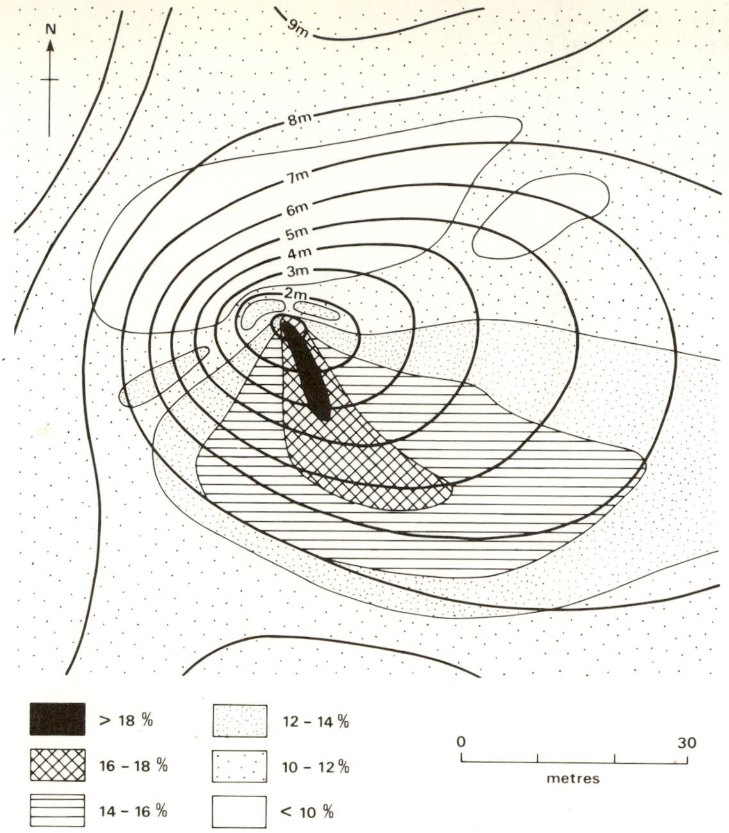

Fig. 3.31 Topographic influences on soil conditions in a doline with uniform, ploughed soil cover: per cent values refer to soil moisture by weight. Isolines are 1 m contours above the doline base. Note both the influence of topography (drainage) and aspect on soil moisture (modified from Jakucs, 1977).

Table 3.5 Rates and location of limestone erosion (modified from Atkinson and Smith, 1976)

Area	*Overall rate* m^3 $km^2 . a^{-1}$	*Location*
Fergus R., Ireland	55	60% at surface, up to 80% in the top 8 m
Derbyshire	83	Mostly at surface
North-west Yorkshire	83	50% at surface
Jura Mountains	98	58% at surface, 37% in percolation zone, 5% in conduits
Cooleman Plains, NSW, Australia	24	75% from surface and percolation zone, 20% from conduit and river channels, 5% from covered karst
Somerset Island, NWT, Canada	2	100% above permafrost layer

the disposition of limestone erosion within the landscape and they provide little information on landform evolution. Indeed data on which portions of the landscape are being eroded more or less than others, i.e. the spatial distribution of erosion rates, are rare. Until attention is focused upon this topic, little progress can be made in the detailed understanding of differential landscape evolution.

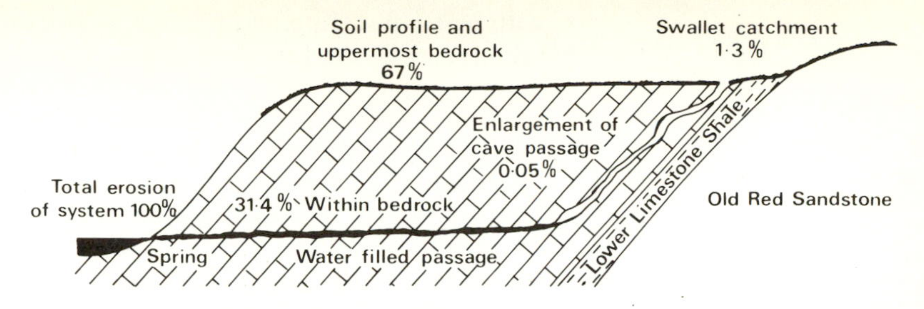

Fig. 3.32 Distribution of erosion rates in a limestone landscape, based on the Mendip Hills, Somerset, UK (modified from Atkinson and Smith, 1976).

3.5 Global variations

3.5.1 Climatic geomorphology and solution processes

In seeking to provide explanations for landforms in different climates, climatic geomorphologists have looked to climatically influenced variations in soil carbon dioxide levels. Williams (1949) suggested that carbon dioxide is more soluble at low temperatures; thus limestone dissolution could be greater in cold regions. This is however an erroneous proposition as biological activity is low and soil covers are thin in cold regions. Harmon *et al.* (1972) noted little climatic consistency in water calcium content in the USA, although there was a marked variation of water temperature with latitude. Smith and Atkinson (1976) have compiled water hardness values and erosion rates for climatic zones, noting an overlap in all the data (Figs 3.33, 3.34). There are thus no simple climatic trends to be seen in the data for solute concentrations or erosion rates (this topic is discussed further in Ch. 8).

Fig. 3.33 Solute concentrations (calcium carbonate): (a) tropical; (b) temperate; (c) arctic-alpine (from Smith & Atkinson, 1976).

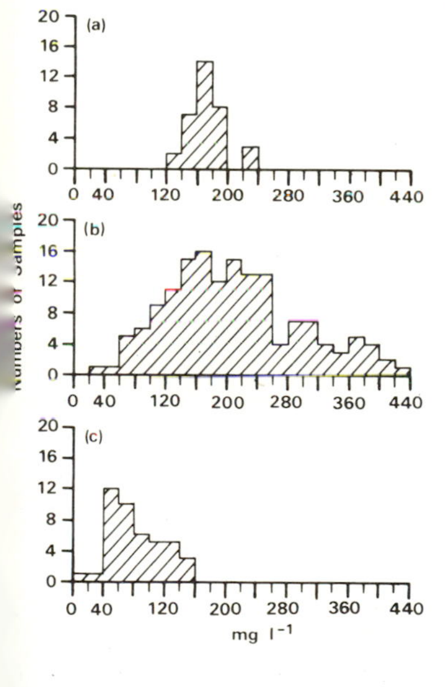

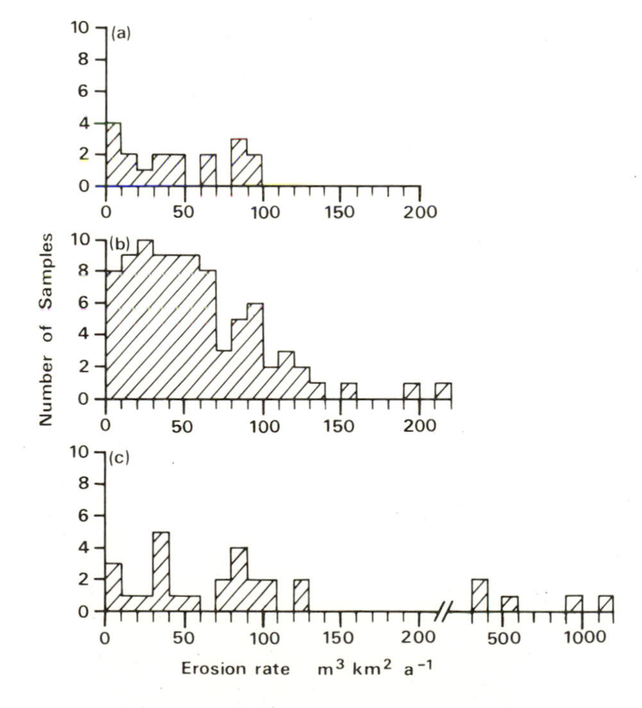

Fig. 3.34 Erosion rates in contrasting climates: (a) tropical; (b) temperate; (c) arctic-alpine (from Smith & Atkinson, 1976).

3.5.2 Conclusions: a balance of factors

Any proposition that seeks to provide explanations of variations in limestone landforms in simplistic terms, such as the study of single factors like soil carbon dioxide levels, is bound to be inadequate in what is a multivariate situation. Specifically, dissolution potential in combination with water-flow rate and volume will provide the chief control upon solution processes in limestone areas. Combining this with lithological variations in rates of uplift and length of time needed to produce features at various scales will provide tremendous variations of combinations of situations and balances of factors.

However, general propositions can be put forward. First, a high volume of run-off will increase limestone dissolution rates. Second, high soil acidity and high carbon dioxide productivity will also increase dissolution rates in permeable soils.

It is the balance of these two factors, dissolution potential and water flow, which controls the overall limestone erosion rates. The investigation of the spatial distribution of erosion rates in relation to historical legacy of landforms, time allowed for adjustment and present processes, still await full attention.

4 Limestone pavements and karren forms

4.1 Introduction

Limestone pavements are areas of bare rock dissected by opened joints (Fig. 4.1) and 'karren' forms are subaerial ones in the shape of pinnacles, small hollows and other morphologies characteristically regarded as being produced by the dissolution of limestones. 'Karren' is the German term for such features and 'lapiés' a French descriptive term, but they are both now incorporated into the general geomorphological literature.

Limestone pavements have been the focus of much attention by geomorphologists (for example, Williams, 1966) largely because of their

Fig. 4.1 Limestone pavement on near-horizontal glaciated bedding plane surface, County Clare, Eire. The block structure (clints) of the pavement along opened joint lines (grikes) is extremely marked. (Photo: S. T. Trudgill.)

distinctive features, thought to be a characteristic of a limestone bedrock. However, they are in essence glacially scoured rock platforms such as can be found on other rocks as sandstones and igneous rocks. What gives them their distinctive appearance is first of all the fact that they are developed on a rock where distinct bedding planes are present. This has meant that glacial erosion has been able to remove pre-existing weathered material down to a well-marked bedding plane, revealing a flat, uniform surface. Second, limestone pavements are distinct from many other rock pavements because of the jointed nature of the limestone. This is especially true of the Carboniferous Limestone in Britain (as shown in Fig. 4.1). The joints act as lines of weakness which have been exploited by preglacial, glacial or postglacial weathering and erosion, leaving deep clefts between the tabular blocks of limestone. Third, a distinctive feature of limestone pavements is that they are often free of soil and are highly visible in the landscape. This contrasts with some other glacially scoured rock platforms which are frequently covered with drift and soils. This has given rise to the speculation that many limestone pavements were postglacially covered more extensively with drifts and soils and that these covers have been subsequently eroded. This erosion could have been by surface wash induced by the felling of trees by man in order to utilise the calcareous soils for pasture or by the subsurface abstraction of soil down progressively widening joints. Dissolution of limestones and the opening of joints will obviously be enhanced under acid soils, but this can result in the lowering of the soil surface down the developing cleft, possibly also with soil loss into near-surface cave systems (Fig. 4.2). Also, in contrast to the case of sandstone, weathering of a pure, mostly soluble rock leaves little residue, and it can therefore be argued that little soil could develop on limestone in terms of accumulated weathering residue. Thus, an alternative view is that soil covers may never have formed postglacially. Controversy thus exists about the role of a soil cover in limestone pavement evolution and the evidence for the various possible roles are discussed in section 4.3.

Fig. 4.2 Soil loss down solutionally opened joints in limestone, sequence (a) to (d) showing progressive soil loss. Soil surface lowering may simply occur because of solutional loss of the bedrock, with no actual loss of soil material; alternatively, physical erosion may occur, with loss to subsurface cave systems.

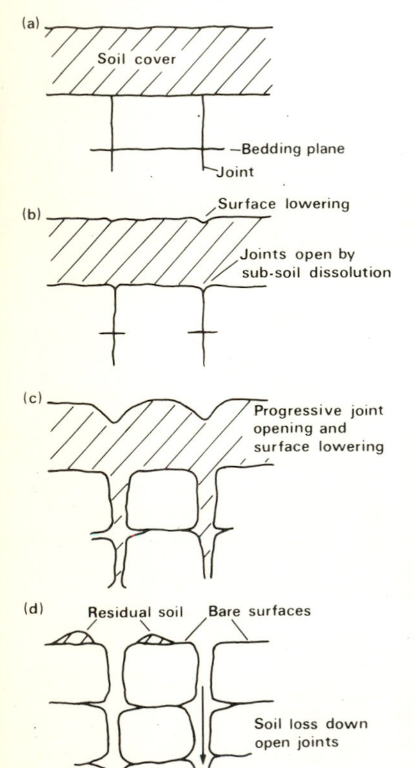

While the production of karren is a result of the contact of water charged with carbon dioxide and organic acids causing the hydrolysis of calcite, the drainage features of the rock surface are the most important factors influencing the evolution of the surface morphology. On bare rock surfaces rapid run-off occurs on sloping surfaces but water will be resident on the surface much longer on flat surfaces or in basins. Thus, on sloping surfaces, forms will be related to dissolution kinetics and, with rapid transport, the overall chemical reactions will be rate limited. On flat surfaces, in basins and under soils, where water–rock contact will be prolonged, the reactions will be transport limited and the features produced will be interpretable in terms of equilibrium rather than kinetic processes. Soil moisture distribution, drainage rates and water-flow routes will have a strong influence on the distribution of subsoil solutional erosion and landforms.

On bare, sloping surfaces rills often occur, especially under intense rainfall regimes. These rills are similar to, but not identical with, rills forming on unvegetated loose earth surfaces such as gullies in semi-arid areas or rills on mining spoil heaps. Their distribution relates to the way that rainwater falling on to a sloping surface becomes organised into thick and thin layers due to surface tension effects, with preferential flow paths forming on the surface. The spacing of these ribbons of water relates to the discharge of water and the slope angle (Woo and Brater, 1962). With unconsolidated material, rills and gullies form in response to erosion caused by greater turbulence and entrainment of sediment in the ribbons of water. The existence of water ribbons on a rock surface formed by the coalescence of sheets of water means that there will be a jump from laminar flow in the sheets to turbulent flow in the ribbons and a corresponding facility for the removal of dissolution products in the ribbons of water by disruption of the surface diffusion layers. This disruption is also effected by rain splash, as discussed more fully below. Where dissolution is rate-limited, this enhanced removal of dissolution products will increase the rate of solid to solvent transfer of ions, facilitating the deepening of the turbulent flow path forming a solutional channel or rill. It is thought that an equilibrium process–form relationship may be established between rainfall characteristics, slope and rill depth on the soluble rock.

Small channels and rills are not confined to limestone rocks but also occur on other soluble rocks, such as basalt. The German term 'rillenkarren' has become applied to the form on limestone (Fig. 4.3). Dissolution processes under soils and on bare rock surfaces are discussed in section 4.2.

4.2 Morphology and dissolution

The occurrence of differentially eroded surfaces can only be explained in terms of an unequal distribution of erosion rates in relation to either lithological inhomogeneities or spatially unequal process distribution. Karren forms are so widely distributed across a range of microlithological differences that it is difficult to provide consistent explanations for them in terms of lithological variations. However, several process patterns can be modelled in order to provide explanations for major differences in karren forms, though some of the reasons for detailed differences are far from clear. On bare surfaces three major types of karren are formed: **rillenkarren** (Fig. 4.3), (Fig. 4.4), **trittkarren** which are smooth surfaces with small steps, and solution basins termed **'kamenitza'** by Bögli (1960) (Fig. 4.5). Under drifts and soils, smooth glacial surfaces may be found, frequently with striations preserved;

Fig. 4.3 Sharp-edged rillenkarren forms at Wee Jasper, near Canberra, ACT, Australia. The grooves are approximately 1–2 cm across and up to 1–2 cm deep. (Photo: S. T. Trudgill.)

Fig. 4.4 Rillenkarren in Chillagoe karst, North Queensland, Australia. Note that most surfaces are covered with vertical rills. (Photo: S. T. Trudgill.)

Fig. 4.5 Solution basins, termed 'kamenitza' on a limestone surface, County Clare, Eire. Lens cap 55 mm diameter. (Photo: J. D. Hansom.)

Fig. 4.6 Limestone surface beneath acid till, County Clare, Eire. Note the deep incision (the 1 m long auger is placed down a soil-filled joint) and sharp-edged solutionally etched forms revealed by excavation (Photo: S. T. Trudgill.)

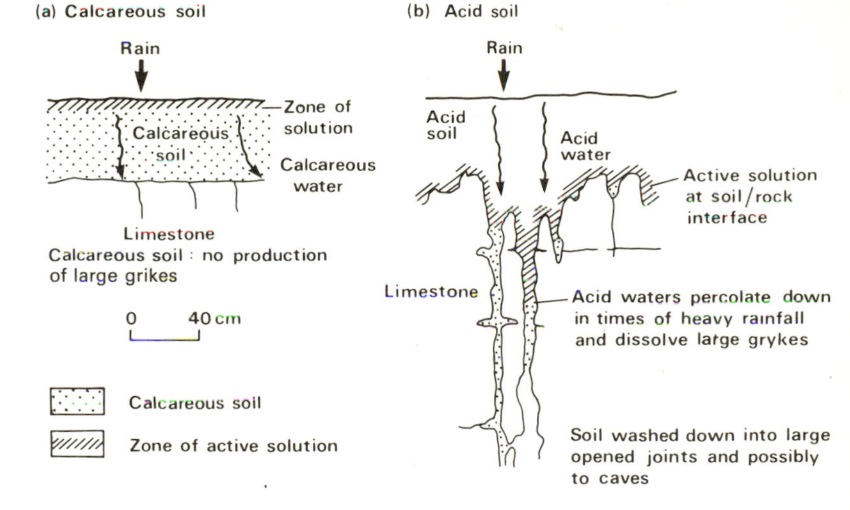

Fig. 4.7 Diagrammatic representation of subsoil weathering under different soil types. (a) Calcareous soil protects the bedrock beneath, with no opening of joints. (b) Under acid soil, subsoil limestone is extensively weathered, leading to the formation of solutionally opened joints, often termed grikes (from Curtis, Courtney and Trudgill, 1976).

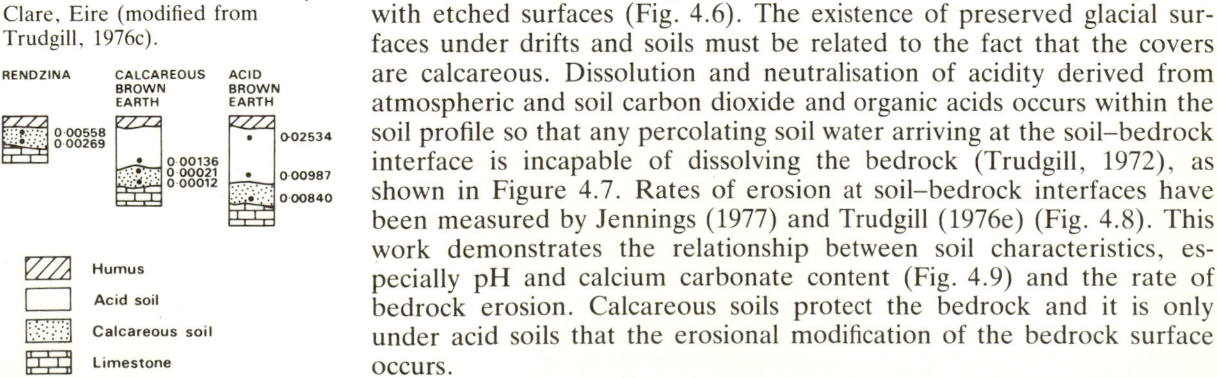

Fig. 4.8 Rates of subsoil erosion measured using a limestone tablet weight loss technique (Trudgill, 1975). Results in mm a^{-1}, showing that rates are low in calcareous material; limestone soils, County Clare, Eire (modified from Trudgill, 1976c).

rounded, runnelled and pinnacled forms are also common, frequently with etched surfaces (Fig. 4.6). The existence of preserved glacial surfaces under drifts and soils must be related to the fact that the covers are calcareous. Dissolution and neutralisation of acidity derived from atmospheric and soil carbon dioxide and organic acids occurs within the soil profile so that any percolating soil water arriving at the soil–bedrock interface is incapable of dissolving the bedrock (Trudgill, 1972), as shown in Figure 4.7. Rates of erosion at soil–bedrock interfaces have been measured by Jennings (1977) and Trudgill (1976e) (Fig. 4.8). This work demonstrates the relationship between soil characteristics, especially pH and calcium carbonate content (Fig. 4.9) and the rate of bedrock erosion. Calcareous soils protect the bedrock and it is only under acid soils that the erosional modification of the bedrock surface occurs.

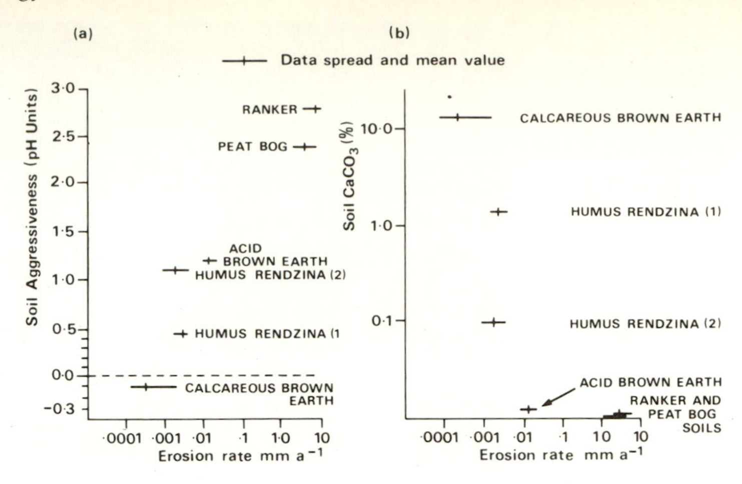

Fig. 4.9 Rates of erosion for limestone bedrock surfaces under a variety of soils, data from tablet weight loss technique. (a) Weight loss data show an increase from calcareous soils to acid soils; 'aggressiveness' units are the increases or decreases in soil pH when powdered calcium carbonate is added (see section 2.5.5, p. 17). (b) Weight loss decrease with increasing soil calcium carbonate content. Limestone soils, County Clare, Eire; ranker is an acid humus soil occurring over limestone in high rainfall areas; two samples of rendzina (alkaline humus soils over limestone) were used (modified from Trudgill, 1976e).

The essential difference between erosion on surfaces covered with acid soil and those which are soil free, is that the former suffers a corrosive soaking, while the latter suffers only episodic dissolution events under the action of rainwater. Not only is the action episodic, rainwater has considerably less dissolution potential than acid soil water (see Table 3.3, p. 34). Under the soil, surfaces can be attacked from many directions, whereas the action on subaerial surfaces may become more disparate (Fig. 4.10). Thus, under conditions of rapid, transient flow on shedding surfaces where water depths are limited, only those constituents of the rock which are rapidly soluble will be removed. This may give rise to a dissected surface (termed **'spitskarren'** by Bögli, 1960) if micro-inhomogeneities exist, such as between dissolution rates of grains and cements (Fig. 4.11) or in relation to preferential attack by algae (Folk *et al.*, 1973). Spitskarren can be found in tropical environments where lithology is very heterogeneous (Fig. 4.12) but also occur on more homogeneous calcarenites. Even on these rocks, however, Trudgill (1979a) has demonstrated that differential solubility may exist (Fig. 4.13), giving rise to spitskarren surfaces (Fig. 4.14).

Where laminar flow exists on slopes of low angle, stepped trittkarren may be identified. It can be argued that the steps are formed by unsteady flow, with transitory turbulence occurring as depths of water flow momentarily increase or where a slight increase in slope occurs. Once the steps are initiated, they will promote the existence of turbulent flow and perpetuate the step.

Rillenkarren are associated with steeper slopes under intense rainfall, usually without a tree cover which would break the fall of raindrops, but their climatic distribution is not fully understood.

Naturally occurring declivities in rock surfaces will cause water to collect where dissolution can proceed to equilibrium. The erosion process may also be enhanced by the presence of algae, such as the blue-green alga *Nostoc*. These will increase the carbon dioxide content of the pool by respiration at night, and also extra-cellular production of organic acids may occur. During dry periods precipitation of carbonates, or organic-carbonate granules, can occur and these may be removed by deflation.

Under soils, attack on tabular blocks from three sides may give rise to a pinnacle form (Fig. 4.15). Arcuate and cuspate forms also occur under soils (Fig. 4.16). The precise process-form relationships of these

Fig. 4.10 Diagrammatic representation of the differences of dissolution distribution. (a) on bare surfaces and (b) under acid soils.

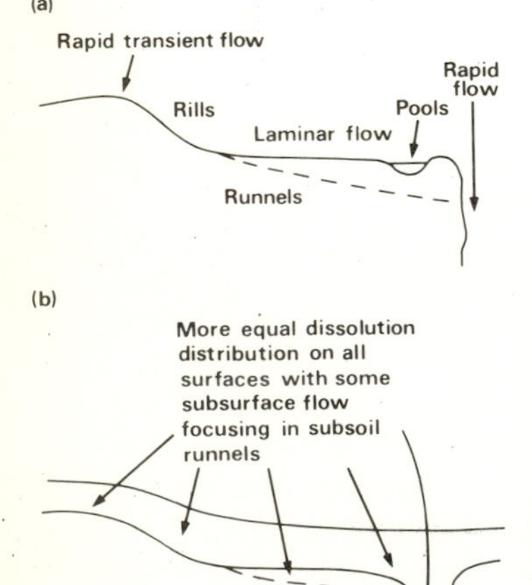

Fig. 4.11 Differential dissolution rates of rock constituents 'A' and 'B' of comparable final solubility (S) but different rates of achieving 'S' over time (T). At short solvent–solid contact, e.g. T_x, 'A' will be dissolved more than 'B', giving a dissected surface if 'A' and 'B' are juxtaposed, e.g. if 'B' are grains or fossils and 'A' is cement or matrix.

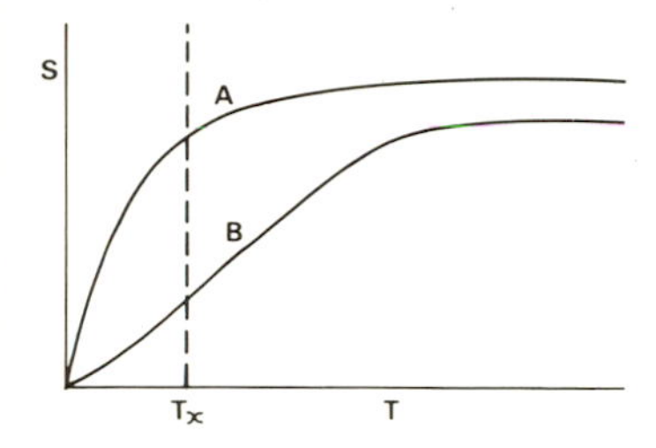

are unclear, but the existence of preferential flow paths in soils will, it can be argued, lead to the evolution of subsoil runnels.

Attempts have been made to quantify the qualitative assessments of process–form relationships outlined above. Fundamentally, water flow on inclined surfaces has been reviewed by Emmett (1970): spatially varied flow may be produced from controlled rainfall under experimental conditions (Woo and Brater, 1962) and incident rainfall has the effect of disrupting laminar flow. For a slope, Emmett describes the relationships specified by Horton (1945) where areas of turbulent flow are interspersed with areas of laminar flow. For turbulent flow the relationship holds:

$$q = KD^{5/3} \qquad [4.1]$$

where:

q = discharge
K = coefficient of run-off related to slope and surface roughness
D = depth of flow.

For laminar flow the equation becomes:

$$q = KD^3 \qquad [4.2]$$

where K is related to slope and viscosity.

These relationships can be expressed in the general form:

$$q = KD^{\mathrm{m}} \qquad [4.3]$$

where m is an exponent reflecting the degree of turbulence (5/3 for fully turbulent flow and 3 for fully laminar flow). Thus, with increased discharge, depth increases more rapidly in turbulent flow than in laminar flow. Surface roughness has a great effect on shallow flows, increasing depth of flow. Raindrop impact has a similar effect. The transition from laminar to turbulent flow is expressed by the Reynolds number (Re) which has the general form:

$$\mathrm{Re} = \frac{4\mathrm{VD}}{r} \qquad [4.4]$$

Fig. 4.12 Dissection and pointed rock surfaces (similar to Bögli's, 1960, spitskarren), dolomite limestone, Cockpit Country 1960, Spitskarren), dolomite limestone, Cockpit Country, Jamaica. (Photo: S. T. Trudgill.)

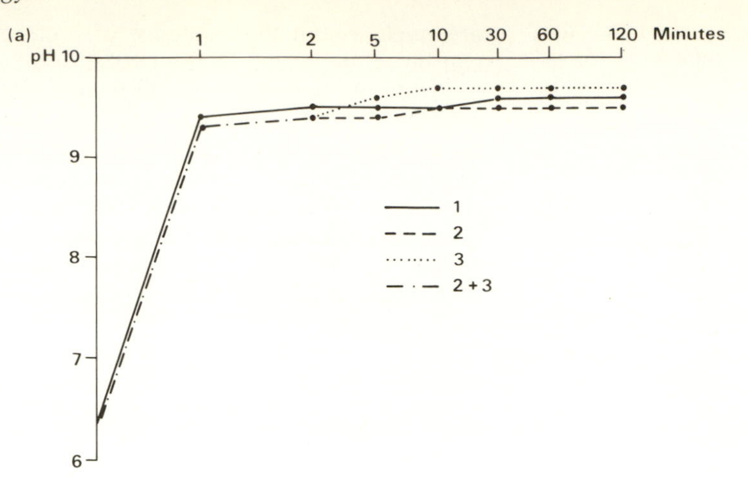

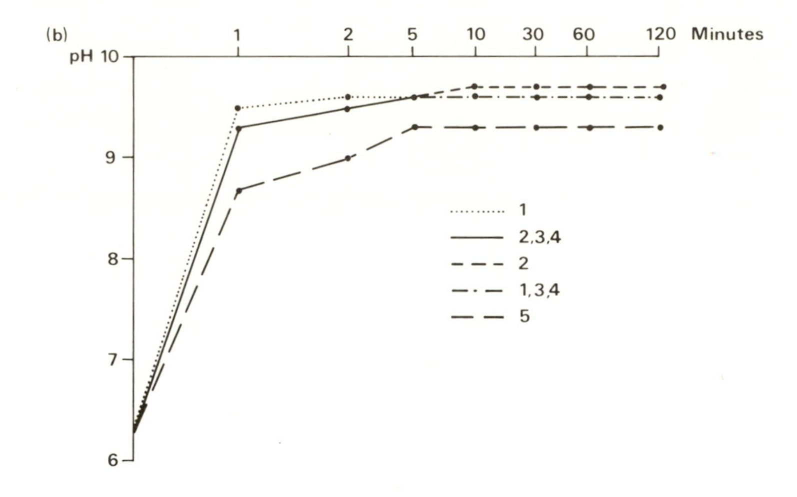

Fig. 4.13 (a) dissolution velocity of calcarenites. 1, aragonitic calcarenite. 2 and 3, high magnesium calcite calcarenites. (b) dissolution velocity of clastic components. 1, aragonite coral. 2, Platygyra coral. 3, calcite and aragonite shells. 4, *Goniastrea* coral. 5, phosphate clasts. Material from Aldabra Atoll, Indian Ocean (from Trudgill, 1979a).

where:

V = mean velocity
D = mean depth
r = viscosity

For flow on sloping surface Re has been shown to have a value of between 2500 and 12,000. Emmett (1970, p. A15) has further quantified the relationship (Fig. 4.17), with a break in the relationship between depth of uniform flow and Re between Re = 1500 and 6000. There is an increase in depth with decreasing slope from 0.003 to 0.08 gradient on a smooth surface and 0.02 to 0.08 gradient on a rough surface. The critical value of Re for a change from laminar to turbulent flow increases with increased slope, indicating that steeper slopes are somewhat more stable against changes to turbulent flow. The morphological model for run-off–sediment erosion relationships produced by Emmett from experimental data on particle entrainment is shown in Fig. 4.18. This fundamental relationship can be translated into a solutional context with reference to the work of Curl (1966) and Ford (1980) who argue that under turbulent flow the chemically saturated solid-solvent boundary layer becomes disrupted, enhancing diffusion and solutional erosion.

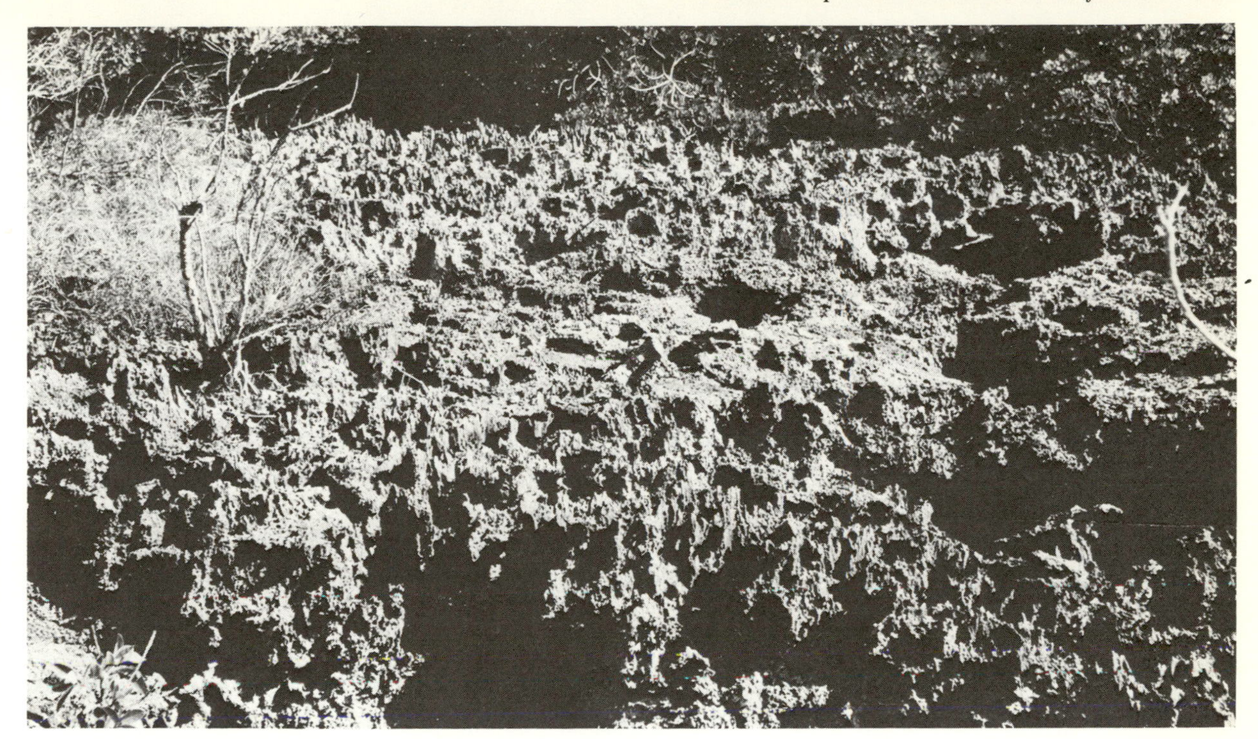

Fig. 4.14 Dissected limestone surfaces on calcarenite, Aldabra Atoll, Indian Ocean. (Photo: S. T. Trudgill.)

Fig. 4.15 Formation of a pinnacle from a rectalinear joint block, (left). Erosional attack from all sides reduces block to an upright linear form.

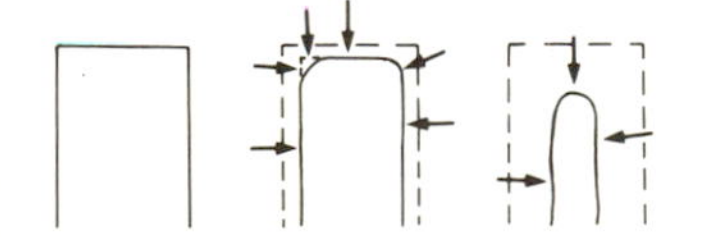

This can be understood with reference to the Nernst equation (Nernst, 1904):

$$\frac{dC}{dt} = kd\,(C_s - C) \qquad [4.5]$$

where:

$\frac{dC}{dt}$ = transfer rate of solute ions from the solid phase to the solvent.
kd = transport rate constant, related to surface area thus:

$$kd = \frac{D}{\delta} \quad \frac{As}{V} \qquad [4.6]$$

where:

D = the molecular diffusion coefficient
δ = the transport distance
V = Volume of solution which is in contact with As
As = Surface area of the dissolving mineral
Cs = concentration at saturation
C = concentration of the solute in the solvent.

As C approaches Cs, dC/dt decreases. Disruption of the boundary layer, where C tends to approach Cs, brings in new low solute concentration water and increases the differential between C and Cs, thus increasing dC/dt.

The work of Glew (1977), Glew and Ford (1980) and Ford (1980) challenges the analogy of rillenkarren formation with the overland flow model based on Horton's work and described by Emmett (1970, Fig. 4.18). In long profile, rillenkarren do not exhibit the upper convex

Fig. 4.16 Diagrammatic cross sections of subsoil limestone pavement types observed by the author after excavation and removal of the soil. (a) smooth glacially striated form where a glacially eroded surface is found preserved under a calcareous drift cover. (b) undulating surface under a thin turf mat, the surface having been produced glacially and showing some signs of etching. (c) cuspate form developed under acid mineral soils. (d) arcuate form found under wet, acid peat soils. Both (c) and (d) show marked signs of surface etching. (e) platy or lamellar form and (f) rubbly surfaces, both produced in relation to the presence of non-massive limestone lithology (e) with close bedded limestone and (f) in incoherant limestones. Observations from Country Clare, Eire and North Yorkshire, England (modified from Trudgill, 1976e).

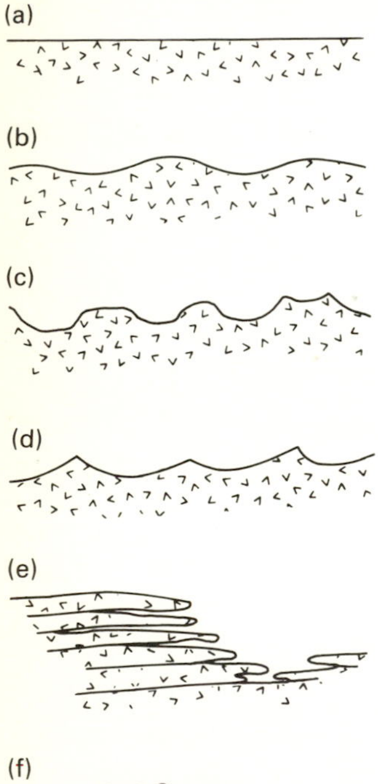

(a)

(b)

(c)

(d)

(e)

(f)

profile characteristic of laminar flow (Fig. 4.18) but are straight channels, heading at the crest of the slope, extending directly down the fall line of the slope, extinguishing downslope into a planar solution surface (Glew and Ford, 1980; Fig. 4.19).

Glew and Ford have shown that this is a characteristic form produced under the influence of raindrop impact. The rills do not have a semicircular cross section as would be characteristic of a river channel, but a quadratic parabolic cross section (Fig. 4.20) which focuses parallel forces through a point; that is, it focuses parallel raindrops at the trough base. Rillenkarren are produced where the dominant erosion process is solution by droplet impact on to an inclined surface, because this form maximises energy conservation. They extinguish downslope when water attains a critical depth which prevents focused rainsplash from pen-

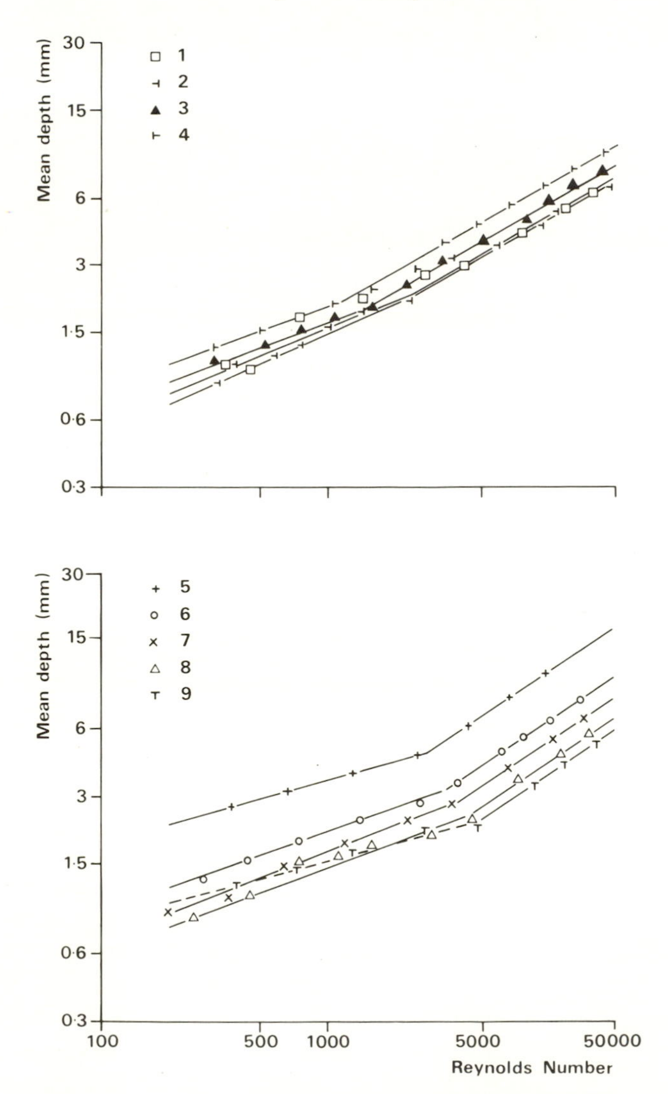

Fig. 4.17 The relationships between the depths of uniform flow and Reynolds number (Re, equation 4.4). Upper diagram: rough surface: lower: smooth surface. Slope values – 1: 0.0775, 2: 0.0550, 3: 0.0342, 4: 0.0170, 5: 0.0033, 6: 0.0170, 7: 0.0342, 8: 0.0550, 9: 0.0775. The breaks in the relationships between Re 1500–6000 marks the change from laminar to turbulent flow; there is an increase in depth with decreasing slope (modified from Emmett, 1970).

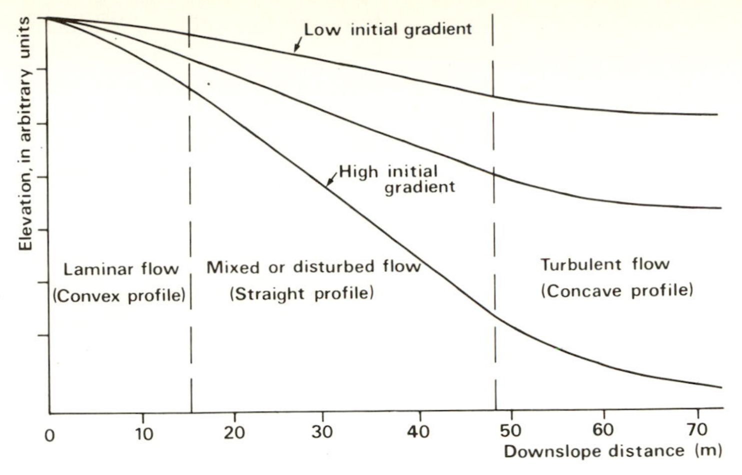

Fig. 4.18 Diagrammatic representation of the changes in flow regime downslope with a transition from laminar flow to turbulent flow, producing a transition from a convex to a concave profile (modified from Emmett, 1970).

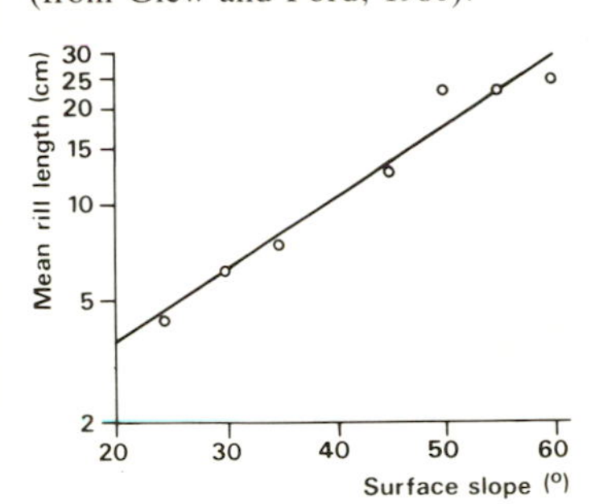

Fig. 4.19 Consistency of surface slope with varying rill length. (from Glew and Ford, 1980).

etrating directly to the rock surface and disrupting the boundary layer. For a rainfall of 35–40 mm hr^{-1} Glew (1977) found this depth to be 0.15 mm. Rills appear at slopes of 8–12° and lengthen to slopes of 70° where irregular scallops appear.

4.3 The role of a soil cover

A paradox in the literature can often be a fruitful source of ideas for further research. On the topic of the solution of limestone beneath a soil cover, one such paradox is provided by a juxtaposition of the works of Dakyns (1890) and Hughes (1901) with that of Jones (1965). Dakyns and Hughes both observed that joints in limestones were unopened and unweathered near and beneath soils developed from glacial tills. Away from a soil cover, bare limestone was dissected by rainwater weathering and the joints were well opened, forming grikes. On the other hand, Jones asserts that 'observation does not confirm this view' and discusses Dakyns's and Hughes's 'acid raindrop hypothesis' of limestone weathering, substituting his own 'biological weathering hypothesis'. He proposes that the solution of limestone by soil water is of far greater importance than the action of rainwater and that soil-covered surfaces are more deeply eroded than bare limestone. Since all three authors rest their arguments upon the same types of evidence, we must conclude that either one set of authors were not accurate in their observations, or that this is a case in which conflicting attempts are being made to propose general theories from local observations. It is probable that both hypotheses are tenable, each under its own conditions of operation, and that in some situations bare limestone is eroded faster by rainwater than is buried limestone by soil water, while in others the reverse is true.

It was Williams (1966) who resolved this paradox by careful observation of the nature of the superficial deposits. Working in the Carboniferous Limestone area of County Clare, Eire, he concluded that 'grikes are not often present beneath boulder clay. However, it is frequently quite obvious that . . . enlarged joints exist beneath a shallow till or vegetation cover. This apparent contradiction can be resolved by closer examination of the superficial cover. Calcareous tills . . . protect the underlying limestone if water becomes saturated with bicarbonate on passing through the till, but if the percolating water is not saturated,

Fig. 4.20 Rill cross sections. (a) cross sections of limestone at 70° and simulated plaster (Gypsum) surface at 60° produced under simulated rainfall. (b) evolution of simulated surfaces, with cross sections (from Glew and Ford, 1980).

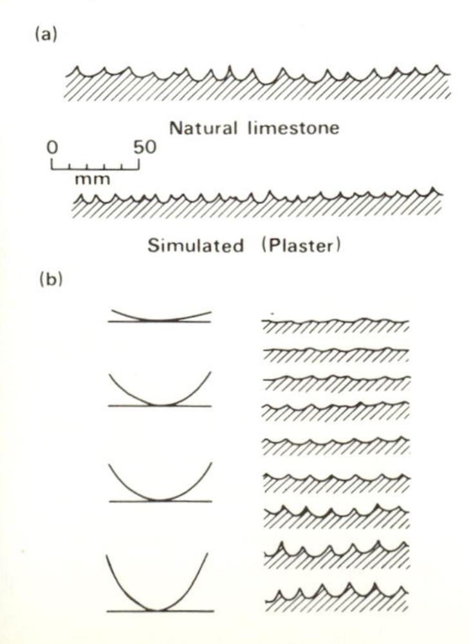

Fig. 4.21 Soil type and subsoil erosion of Carboniferous Limestone. (a) acid carbonate-free drift and eroded limestone. (b) carbonate drift and protected limestone.

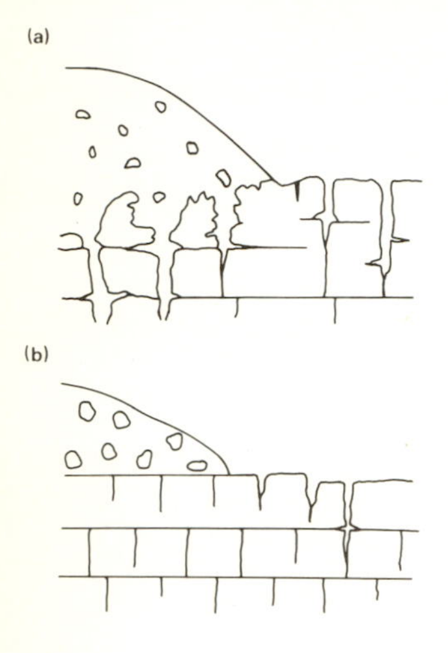

then solution will take place'. William's ideas have been amplified and quantified by Trudgill (1972) as follows.

In general, calcareous tills with high pH values protect the underlying limestone from erosion, whereas acid tills containing no limestone material usually overlie deeply corroded bedrock. Erosion of limestone is most severe beneath deposits supporting an acid vegetation and with a pH between 4 and 7 and a calcium carbonate content of 0 to 1% (Fig. 4.21a). Deposits with a fescue–herb vegetation, a pH of 7 to 9, a calcium carbonate content greater than 10% protect the underlying limestone almost completely from erosion (Fig. 4.21b).

However, dissolution of carbonates will take place within a calcareous soil profile. This will provide calcium in soil drainage waters but will find no expression in bedrock morphology. It will contribute to surface lowering within the soil but not in terms of bedrock lowering. The weathering of bedrock beneath the soil thus depends upon the aggressiveness (see Ch. 2) of percolating soil waters. This could be measured repeatedly in order to assess variations of solution potential throughout the year but it is easier, and equally useful, to assess the 'aggressiveness' of the soil material. A soil sample can be air dried, weighed and distilled water added until the soil is moist and plastic; this is known as the 'sticky point'. The pH of the soil can then be measured using a spearhead glass electrode carefully inserted into the soil paste. Calcium carbonate powder can then be added (1 g per 10 g soil) and the sample remixed. The pH can then be remeasured. A rise in pH indicates that calcium carbonate has gone into solution in the soil solution and thus that the soil material can provide aggressiveness. A fall in pH or no change in pH indicates that the soil material is already saturated with respect to calcium carbonate. This measurement is analogous to buffer capacity (see Brady, 1974, p. 384) but is specific to calcium carbonate; ground samples of country rock can also be usefully used in order to assess the aggressiveness of soil material to the local rock in question.

Fig. 4.22 Changes in soil pH on adding powdered calcium carbonate ('aggressiveness'). Soil types: 1 and 2, peat rankers over limestone. 3, calcareous brown earth. 4, clay over limestone, County Clare, Eire.

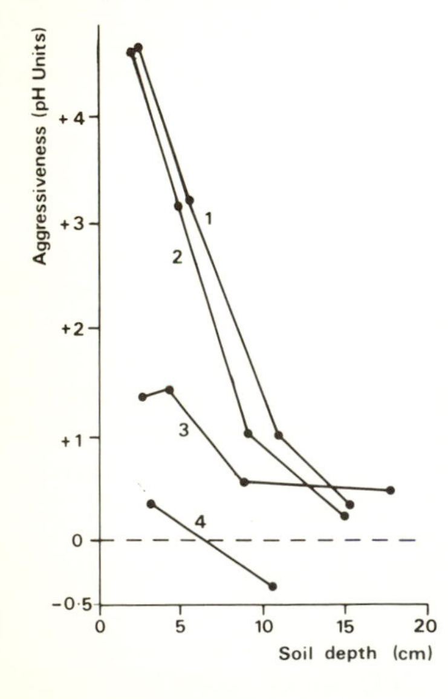

Figures 4.22 and 4.23 show the results of aggressiveness determinations on soil water from four soil profiles in County Clare, Eire; aggressiveness is scaled in terms of the change in pH on adding calcite. Both figures show that the aggressiveness increases with increasing organic matter content of the soil, with Fig. 4.22 indicating a decrease in aggressiveness with depth and Fig. 4.23 a decrease with calcium carbonate content of the soil. As calcium carbonate content increases with depth in these soils, and organic matter decreases, these trends are not independent. However, Figs 4.22 and 4.23 do suggest causal connections between these soil characteristics and aggressiveness.

Data on soil properties and the subsoil morphology of limestone have been collected from a number of sites in County Clare and the Mendip Hills (Trudgill, 1976c,d). Both areas are composed of Carboniferous Limestone which is massive and crystalline and does not break up easily into small fragments. At each site, data were collected on soil pH and carbonate content at the surface, in the mid-profile and at the base of the soil, and also on vegetation, soil depth, texture, parent material and slope. At the same sites the micromorphology of the subsoil limestone surface was examined and the degree to which it was eroded assessed. Soils evolve, in general, over periods of tens to thousands of years, and it is clear from the work of Williams (1966) and Pigott (1962) that many features of the bedrock morphology show the marks of a glacial legacy, and have therefore evolved over a period of time which is not necessarily

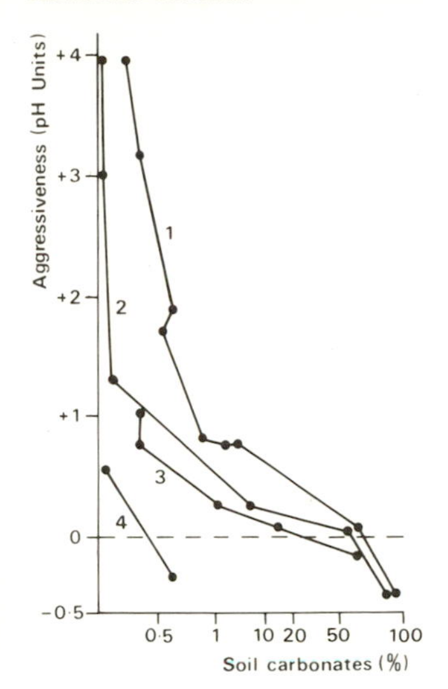

Fig. 4.23 Data from Fig. 4.22 replotted against soil calcium carbonate content.

represented by present-day measurements of soil properties. Therefore only the micro-morphology of the limestone surface was employed in classifying the degree to which the surface had been eroded, as micro-morphology is more likely to have evolved in the recent past and is thus more likely to be related to soil properties measurable at the present day (Trudgill, 1976c). The term 'micromorphology' arbitrarily was taken to include all irregularities less than 1 cm in height. An index of erosion based upon the texture of the surface and the differential weathering of crinoid fossils, chert, and other siliceous material was defined as follows:

Index number	*Characteristics of micromorphology*
1. *Very eroded*:	Etched and pitted: fossils and chert stand out more than 5 mm
2. *Moderately eroded*:	Moderately etched and pitted: fossil relief 2 to 5 mm
3. *Slightly eroded*:	Slightly rough surface: fossil relief 0 to 2 mm
4. *Not eroded*:	Smooth surface: glacial striae preserved.

While the finer distinction between indices 2 and 3 may involve some subjective judgement, the difference between index 1 and index 4 is unequivocal.

Erosion indices and soil characteristics are compared in Fig. 4.24. It can be seen that the degree of erosion varies inversely with pH and carbonate content. Similarly, it is greatest under plants adapted to acid conditions and least beneath grasses and herbs. These factors have already been noted to be of importance in controlling the aggressiveness of soil water. The degree of erosion is greater under deeper soils, which may reflect the increase in soil air carbon dioxide values with greater depth, reported by many authors (see p. 29). The nature of the parent material and the soil texture appear to exert relatively little influence, except in so far as calcareous parent material tends to protect the underlying limestone, whereas non-calcareous or organic parent materials give rise to acid soils and greater erosion. Hypothetically, a light textured soil should enhance leaching with a concomitant decrease in pH and increase in erosion but this effect, if present, is masked by other effects. Finally, there is an overall decrease in erosion beneath soils on steep slopes compared with those on flat or gently sloping ground. This may arise from several causes. Soils on slopes tend to be thinner, with a higher rate of downslope removal of material, than those on flat ground. On steep slopes the top of the soil profile is often truncated, as the rate of mechanical erosion is such that a true A horizon cannot become established. In theory, bedrock erosion rates should vary with position on the slope since soils on the steeper upslope sites tend to be thinner and less weathered.

In general, the data on the micro-morphology of the subsoil surface tend to support the conclusions reached regarding the nature of soil water aggressiveness. It should be remembered, however, that these data refer only to massive limestones and not to more coarse textured, friable rocks. As noted by Syers (1964), Gagarina (1968) and Ciric (1967), the nature of the limestone is important in determining the rate of its erosion beneath a soil. Hard, massive limestones do not break down easily and the soils above them tend to be easily leached. More friable rocks, such as the Jurassic Oolites of the Cotswold Hills, Eng-

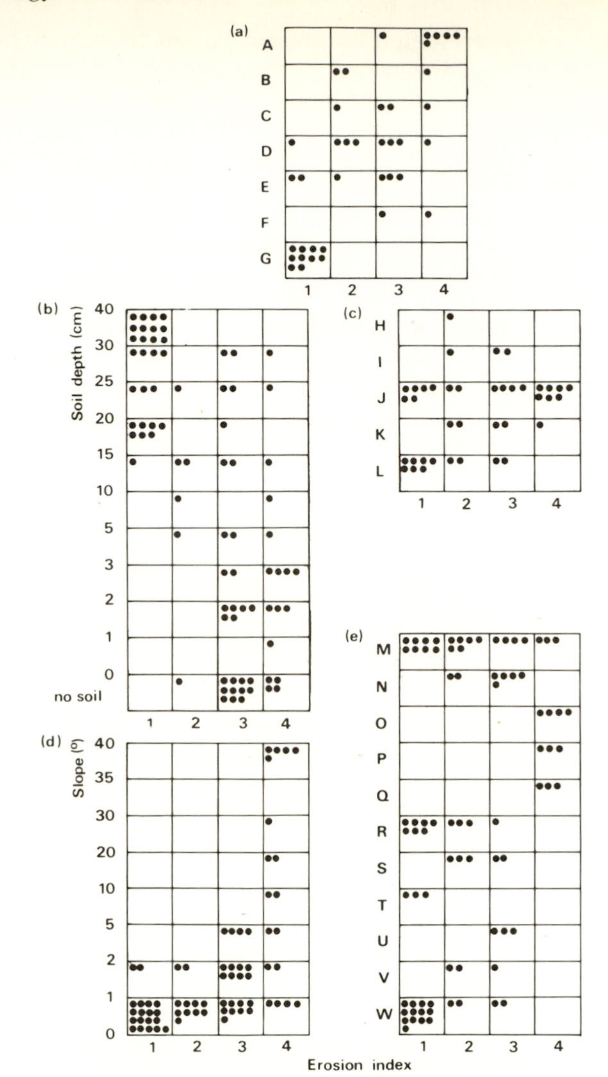

Fig. 4.24 Observations of micromorphological types, County Clare, Eire (see text for explanation of erosion index; 1 is the most eroded). (a) with vegetation type. Key: A, turf mat. B, *Thymus drucei*. C, *Lotus corniculatus*. D, *Agrostis tenuis – Festuca ovina*. E, *Dryas octopetala*. F, *Molinea caerulea – Erica cinerea*. G, *Calluna vulgaris*. (b) soil depth (c) soil texture. Key: H, silt loam. I, Loam. J, silty clay loam. K, clay. L, peaty loam. (d) slope angle (e) soil type. Key: M, silty clay loam. N, thin humus. O, calcareous drift. P, calcareous sand. Q, calcareous clay. R, sandy clay drift. S, shale drift. T, sandstone drift. U, colluvium. V, clay. W, peat.

land, fragment more easily and become incorporated in the soil, making it less acid. This, in turn, affects the vegetation, and whereas hard limestones may support a calcifuge vegetation, such as *Calluna*, which has the effect of increasing the erosion rate, softer rocks typically support a calcareous grassland flora (Shimwell, 1971; Hope Simpson and Willis, 1955).

Time is a factor of great importance in determining whether or not soil characteristics will tend to encourage weathering or protect the rock beneath it. In this context it is appropriate to examine briefly the development of soils on a limestone terrain. Various aspects of this topic are discussed more fully by Bryan (1967), Bullock (1964, 1971), Syers (1964), Grime (1963), Pigott (1962), Gardiner and Ryan (1962), and Curtis, Courtney and Trudgill (1976, Ch. 6).

The process of soil formation from a bare limestone surface is summarised in Fig. 4.25. Endolithic algae will colonise any fresh surface within one or two years and are followed by fungal symbionts, forming a crustose lichen colony. Unless the rainfall is too low, or the surface

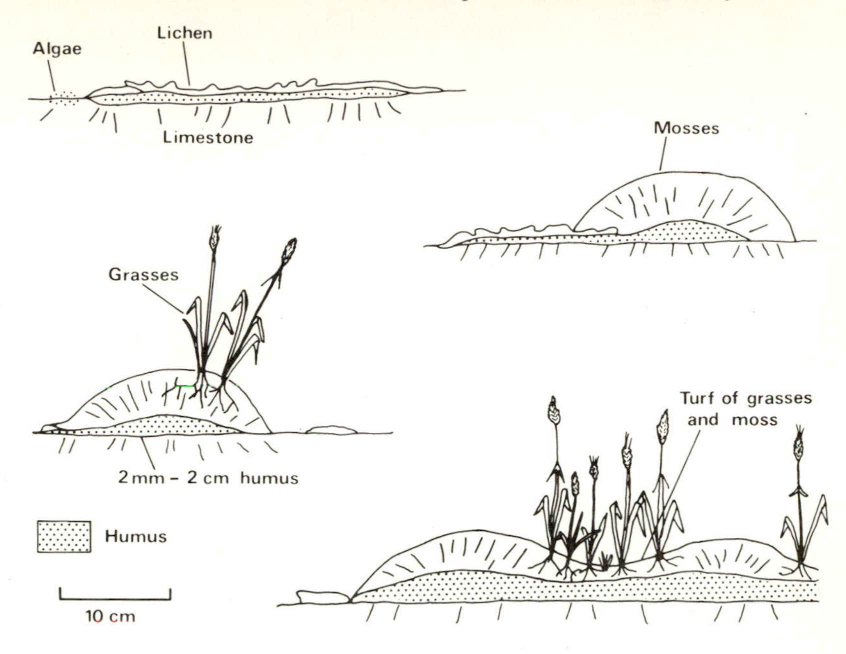

Fig. 4.25 Diagrammatic representation of soil formation on an initially bare limestone surface during colonisation by vegetation. (from Curtis, Courtney and Trudgill, 1976).

too steeply sloping to keep the lichen moist, the next stage of development is the colonisation of the lichens by mosses. Humus accumulates beneath the moss, but unless the rainfall is very high it remains alkaline because of the upward transport of solutes from the underlying rock. Colonisation of the moss by grasses and herbs further protects the humus from wind erosion and enhances the moisture-retaining ability of the soil. Further accumulation of litter and faunal droppings lead to the formation of the climax soil, which is a thin, mull-like rendzina with a neutral or alkaline pH, except near the surface where it may be slightly acid. Erosion rates beneath such a soil are low and limestone fragments incorporated within it are usually almost uneroded, except near the surface. Micro-erosion meter results are available for sites under this type of soil in the Mendip Hills, England. The soil had a pH of 6.5 to 7.0 and a carbonate content of up to 2%. Erosion rates were measured directly (High and Hanna, 1970; Trudgill, High and Hanna, 1981) and ranged from 0.003 to 0.005 mm per year. Beneath lichens, however, erosion rates are much higher than this, so that a thin mull-like rendzina type of soil tends to develop quite quickly. A single lichen-covered site close to that just mentioned gave an average erosion rate of 0.112 mm a^{-1}, about thirty times greater than that beneath the rendzina.

Only if the rainfall is very high will a bare limestone surface give rise to an aggressive soil. In areas of high rainfall, humus accumulating beneath a mat of mosses may be continuously leached and kept acid. This promotes the rapid growth of the moss and the formation of an acid peat ranker soil (Kubiena, 1953) which may be colonised by *Calluna* and other acid vegetation. These soils are common on the Carboniferous Limestone in County Clare where the annual precipitation of about 2500 mm is high enough to prevent drying and oxidation of the humus. Erosion rates beneath these soils are high (up to 5 mm a^{-1}).

In theory, the continued solution of limestone beneath an organic soil should lead to an accumulation of acid-insoluble residue from the rock, and the gradual development of a mineral soil. In fact, the rate of erosion beneath humus rendzinas is so low, and most hard limestones contain so little insoluble residue, that there has not been enough time since the last glaciation for this to occur in temperate areas. The average acid insoluble residue of nine samples of the Carboniferous Limestone, for example, was only 7.55% so that it would take 264,000 years for a mineral horizon 10 cm thick to form if the limestone were eroding at 0.005 mm/yr. However, mineral soils are common on limestones in Britain and in other parts of the world that were also glaciated or nearby glaciated areas during the Pleistocene. In the former case, soils are formed on drift, and in the latter case Pigott (1962) has emphasized the role that Late Glacial and Early Post-Glacial loessial deposits may have had in the formation of mineral soils. In the Mendip Hills, Findlay (1965) has examined the heavy mineral suite of the Nordrach Series, a brown earth soil, which overlies Carboniferous Limestone to a depth of over a metre. The mineral horizons of the soil contain heavy minerals which are rare or not found at all in the limestone beneath.

If a mineral horizon is present at the start of soil development, the soil which subsequently develops may be a great deal more aggressive than a humus-rendzina. If the bedrock is massive and does not become incorporated in the profile, leaching will remove bases from the soil, rendering it more acid and aggressive. The humus which accumulates in the A horizon will be of low pH because it is separated from the supply of bases from the limestone. Water percolating through the A horizon is acidified by it, and is thus more effective in leaching bases from the mineral horizons. The effect of a mineral horizon in increasing erosion rates is shown by the results from a micro-erosion meter site in County Clare, Eire. Here the mineral soil is 20 cm deep and has an acid pH down to 15 cm. The rate of bedrock lowering was measured as 0.025 mm a^{-1}, which is five to eight times faster than beneath an adjacent humus-rendzina. Table 4.1 shows erosion rates measured in the Mendip Hills, Somerset beneath both types of soil by a different method. Weighed tablets of limestone with a known surface area are buried in the soil, excavated after a suitable period of time and reweighed. The erosion rate is expressed in mm a^{-1} over the whole surface (Trudgill, 1975). The results shown agree well with those obtained by micro-erosion meter, and show the clear difference between soil types.

Table 4.1 Erosion rates beneath mineral and humus rendzina soils, Mendip Hills, Somerset

Soil	*Vegetation*	*Depth*	*pH*	*Carbonate (%)*	*Rate (mm a^{-1})*
Nordrach Series (Brown Earth)	Grassland	15 cm	6.0	0.02	0.0253
Humus Rendzina	*Thymus drucei*	5 cm	6.5	0.1	0.0027 – 0.0056
Humus Rendzina	Herb rich mat	4 cm	7.0	2	0.0047 – 0.0050

It is clear that humus-rendzinas are not at all active in producing erosion. Only in areas of high rainfall will acid organic soils (rankers) form over limestone. Where a mineral horizon is present the soil, unless calcareous, will promote erosion of the underlying rock, the erosion rate depending in the long term upon the efficiency of the soil leaching system, and in the short term upon such factors as the pH, carbonate content, texture, vegetation and slope. It is clear from the available measurements, which are summarised in Table 4.2, that solutional erosion rates are faster beneath a mineral or acid peat soil cover than they are on bare limestone. These values were obtained by a variety of meth-

Site	*Erosion rate mm* a^{-1}	*Location*	*Mean annual rainfall mm* a^{-1}	*Source*	*Method used*
Bare limestone	0.04	NW Yorks	*c.* 1500	Sweeting, 1966	2, 3
Bare dolomite	0.25 – 0.30	NW Scotland	1550–2550	Newson (pers. comm.)	1
Bare limestone	0.009	Co. Clare, Eire	*c.* 2500	High (pers. comm.)	2, 3
Bare limestone	0.003–0.004	Co. Clare, Eire	*c.* 2500	Author	2, 3
Bare limestone	0.025–0.075	S Wales	*c.* 1500	Thomas, 1970	2, 3
Beneath peat	5.0 – 8.2	NW Yorks	*c.* 1500	Sweeting, 1966	2, 3
Runnels from peat	6.3 – 11.5	Co. Clare, Eire	*c.* 2500	Sweeting, 1966	2, 3
Beneath mineral soil	0.43 – 0.50	NW Scotland	1500–2500	Newson (pers. comm.)	1
Runnels from mineral soil	0.010–0.015	Co. Clare, Eire	*c.* 2500	Author	2
Pools on limestone	0.02	Co. Clare, Eire	*c.* 2500	Author	2
Grikes	0.1 – 0.2	Co. Clare, Eire	*c.* 2500	Author	2
Beneath mineral soil	0.045	Mendip Hills, Somerset	*c.* 1100	Author	

Methods used:
1. Micro erosion meter
2. Extrapolation from an assumed glacial surface
3. Water sampling

Table 4.2 Erosion rates on bare limestone and beneath soils

ods, including water sampling, direct measurement and extrapolation from unweathered glacial surfaces. Where two or more methods have been used at the same site they are in substantial agreement. However, there is also some evidence to suggest that soils are in fact being removed from areas of pavement. Jones (1965) summarises this evidence, stating that bare areas adjacent to soil are not colonised by lichens and suggesting that the soil may have been lost down grikes. Since these bands of lichen-free rock are sometimes 50 cm wide, and Syers (1964) shows that lichens will generally colonise a bare surface within five years, the rate of retreat of the soil cover may be as much as 10 cm per year. It is thought that grazing pressure may be an important cause of vegetation retreat, with subsequent loss of unprotected soil during high pressure, but colonisation of bare areas where pressure is lower.

A further piece of evidence in favour of soil loss is provided by Gosden (1968), who demonstrates that pollen sequences are truncated at the base of peats occurring over limestone in Yorkshire, when compared with adjacent peats on non-limestone rocks. This could be due to the oxidation of the peat at the limestone surface, but the evidence of peat seen washed down the walls of shallow caves demonstrates that peats may be eroded from below (Standing, 1969; Burke 1970). Trudgill *et al.*, (1979) showed that both retreat (Fig. 4.26) and advance of soil and vegetation could occur in the same area.

Fig. 4.26 Soil retreat and lichen colonisation on a limestone surface, County Clare, Eire. Above the ruler a large white colony of *Lecanora albescens* is well established. Lichenometry work (Trudgill *et al.*, 1979) suggests that colonies of this size are around 50 years old. Down the left-hand side of the ruler progressively younger and smaller colonies are seen, colonisation presumably occurring as the soil retreated. Fresh soil retreat is seen at the base of the ruler, with a pale lichen-free zone. (Photo: S. T. Trudgill.)

A further possible hypothesis, put forward by Parry (1960), is that the amount of erosion represented by runnels on limestone pavements is too great to have occurred under present climatic conditions. It is suggested that they may have been formed by periglacial meltwaters with a high level of carbon dioxide in snowbanks and a greater run-off than occurs at present. This hypothesis may be qualified by considering the measured rates of erosion in periglacial areas today. Smith (1969), in a study of rates of erosion in Somerset Island, Canada, finds that water samples rarely exceed a concentration of 60 mg l^{-1} $CaCO_3$ and are never greater than 95 mg l^{-1}. Rates of precipitation and run-off in this area are low and this fact, combined with the low hardness of the water, produces an erosion rate far smaller than those found in temperate areas. Moreover, solution runnels are unknown. In the Rocky Mountains of Alberta water hardness is 50 to 90 mg l^{-1} $CaCO_3$ above the treeline and 100 to 265 mg l^{-1} below it, where there is a soil cover (Ford, 1971a).

Fresh evidence for the role of a soil cover in pavement evolution comes from a study by the present author. Thomas (1970), Sweeting (1970) and Sweeting and Sweeting (1969) have shown that different rates of erosion may occur on limestones of different lithologies. Rates appear to be highest on biomicrites and lowest on sparites. Therefore, in comparing areas of different morphology to assess the role of the soil in the evolution of each it is necessary to study a single bed where it forms both a bare pavement and disappears beneath an adjacent soil cover. This has been done in County Clare, Eire, where it has been found by digging pits that each of the pavement surface forms shown in Fig. 4.16 (p. 62), is related to different environment. Thus, cuspate and arcuate forms are related to the occurrence of acid soils, whereas an undulating surface always occurs in association with herb rich vegetation. Flat, smooth surfaces may be ascribed to glacial planation, and platy weathering to subaerial solution. If, say, an arcuate form is found on a bare surface, it may be argued that soil removal has occurred.

In summary, the establishment of process-form relationships still represents a challenge in terms of pavements, karren and lapiés. Many of the processes are documented and the erosion rates and forms described, but progress in understanding process-form relationships has been slow. It is often by an experimental approach, adopted, for example, by Glew and Ford (1980), rather than a descriptive approach, that real progress is best made on this topic.

5 Caves

5.1 Introduction

Caves are perhaps some of the most well known karst features but in quantitative terms they occupy a small portion of a limestone mass and are essentially conduits for the transport of water and solutes from near the surface to a lower output point. They are formed along lines of weakness in the rock such as joints and bedding planes and their development is also controlled by hydraulic gradient.

Caves are well developed in Carboniferous (Missippian) Limestone since the rock itself is nearly impermeable and water is focused along joints, that is to say, the rock is pervious rather than porous. In more porous rocks, water can travel through the rock mass and cave development is limited.

In recent years, Uranium-Thorium dating of cave deposits has greatly added to the knowledge of cave formation sequences. In addition, where dated cave sequences are present, it has become possible to date the sequences of formation of whole landscapes, especially in terms of valley downcutting. This is a possibility rare in geomorphological study.

5.2 Cave development

Caves in limestone are water-worn features formed as surface streams become diverted underground by the opening of fissures, joints and bedding planes. In other rocks, such as a porous sandstone, water may be able to move within the rock but an integrated preferential flow net rarely forms because dissolution along joints is limited and therefore flow routes do not become focused. In soluble limestones, zones of preferential flow occur in the aquifer and these become enlarged because the flow leads to the solutional enlargement of the flow-route cavities. Caves thus form as integrated flow networks of water-filled passages in a pervious and soluble bedrock. During the development of a cave system an initial situation may be seen where surface streams exist in valleys and a watertable exists at some depth in the interfluve areas – the groundwater being present in small cracks and fissures. Cave development is then encouraged as the cracks and fissures enlarge into pipe systems abstracting surface flow and integrating subsurface flow. In a pervious rock the relative impermeability of the material between the water-filled cave passages may mean that the passages flow independently, with little mixing of the water bodies in the rock as a whole.

Fig. 5.1 Vertical section through limestones showing cave development sequences. 1, Abandoned, older, dry passage formed in relation to former valley floor level (pecked line). 2, Partly water filled passage. 3, Newer, active passage, completely filled with water, draining to present valley level.

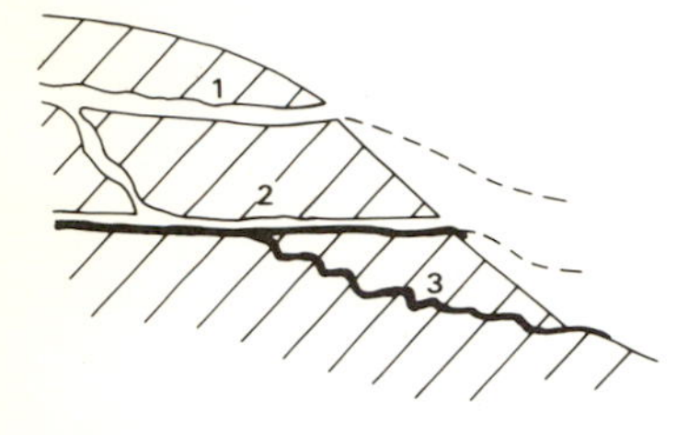

Cave passages progressively cut down into the rock mass, leaving surface streams dry, and at a rate related to the lowering of the base level at the output of the system. Caves which are not normally occupied by water to the roof are a late stage of development: their presence indicates that active water-filled passages are likely to be found at a lower level in the rock. Thus, at any one time, a cave development sequence may often be seen with a lower, younger system, which is water-filled and still developing, and one or more upper systems which are only partially or temporarily water-filled, or which may be dry, and which are essentially fossil. These sequences are outlined in Fig. 5.1.

In temperate regions which have been glaciated, a simple sequence of downcutting has often been modified. Permafrost acts to divert flow back to the surface in periglacial times. Major downcutting of glacial valleys occurs by erosion in glacial times. This erosion leads to the existence of lower base levels in deglacial times and subsequent downcutting of cave levels to the lower base levels (as shown in section 5.11).

Water-filled passages are termed **phreatic** passages. These tend to be circular in cross section. This is because they are dissolved equally in all directions since water contact is present on all sides. As base level is lowered, water tends to be lost from the upper parts of the passages, leaving an air space in the former phreatic tube. Under water-filled phreatic conditions forced flow under pressure may occur or water may flow slowly but, with an air space in the passage, phreatic conditions are replaced by **vadose** conditions, where there is a more 'normal' stream, with an air space above. Here downcutting tends to occur with water contact limited to the lower part of the passage; abrasion by bedload also increases this effect as the tendency for particle suspension in the forced flow phreatic conditions is much reduced under vadose conditions. Thus under vadose conditions vertically elongated cross sections tend to occur (Fig. 5.2).

A transition from phreatic to vadose passages may, then, be seen in terms of hydrological controls, that is, in relation to downcutting and changes in base level to which outflow water is able to move. This is given the qualification that under forced flow phreatic conditions, water need not simply flow under gravity. Water confined in a tube is able to flow upwards provided there is a hydraulic head above the outflow level (Fig. 5.3).

Superimposed on these hydrological controls are the geological controls. The frequency of fissuring, joints and bedding planes and the resistance to chemical and mechanical erosion will combine to influence the shape of passages in cross section and their distribution in plan and long section.

5.3 Joint patterns and caves

The relationships between regional tectonic stresses and joint patterns have been discussed in Chapter 2. While overall cave orientation is largely controlled by hydraulic gradient, joint patterns and other tectonic features, such as faulting and folding, also have a major influence on cavern orientations (Waltham, 1981). Under phreatic conditions, many joints may be opened up, leading to the formation of a phreatic maze. Under vadose conditions, erosion is less evenly distributed and cave alignment often follows one or two major joint systems.

Phreatic network mazes are described by Palmer (1975) as angular grids of intersecting fissures formed by the solutional widening of nearly

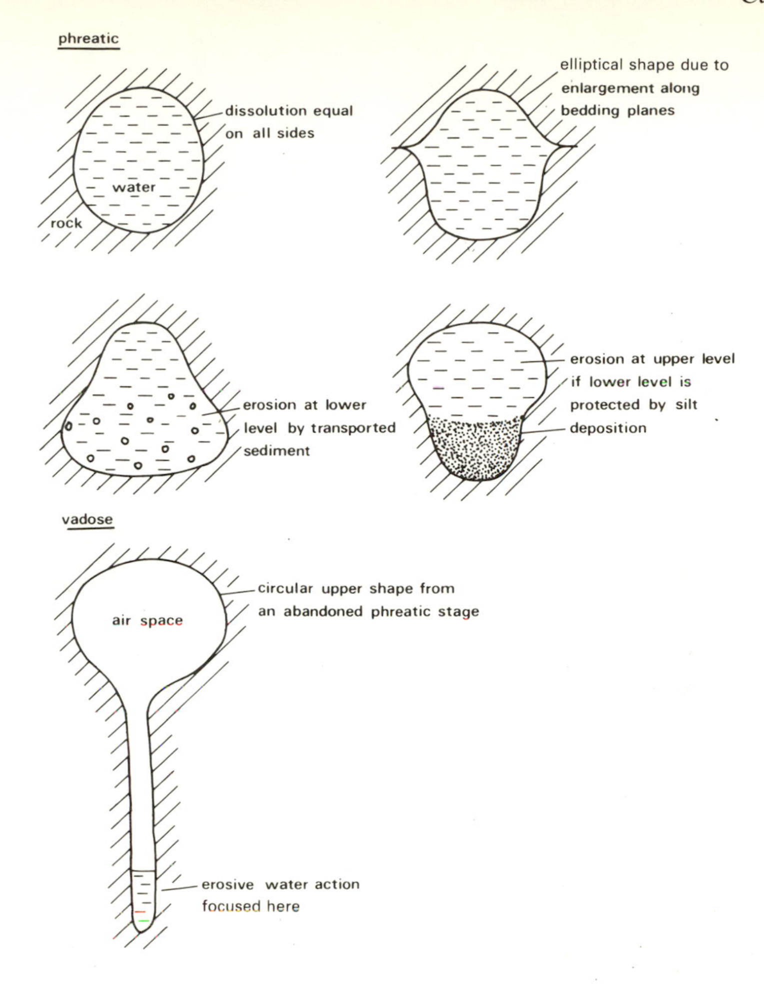

Fig. 5.2 Cave passage shapes under vadose and phreatic conditions. Geological structures may markedly alter the shapes shown.

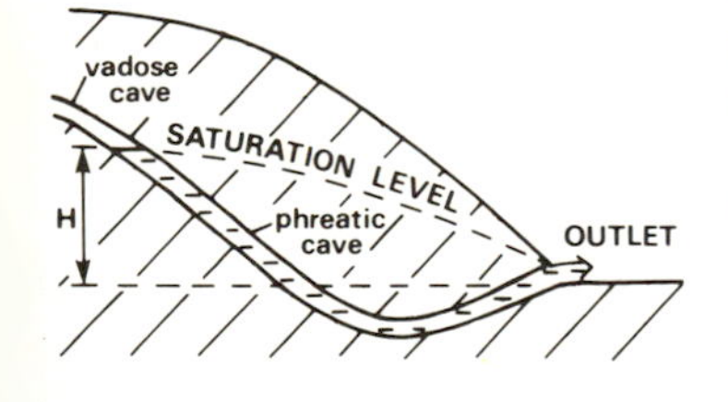

Fig. 5.3 Upward water flow in a confined phreatic tube under the influence of hydraulic head, H, from the saturation (water table) level to the outlet.

all major joints to roughly the same size. The prerequisite for formation appears to be a uniform distribution of water flow. This may be achieved in areas of low hydraulic gradient or where there is widespread seepage from an overlying stratum such as a porous sandstone. An outline plan of a phreatic network cave is shown in Fig. 5.4. The rectilinear joint pattern is clearly seen, with orientations of 68°, 118° and 156°, and a more oblique set in the top left-hand corner of the plan at 20°.

Similar joint control is evident in bigger, more complex systems, as, for example shown by Weaver (1973) in South Wales, UK. Four dominant joint sets are in evidence trending NE–SW, NW–SE, N–S and E–W. The dominance of the NE–SW and NW–SE systems are seen for Dan-yr-Ogof I and II (South Wales) (Fig. 5.5 a). The dominance of the N–S and E–W systems are seen in the Ogof Ffynnon Ddu I cave system in South Wales, both for joints visible at the surface and for joint aligned passages underground (Fig. 55 b). The relationship of both cave systems with regional tectonics is shown in Fig. 5.5(c). Detailed work by Coase and Judson (1977) in Dan-yr-Ogof has shown that while joint-aligned passages dominate the system (Fig. 5.6 a), where synclinal folding is marked, with dip values from 30–80°, then the spread of passage orientations is greater as bedding plane dip acts to influence flow direction as well as joint orientation (Fig. 5.6 b).

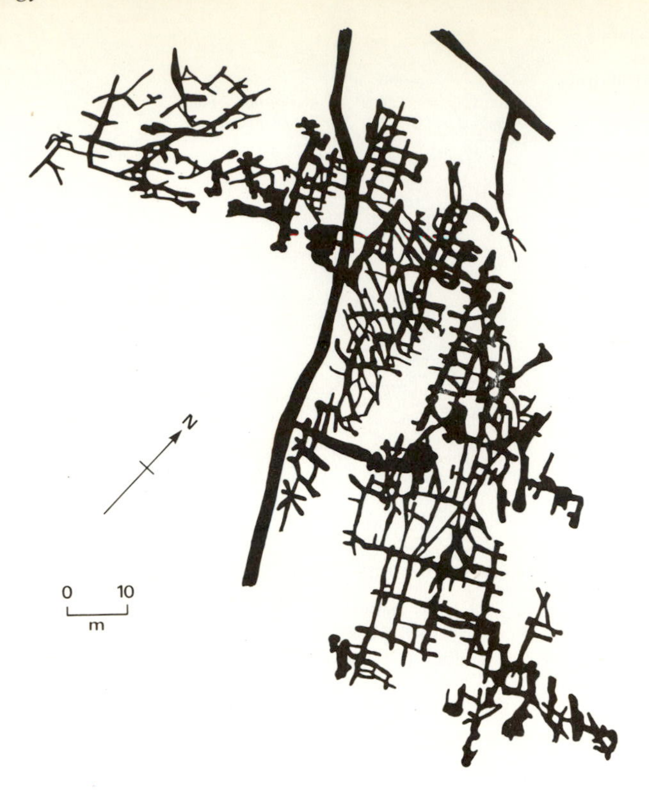

Fig. 5.4 A phreatic network cave, Devis Hole Mine Cave, Swaledale, Yorkshire, UK (modified from Ryder, 1975).

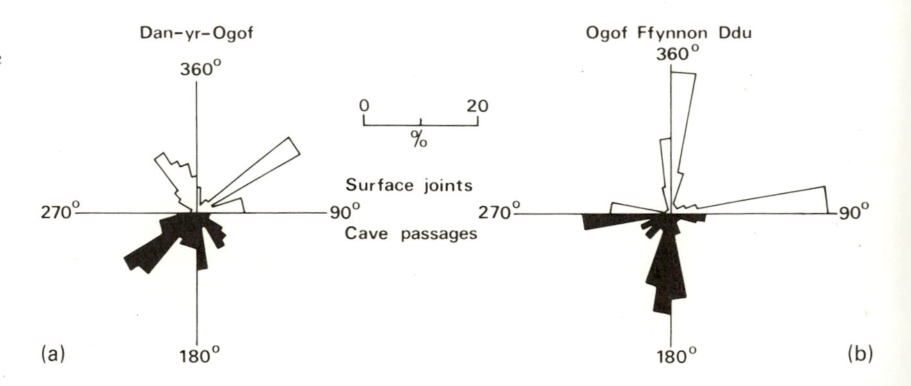

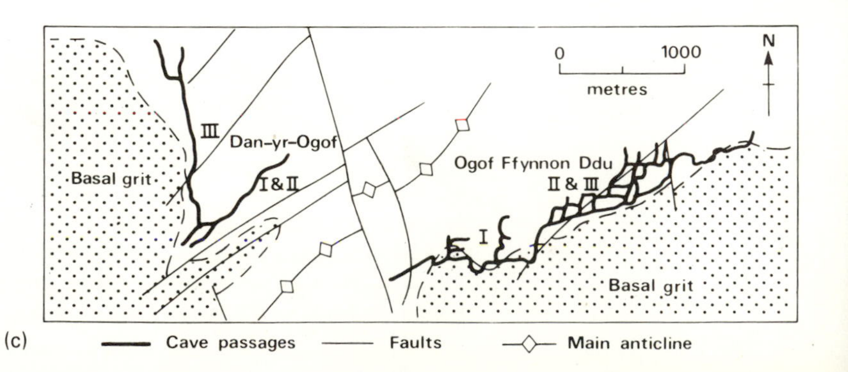

Fig. 5.5 (a) Joint orientations on the surface (unshaded) and in cave (shaded), Ogof Ffynnon Ddu I, South Wales, UK. (b) Dan-yr-Ogof I & II cave system and joint pattern, South Wales. (c) Regional tectonics and cave systems, South Wales (modified from Weaver, 1973).

5.4 Phreatic–vadose transitions

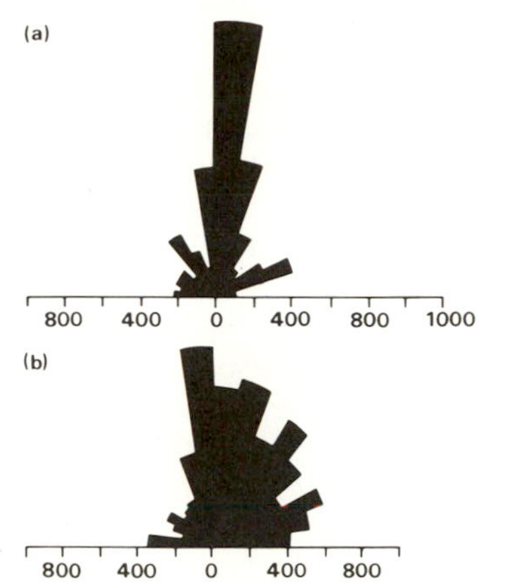

Fig. 5.6 Joint alignment in Dan-yr-Ogof, (a) without and (b) with synclinal folding (modified from Coase and Judson, 1977).

Caves tend to be initiated by seepage along bedding planes, joints and other fissures. Frequently an anastomosing network of water-filled tubes forms the first stages of formation (Fig. 5.7). Eventually, with vadose incision, these networks are abandoned but may be frequently visible on cave ceilings as a series of half-tubes or pendants (Fig. 5.8). In addition, the latter stages of phreatic conditions may be characterised by the deposition of sediments (section 5.9), and these may also be cut into during vadose incision and may be visible as small terraces on the upper parts of old phreatic cave walls.

During the phreatic–vadose transition, tubular passages tend to be basally incised by a vadose stream, often giving rise to keyhole-shaped passages (Fig. 5.2). These features may, however, be substantially masked or modified by lithological variations in the cave wall (section 5.6). The phreatic–vadose transition is not necessarily a sudden jump, but is a transition that may evolve slowly as joints in the phreatic cave floor are gradually exploited and enlarged under the influence of an hydraulic gradient, often controlled by a gradually lowering external base level. Additionally, in glaciated regions, a phreatic system may be adjusted to a preglacial base level; if glacial erosion then deepens the valley external to the cave, deglaciation will reveal a lower base level. This will lead to either an abandoned phreatic system with a new, lower phreatic system or to a deep vadose incision of the preglacial phreatic system.

5.5 Cave walls – scallops

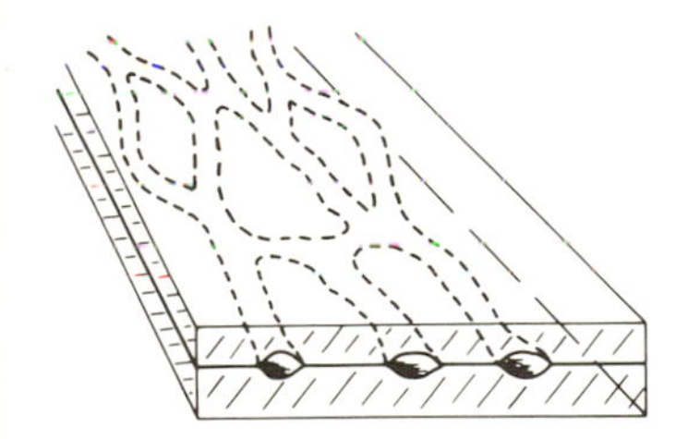

Fig. 5.7 Network of anastamosing phreatic tubes, seen in plan and in section. The tubes are commonly 1–30 cm across.

Cave walls commonly exhibit small contiguous dish-shaped depressions termed scallops; if elongated they may be termed flutes. The scallops evolve because of turbulence in flowing water. They may, in some ways, be compared with ripple marks formed in sand under flowing water. In this case, eddies of greater or lesser turbulence give rise to alternative relative erosion and deposition of sand. In the case of soluble cave walls, eddies in turbulent flow give rise to greater or lesser disruptions of the saturated boundary layer of solutes diffusing outwards from the solid–liquid interface (Fig. 2.4 and 2.12, pp. 14 and 19). This turbulence can be illustrated as in Fig. 5.9.

There is some discussion concerning the initiation and propagation of scallops. They appear to be dependent on the transition from laminar to turbulent flow in water over the rock surface. The occurrence of turbulent flow is related to surface roughness, water velocity and depth of water (as explained in section 4.2, p. 59). Eddies in flowing water create locally high velocities and disrupt the laminar flow occurring close to the cave wall. Under laminar flow, the water shears horizontally, parallel to the direction of flow; if this is disrupted by turbulent flow the dissolution products are locally transported at a more rapid rate, giving rise to a locally greater rate of dissolution. Curl (1966) and Blumberg and Curl (1974) relate the size, spacing and propagation of scallops to water velocity. In addition, Allen (1972) proposes that their initiation and development is closely related to lithological inhomogeneities, such as variations in grain solubilities and small fossil fragments, these factors assisting the formation of locally increased turbulences. Flow velocities for an initial small depression and a developed scallop are shown in Fig. 5.10(a,b). Maximum dissolution is facilitated in the base of the feature where turbulent flow is directed downwards to the rock.

Scallops forming under vadose conditions frequently have their steeper face located on the upstream part of the scallop, with the face pointing downstream. Under phreatic conditions, where larger eddies may occur, this relationship is often obscure because broader, more rounded features are present. In addition, domes often form in the roofs of phreatic caves by the upward eddying motion of water under pressure. Such features can be often used as evidence of former phreatic conditions in a presently dry or vadose cave.

5.6 Cave walls – lithological factors

While micro-lithological variations are implicated in Allen's view of scallop formation, vertical sequences in bedding may also have a major influence on the erosion and shape of cave walls. Frequently, bedding planes may be preferentially exploited, giving rise to 'shelving layers' described by Tratman (1969) (Fig. 5.11). In addition, the presence of insoluble chert layers may often give rise to prominent ledges in caves.

Lithological variation, as described in sections 2.2 and 2.3 (pp. 8–9) can have a strong influence on cave walls because of variations in grain size, cementation and solubility. This is illustrated in Fig. 5.12, from the Ogof Ffynnon Ddu System in South Wales. Here loose, poorly cemented, sandy conglomeratic limestone has been preferentially eroded while the more massive limestones are more resistant. In the section shown the dolomites are more resistant than the calcite limestones. This is also seen in cross section (inset). Such lithological factors may well have a greater influence upon cave passage shape than factors such as the phreatic–vadose transition described above. The inset on Fig. 5.12 owes its cross-sectional shape as much to lithology as it does to its origin as an upper phreatic tube and a lower vadose trench.

5.7 Cave walls – abrasion

Scallops and domes are essentially solutional features, but cave streams carry a suspended sediment load and a bedload, just as surface streams do. Their effectiveness as erosion agents depends on the amount, size and nature of the load carried and the velocity of the water. As with surface streams, the amounts of both mobile bedload and suspended load are smallest at lowest flows; under these situations dissolution is liable to be the dominant erosional process. Under the high flows, abrasion by suspended sediment can be considerable and movement of larger material over the bed is also facilitated. The combined erosional effects of these will attenuate upwards from the cave floor, and thus basal lowering and basal lateral extension of the cave floor are encouraged by these physical processes.

Fig. 5.8 Mode of formation of half-tube and pendants. The half-tubes are commonly 1–30 cm across (modified from Tratman, 1969).

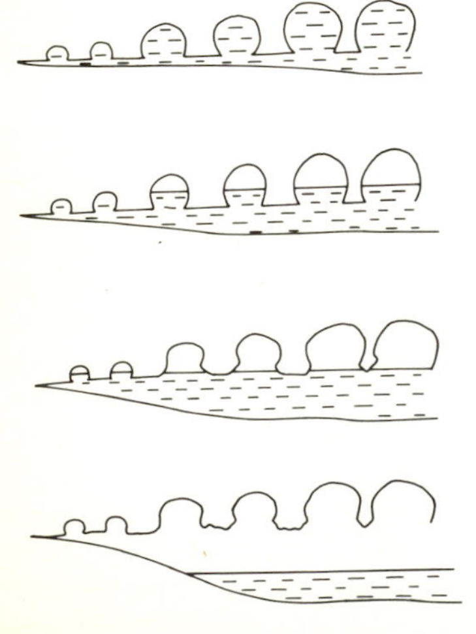

The movement and geomorphological importance of bedload has been little studied in caves. This is possibly because of a relatively greater focus upon solutional processes, to the detriment of an understanding of the mechanical processes operating in caves – and common to all fluvial systems.

A distinction may be made between those limestone conduit systems where the water is derived solely from percolation water and those where water enters the system from a surface stream. In the former case, sources for sediment are limited and, more important, flows are liable to be slow because of the damping effect a percolation feed system will have on intense rainfall events. In a surface stream-fed system, high rainfall can readily lead to flood flows and the entrainment and movement of surface derived sediment.

It has been argued by Newson (1971) that if the effectiveness of abrasion in cave systems is to be demonstrated, then it has to be shown that calcium carbonate sediment exists in water resurging from limestone

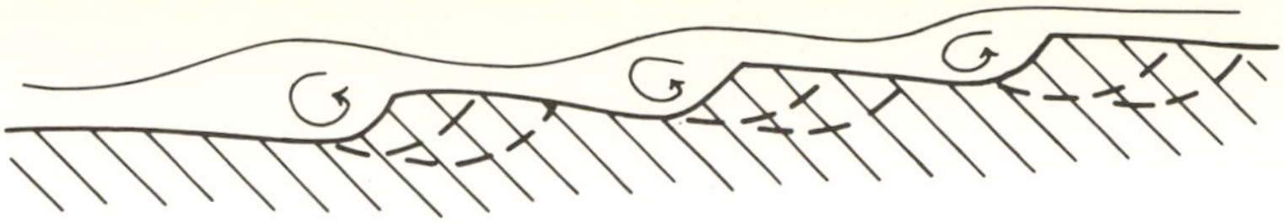

Fig. 5.9 The formation of scallops in turbulent flow (modified from Tratman, 1969).

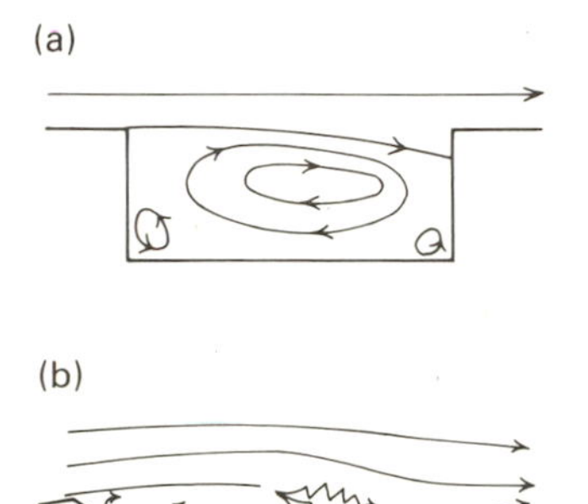

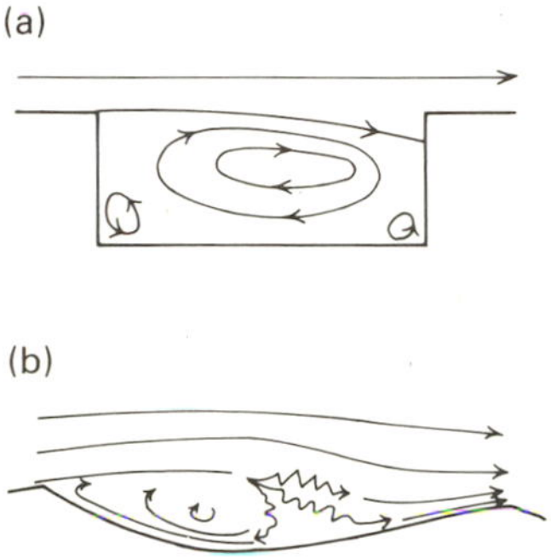

Fig. 5.10 (a) Flow velocities in a small depression (modified from Allen, 1972).
(b) Flow velocities in a scallop (modified from Blumberg and Curl, 1974).

masses. Table 5.1 shows data on the chemical composition of suspended sediment for swallet sites where surface streams enter the limestone and for resurgence sites, both on the Mendip Hills, Somerset. The relative increase of calcareous suspended sediment (acid soluble) is evident. In addition the relative rounding of siliceous sediment can be evident, as is shown in Fig. 5.13 for a South Wales cave system. The argument is that the increase from swallet to resurgence in the percentage carbonate in the sediment represents the particles abraded from limestone walls and bed material by the abrasive action of the siliceous material; the siliceous material itself being rounded in the process. Newson also noted that at the Cheddar resurgence, England, suspended solid calcium carbonate load exceeded the dissolved calcium carbonate load at discharge values of over 5 $m^3 s^{-1}$ when approximately 500 g s^{-1} solid material were transported. In major floods much of this material is, in fact, reworked sediment which may have been resident in the cave system for many thousands of years. Major flood events may occur only infrequently, but can have a considerable effect on cave development, both in terms of transport of previously derived sediment out of the cave system and also in terms of abrasion of cave passages.

5.8 Cave sediments

Caves are repositories for sediments moved in from the surface and also for sediments derived from within the cave. The disposition, sedimentary structures and mineralogy of these sediments can be used to gain interpretations of the environments of deposition of the sediments as well as of the mechanisms of deposition.

Alluvial deposits may be found in caves in similar situations to those

Fig. 5.11 Shelving layers from preferential dissolution along bedding planes (from Tratman, 1969).

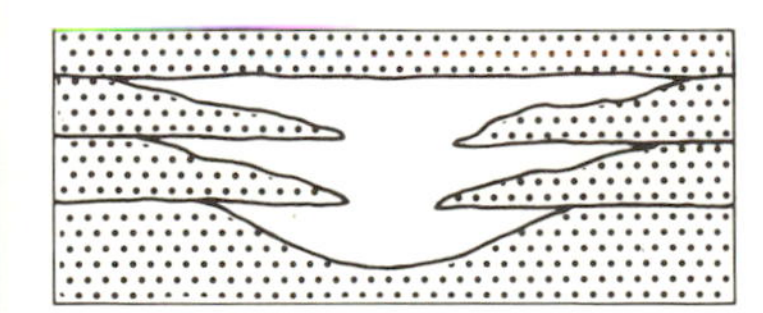

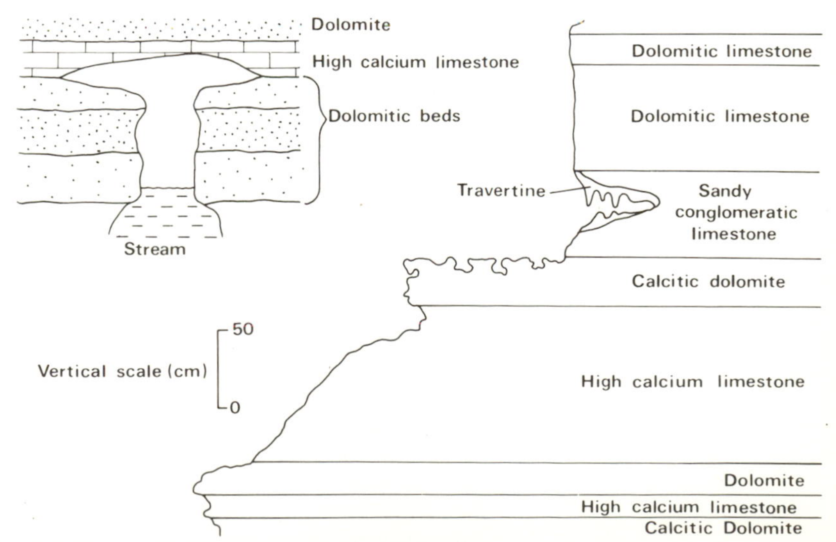

Fig. 5.12 Influence of lithology on cave passage shape, Ogof Ffynnon Ddu, South Wales (modified from Charity and Christopher, 1977).

Table 5.1 Size and mineralogy of suspended sediment in cave systems, Mendip Hills, Somerset, UK (modified from Newson, 1971)

System	*Site*	*Mean % by weight*			
		Size		*Mineralogy*	
		Sand	*Silt and Clay*	*Calcareous*	*Non-Calcareous*
Cheddar	Swallet	82	18	4	96
	Resurgence	14	86	98	2
Burrington	Swallet	79	21	0	100
	Resurgence	5	95	38	62

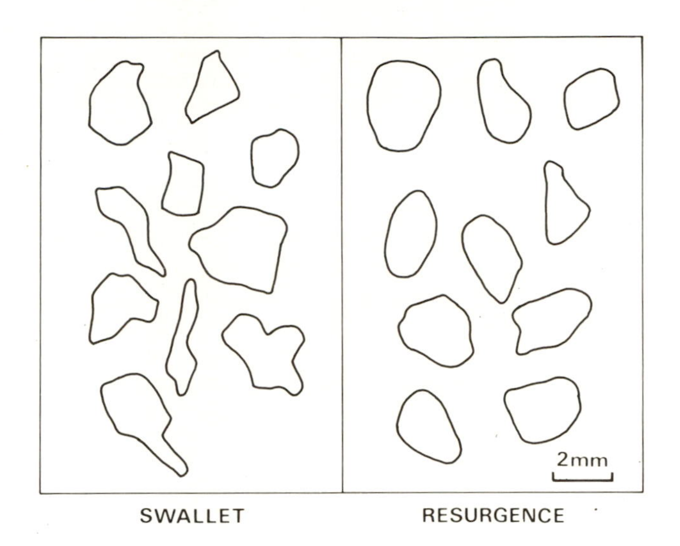

Fig. 5.13 Relative rounding of coarse siliceous sand sampled at the swallet and resurgence of the Dan-yr-Ogof system, South Wales, a distance of 4 km (modified from Newson, 1971).

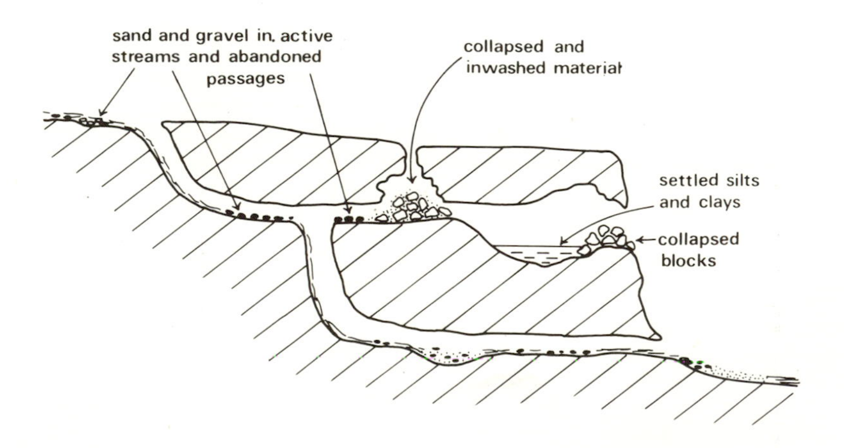

Fig. 5.14 Major types and deposition sites of cave sediments (modified from Ford, 1976).

in surface fluvial systems (Fig. 5.14). In addition, fine grained material is often deposited in still or very slow flowing water. Frequently, major deposits are to be found where rapidly flowing water with a high sediment load is checked in velocity, either by a constriction or where a narrow vadose stream encounters a wider, slower flowing phreatic system (Ford, 1975). Often, in swallet caves, the coarsest material may be deposited in the earlier reaches of the cave while the finest material is

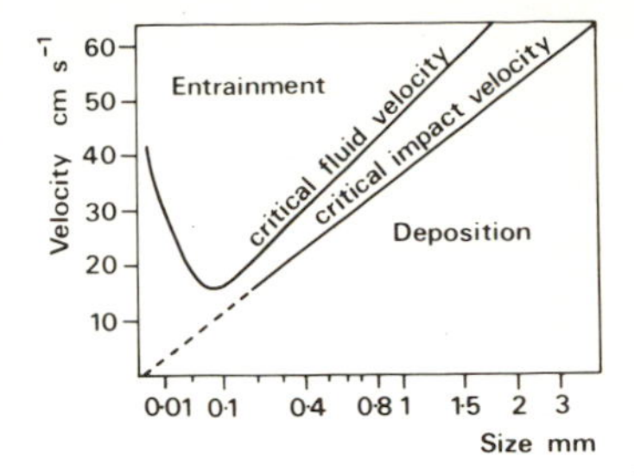

Fig. 5.15 The relationship between sediment size and velocity of entrainment (modified from Briggs, 1977).

carried deeper into the cave. Because of the effects of settling in quiet reaches of water in caves, resurgences tend to have lower sediment concentrations than swallets and this is especially so if comparison is made with an equivalent length of surface streamway; in floods, however, this will tend to be far less marked as velocities will be high enough to mobilise more sediment. The relationship between velocity of entrainment and sediment size is shown in Fig. 5.15.

Scanning electron microscope studies have provided useful information on the sources and weathering environment of cave sediments (Bull, 1976). In a study of Agen Allwed cave, South Wales, many of the grains showed fracture surface characteristics of mechanical weathering during frost shattering. Most of the grains were derived from Millstone Grit outside the cave and, most probably, during one major cold phase of freeze–thaw weathering during periglacial conditions. Smoother material, derived by fluvio-glacial erosion, was also identified.

Also in the same cave Bull (1975) recognised 'birdseye' structures (Fig. 5.16). These are interpreted as resulting from air entrapment during wetting of a previously dry sediment. Such features are thought to be diagnostic of dominantly dry phases in a cave development sequence.

Evidence for deposition in deglacial, periglacial times comes from a paleomagnetic study of cave sediments in the cave (Noel *et al.*, 1979). The magnetic properties of a sediment are related to the polarity (north or south direction) of the earth's magnetic field; changes in polarity have occurred over time which enable sediments to be assigned to periods of contrasting polarity. In the study of Noel *et al.*, sediments could be dated at around 12,000 years ago, when deglacial conditions would have been occurring. Combining the evidence of frost action on grains with the paleomagnetic data strengthens support for a major period at deposition in a periglacial environment in post-glacial times.

5.9 Cave deposits

As distinct from the largely unlithified *sediments*, cave *deposits* are formed by crystal growth, usually of calcite though many other minerals may also be present (White, 1976). **Stalactites** and **stalagmites** are familiar examples, and the term **speleothems** is commonly given to all such deposits.

Fig. 5.16 Birdseye structure in cave sediments (modified from Bull, 1975).

Speleothems are formed in most cases because degassing leads to the precipitation of calcite:

$$\underset{\text{ion species in water}}{Ca^{2+} + 2HCO_3^-} \longrightarrow \underset{\text{calcite deposit}}{CaCO_3} + \underset{\text{water}}{H_2O} + \underset{\text{carbon dioxide degassed}}{CO_2 \uparrow} \qquad [5.1]$$

This process tends to occur when waters with calcium and hydrogen carbonate contents which are equilibriated with high values of carbon dioxide from the soil and bedrock fissures encounter the lower values of carbon dioxide present in the cave atmosphere (Fig. 5.17). A disequilibrium situation occurs when percolation water reaches the cave air and the carbon dioxide content of the water drops, equilibriating with that of the cave air; calcite precipitation results (Fig. 5.18). In some circumstances, the cave air is also dry enough to allow evaporation to occur; frequently cave air is close to 100% relative humidity, but in caves near to the surface and especially in semi-arid climates or in climates with a

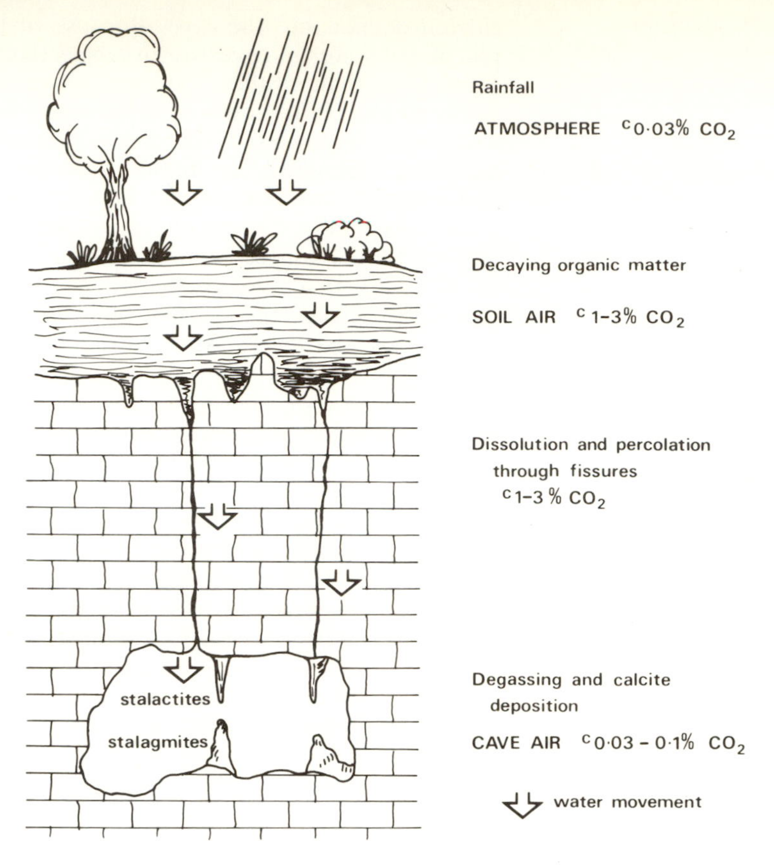

Fig. 5.17 Model for the formation of calcite deposits in cave systems by degassing, the carbon dioxide being primarily derived from soil air (modified from White, 1976).

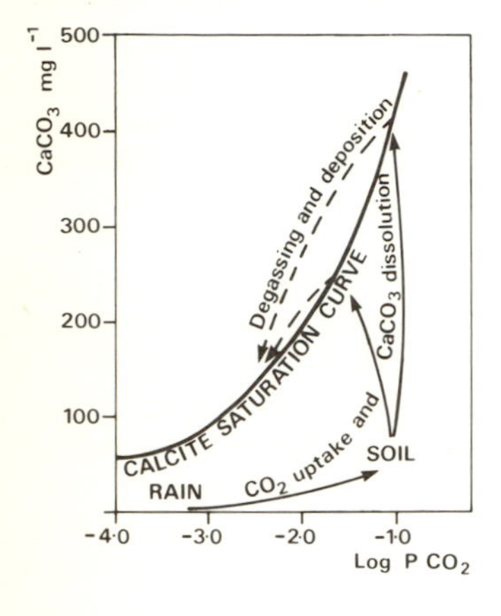

Fig. 5.18 Calcite solubility in the rainwater – soil water – cave water system, as a function of pCO_2 (modified from White, 1976).

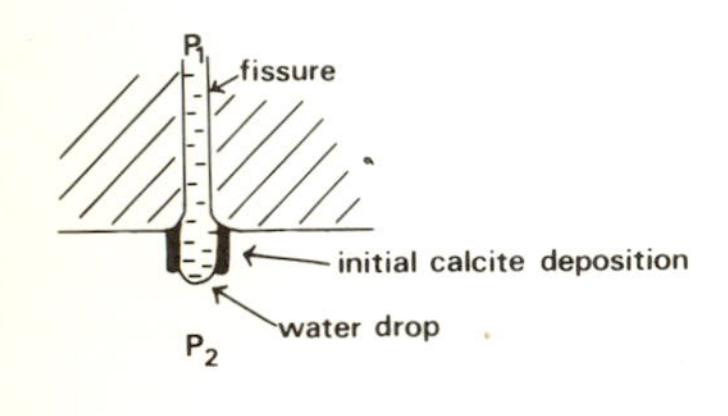

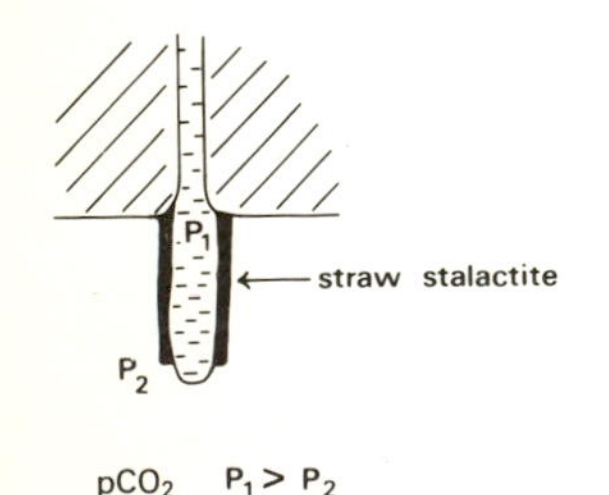

Fig. 5.19 Formation of stalactite around a drop of water in relation to pCO_2 differences, where $P_1 > P_2$ (modified from Roques, 1969).

dry season, this may drop and calcite may be precipitated simply by evaporation of water.

Stalactites may be initiated by a crystal rim around a drop of water (Fig. 5.19), the rim enlarging to form a straw stalactite. More massive stalactites are formed during seepage round the outside of a stalactite. The position of most stalactites is related to that of a joint or other fissure in the rock, down which percolating solutions move. Combinations of flow rate, calcium content and local conditions of topography, climate, trace element composition and flood frequency combine to produce a large variety of speleothem forms. Two basic divisions are **flowstones**, produced by sheet flow over an inclined surface, and **dripstones**, produced by point sources of water. **Helictites**, or erratic stalactites, grow in curved shapes under conditions when flow rates are small. Instead of being orientated gravitationally by the weight of dripping water, crystals form in thin films of water and can grow in a variety of directions.

5.10 Cave dating

The major geomorphological importance of speleothems lies in the fact they can be dated. First, stalagmite cannot be deposited on a cave floor while an active stream is flowing over the floor; therefore, deposition must be subsequent to the drainage of the cave by the diversion of flow to a route lower down in the rock mass. This gives a relative indication of a cave sequence. More important, uranium series dating of stalagmites can be used to give a minimum date for the abandonment of that passage. In this way, not only can the passage sequence be dated, but also, if it is assumed that passage abandonment sequences occur in relation to base level lowering sequences (Figs. 5.1 and 5.3), then the cave abandonment data can be used to date the surface lowering sequence occurring outside the cave, e.g. Williams, (1982).

Dating by uranium series is limited to 350,000 years BP and is based on the radioactive decay of ^{234}U uranium isotope to the thorium isotope ^{230}Th, the former having a half life of 2.47×10^5 years. Age can be determined from measurements of the $^{230}Th/^{234}U$ ratio and a knowledge of the decay rate (Gascoyne *et al.*, 1978).

The technique is applicable only with certain qualifications and it is based on certain assumptions. First, no clay or other insoluble detritus should be present in the speleothem because thorium is strongly adsorbed on to clays in the form of the Th^{4+} ion. Clay-free speleothems are therefore also initially free of thorium, but if clay and associated thorium are initially present then the assumption that all thorium present is derived from uranium decay is invalid. Samples containing more than 1% acid-insoluble residues are rejected for analyses. Second, uranium is derived from the rock through which the speleothem-forming percolate travels and the speleothem must contain uranium concentrations in excess of 0.05 ppm in order that determination resolution may exceed analytical errors. Third, samples where partial dissolution and/or recrystallisation are evident are rejected since fresh material of a later date may have become incorporated into the material. A date sequence is determined from samples from the entire length of the speleothem and the oldest date should be found at the base; where this is not the case, disturbance is evident and the sample is rejected.

5.11 Sequences of cave development and surface landforms

Published speleothem dates are largely coincident with dates commonly regarded as interglacial (Atkinson *et al.*, 1978) (Fig. 5.20). The data for the United Kingdom show (1) a late Devensian and Holocene deposition period from 17,000 years BP to the present; (3) a period from 90,000 to 140,000 years BP and (4) from before 170,000 BP to the resolution limits of the method; in addition there is a short period (2) around 60,000 BP. Similar dates are presented by Harmon *et al.* (1977) for the Canadian Rocky Mountains (Fig. 5.21).

During interglacial periods, sea-levels were higher than during glacials

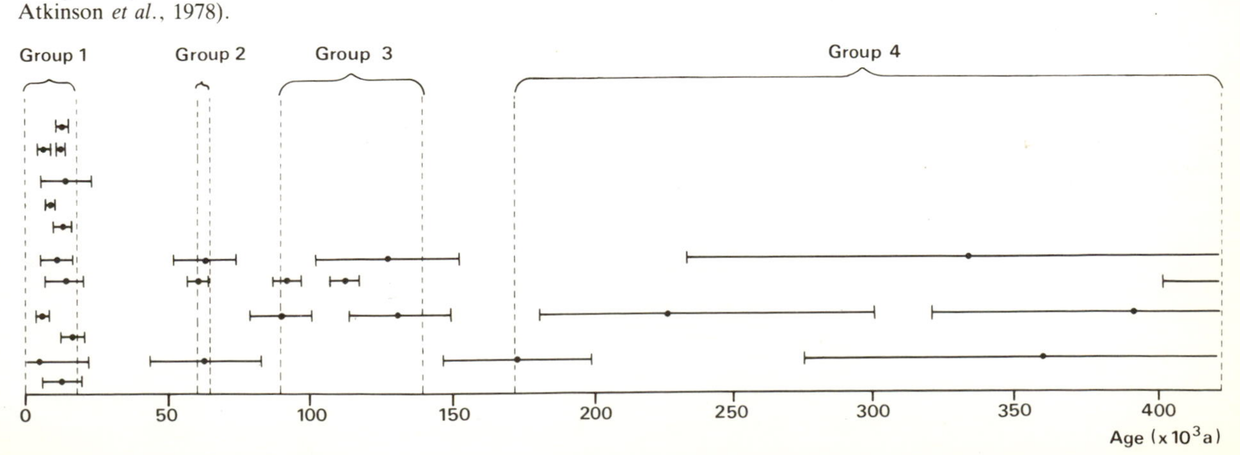

Fig. 5.20 Distribution of age of British speleothems. ├──•──┤ = error bars; brackets = interglacial periods (modified from Atkinson *et al.*, 1978).

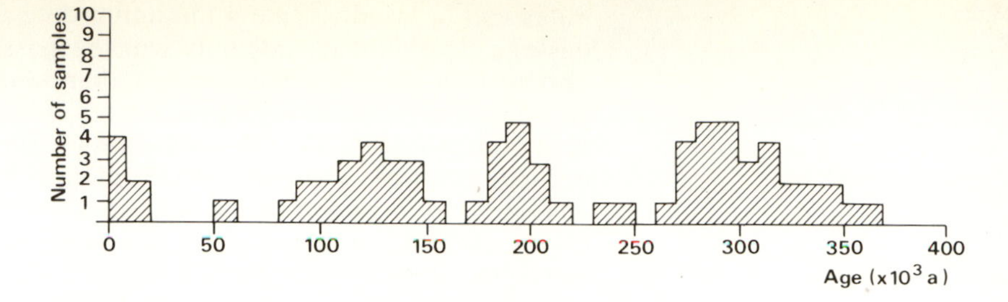

Fig. 5.21 Histogram of speleothem ages determined from the Canadian Rocky Mountains. Data plotted without showing error bars on age determinations (modified from Harmon *et al.*, 1977).

as, in the glacials, much of the earth's water volume was present as ice. Thus, for caves in Bermuda which are presently under sea, but which were terrestrial during low, glacial sea levels, Harmon *et al.* (1978) show that speleothem dates coincide with glacial periods, giving a mirror image of dates for glaciated areas (Fig. 5.22).

Dating work can also be corroborated by analyses of $^{18}O/^{16}O$ ratios in speleothems. The amount of ^{18}O decreases with increasing temperatures and therefore lower values of ^{18}O in terrestrial speleothems in northern latitudes correspond with warmer interglacial periods of speleothem deposition (Gascoyne *et al.*, 1978).

In geomorphological studies sequences of cave development have long been recognised in several areas, with upper, older series and lower, younger series, as, for example, described by Miotke and Palmer (undated) for Mammoth Cave, Kentucky, USA. Here the successively lower cave levels are related to valley incision, with evidence for glacial terraces (Fig. 5.23). A similar, but simpler, sequence of downcutting is

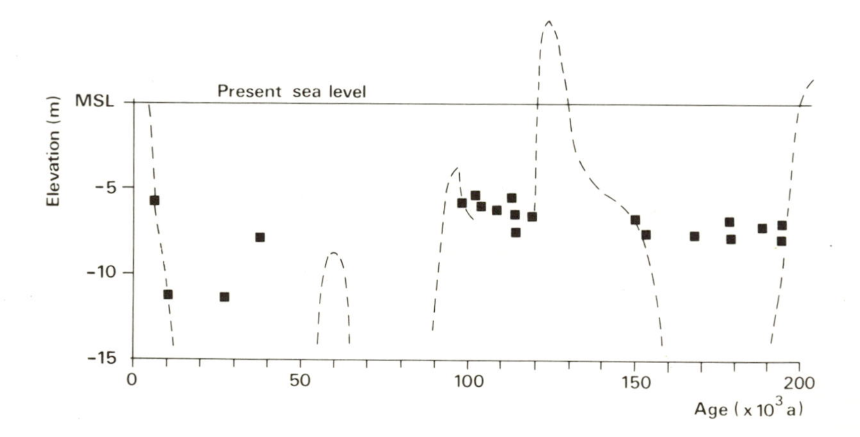

Fig. 5.22 Ages for speleothems now submerged in Bermuda. Deposition is inferred to be interglacial, at times of low sea level (modified from Harmon *et al.*, 1978).

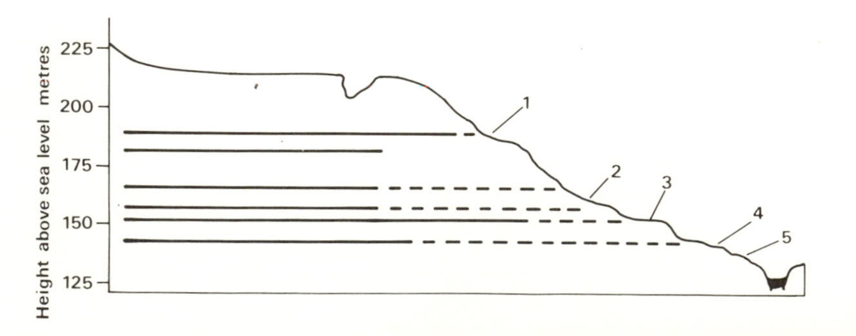

Fig. 5.23 Generalised profiles of cave levels, Mammoth Cave, Kentucky, USA. 1, Upper terrace (?pre-glacial). 2–3, ?pre-last glacial terraces. 4, Upper last glacial. 5, Lower last glacial (modified from Miotke and Palmer, undated).

Fig. 5.24 GB Cave system and Cheddar Gorge, Mendip Hills, Somerset, UK. 1, GB Cave, a swallet stream cave. 2, Great Oones Hole, a dry upper level. 3, present resurgence. . . . . = inferred upper level drainage line. ——— = present floor of Cheddar gorge (modified from Atkinson *et al.*, 1978).

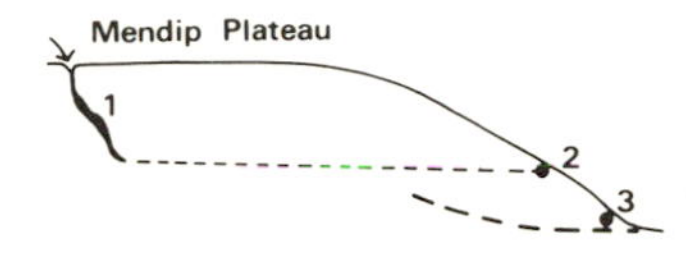

seen in Fig. 5.24 for the Mendip Hills, Somerset. Here, however, speleothem dates are available which can be used to indicate the rate of downcutting and the long length of time of the evolution of the system (Atkinson *et al.*, 1978). On Fig. 5.24, the upper level (2) was abandoned at around 350,000 years BP, the date of initiation of speleothem deposition. Karst scenery with underground drainage had thus been well developed prior to this time. This is also attested to by the presence of abandoned phreatic passages at levels higher than level (2). Surface erosion and valley downcutting is thought to have been active in the area because although the area was unglaciated during the Pleistocene it was close to the ice margins and subject to periglacial activity with attendant permafrost and surface meltwater erosion. Downcutting from the upper, dated, level to the present, lower, level of resurgence can be calculated to have occurred at a mean rate of approximately 0.2 m per thousand years.

In the Ingleborough district of North Yorkshire, UK, an upper level of dated stalagmites, post-dating the draining of the cave passages, is given at around 350,000 years BP. This is down to the levels (1) and (2) shown on Fig. 5.25. Cave levels at up to nearly 100 m above present levels are indicated at this time, with an intermediate stage some 50 m above the present level at 250,000 years BP. An overall, mean erosion rate from the upper to the present level is 0.12 m per 1000 years. However, this area, unlike the Mendips, was glaciated and it is likely that valley incision was largely a result of glacial erosion. If the incision was confined to glacial epochs alone, the rate for those periods alone would be 0.21 m per thousand years (Waltham, personal communication). It is inferred from this work that two thirds of the evolution of the Yorkshire Dales landscape pre-dated the glacial erosion, with major valley incision before that time. Glacial action in the valleys deepened them, leading to a further downcutting of the caves of up to about 100 m.

Fig. 5.25 Inferred landscape development, Ingleborough region, North Yorkshire, UK. 1, pre-glacial valley floor with 2, level of maximum downcutting at maximum depth of cave development; 3, present maximum depth of cave development (modified from Atkinson *et al.*, 1978).

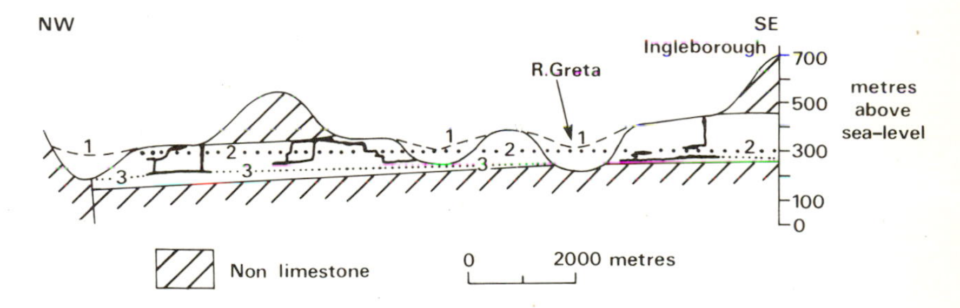

The antiquity of some cave systems in general is indicated by the fact that many isolated upper level remnant caves studied have dates beyond the limit of the uranium-thorium dating method, in excess of 350,000 years. In addition, there is evidence of pre-last glacial caves cut by glacial erosion, for example, in County Clare, Eire (Fig. 5.26) where there are clear striations on cave walls.

In tropical areas, where glacial deepening of valleys has not occurred, uplift and deep incision have been important features, as, for example, inferred by Brook (1977) for the Finim Tel caves in the Hindenburg Ranges of New Guinea (Fig. 5.27). Here, uplift and tilting since Miocene times has been responsible for the extent and depth of cave development. The caves increased with depths as the height above sea level increased in relation to uplift.

Very old, large and spectacular caves exist in other tropical areas, notably the Mulu area of Sarawak. Here, Deer Cave (Fig. 5.28) is of vast size, indicative of a former very active system, but it is now an isolated cave fragment, following subsequent erosion around the cave system.

5.12 Conclusion

Caves are not only worthy of attention because of their intrinsic interest through exploration, physical challenge, natural beauty and geomorphological features; they also present an ability to estimate dates for the downcutting of entire landscapes. This gives an opportunity for the discussion of the relationships between processes, forms and rates of erosion rare in geomorphology. While these kinds of relationships are still not fully understood, it is evident that many landscapes visible at the present day are old, and that many of the major features evolved preglacially and glacially.

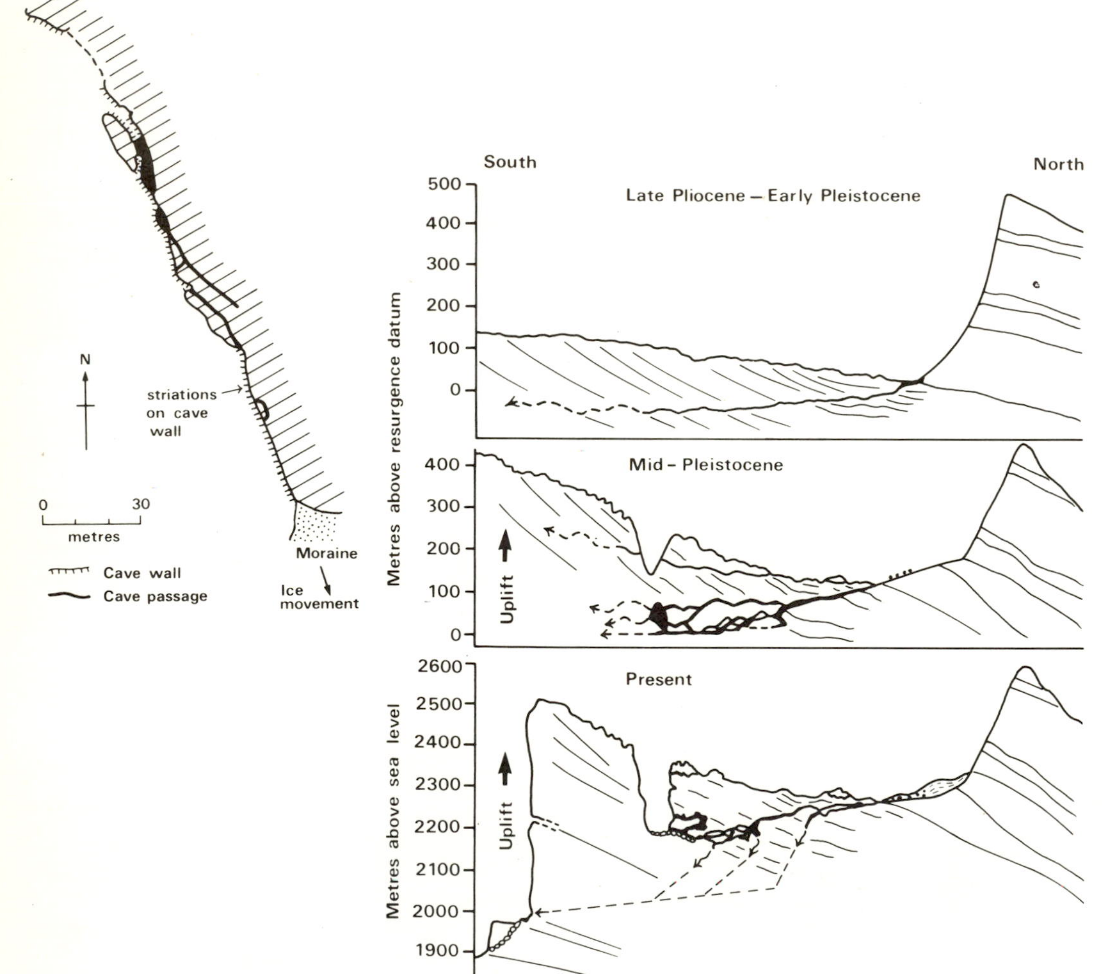

Fig. 5.26 Longitudinally eroded cave section, Poulcraveen, County Clare, Eire. Glacial striations are found superimposed over cave scallops at the south end of the cave wall (modified from Trudgill, 1971).

Fig. 5.27 Inferred evolution stages of caves in the Hindenburg Ranges of New Guinea (modified from Brook, 1977).

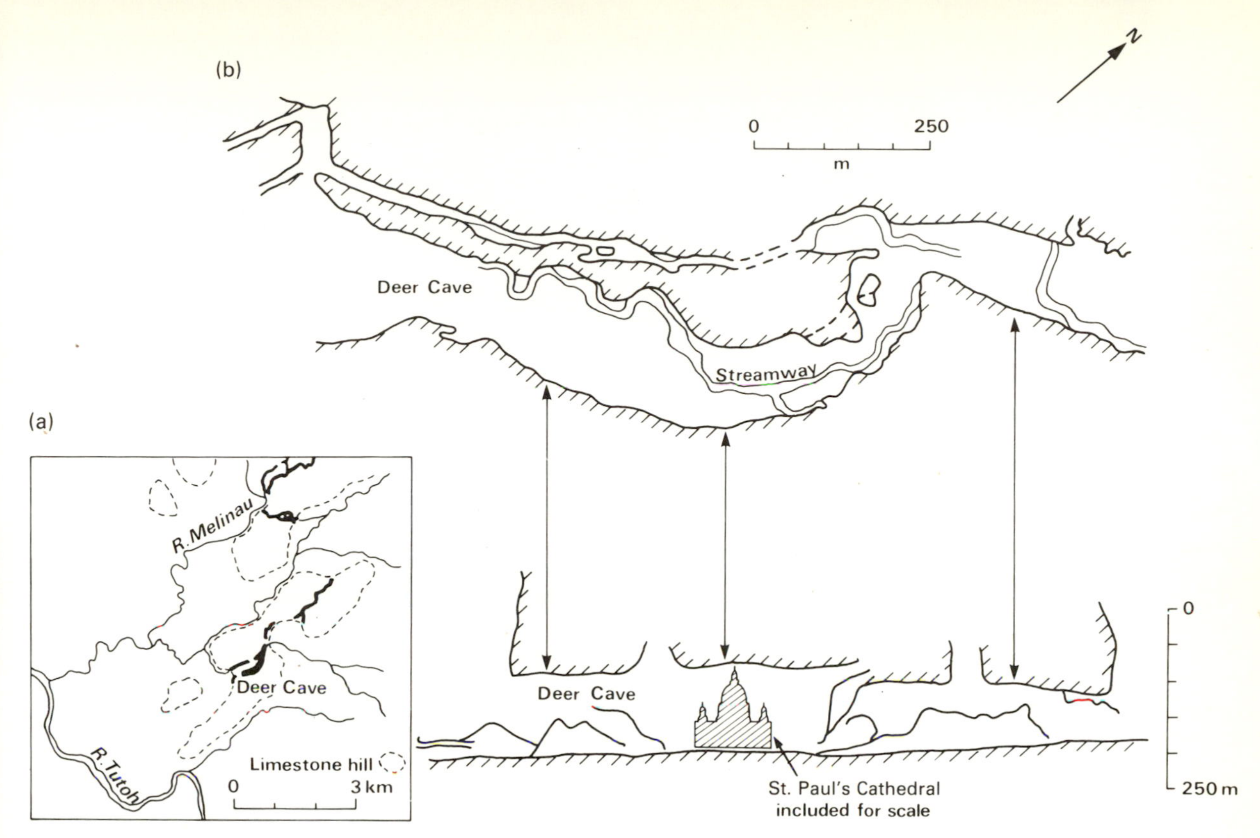

Fig. 5.28 (a) location and (b) plan and section of Deer Cave, Mulu, Sarawak. Note the large size of the cave passage and its isolation, evidence of former active cave formation and subsequent erosion of surrounding land (modified from Brook and Waltham, 1978).

6 Limestone landforms in a fluvial system

6.1 Introduction

Integrated fluvial systems develop on the land surface as networks for the transport of run-off waters to the sea; in addition they carry solutes and sediments and thereby facilitate erosion in drainage basins and the progressive downwearing of the landscape. Fluvial systems in limestone landscapes are no different from those in non-limestone landscapes. They perform the same functions and are merely located partly or wholly below the surface (Williams, 1978a, p. 260).

While the general functions of fluvial systems may be similar in limestone and non-limestone areas, their geomorphological effects may be different. This is because the diversion of fluvial systems underground acts to short-circuit much of a rock mass when viewed in the vertical dimension. In a non-limestone area, retreat of a river profile upstream proceeds from a base level with significant lateral as well as vertical erosion. The balance of vertical and horizontal effects is determined by the rate of downcutting relative to the rate of any uplift of the land mass. In a limestone system, the same is true in essence, but the balance point is somewhat different since much of the rock mass may be bypassed by the diversion of erosional activity underground: this gives the system a greater vertical development and encourages the formation of relatively dry, upstanding masses. The relative relief from interfluve to stream bed is thus greater in a limestone system. The upstanding masses are attacked basally by fluvial systems which have adjusted to base levels more rapidly than would be the case in a non-limestone system. In non-limestone areas the river has to erode a bed through all the rock mass occurring between any given initial level and base level; in limestone areas this stage may be shortened by the integration of underground flow (Fig.6.1).

Fig. 6.1 By-passing of rock mass by karstification of fluvial systems. (a) initial profile of river bed and base level. (b) Non-limestone system, successive lowering of profile by erosion of the river bed. (c) limestone system, rock mass between 1 and 2 is bypassed by development of underground system at level 2.

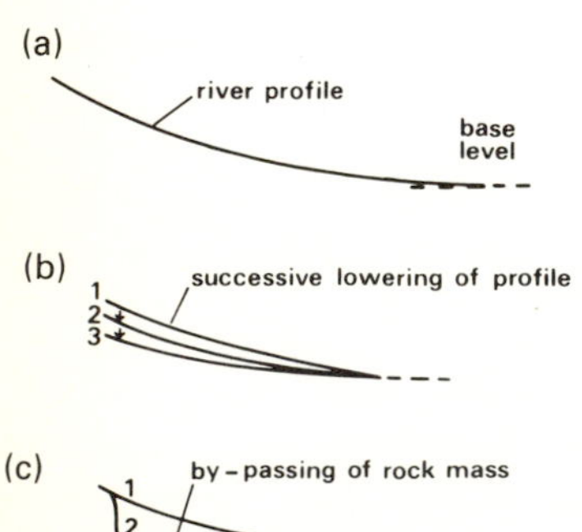

Limestone landforms can thus be seen in the context of a fluvial system. Both relict and active features may be seen. First, relict fluvial features occur, principally dry valleys, abandoned because of the underground diversion of fluvial action. Second, vertically orientated features form actively in relation to the subterranean drainage levels. These include closed depressions which have no surface drainage outlet and depend on subsurface vertical percolation for their drainage and on subsurface abstraction of weathering and erosion products for their propagation. The drainage from these features collects and is transported away by the underground fluvial system. Closed depression drainage systems thus operate in a similar fashion to hillslope throughflow and tributary streams in a surface fluvial system. Third, as base level is approached, or an underground fluvial system reaches an impermeable non-limestone basement rock, lateral erosion becomes more important. Limestone masses become isolated as basal fluvial erosion proceeds. The relative rates of fluvial erosion and surface erosion will, together with relative rates of uplift, act to determine the shape and height of any landforms existing at any one time. These general outlines will be discussed in specific contexts below.

6.2 Abandoned and reorganised fluvial systems

Dry valleys occur as a consequence of the development of an underground fluvial network in relation to a given base level. It follows that the prerequisites for the abandonment of dry valleys are the opening of joints and other fissures in an aquifer and the integration of flow in a network to an output point whereby the network has sufficient capacity to carry all the water previously carried on the surface. This may occur with or without active base level lowering, but base level lowering and land mass uplift will encourage the process. This developmental sequence is common to all parts of the world. In addition, where glaciation occurs, the existence of permafrost conditions under periglacial conditions adjacent to ice masses, whether at their maximum extent or retreating, will tend to discourage subsurface flow, returning flow to the surface. This will encourage deepening and widening of valleys which would otherwise have been dry. Glacial and periglacial meltwaters will be the major source of run-off under these conditions. In areas which have undergone glaciation, much of the abandonment of dry valleys will have been facilitated by glacial erosion acting to lower local base levels, encouraging steep hydraulic gradients in the aquifer and the diversion of streams underground.

Fig. 6.2 Variation in underground flow route according to input conditions, Bishopston Valley, Gower Peninsula, Wales, UK. The basic system, (a) operates progressively as input increases from (b) to (f) (modified from Ede, 1975).

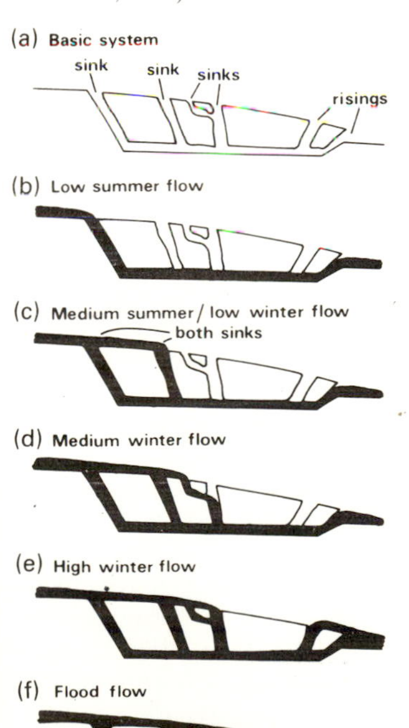

The hydrological links of dry valleys with subsurface systems can be seen by the fact that they may take water flow during high flood events when the capacity of the underground system is exceeded. A system of progressive diversion of water flow towards the surface is recognised by Ede (1975) (Fig. 6.2). Similarly, the Cheddar Gorge system on the Mendip Hills, Sometset, UK, was flooded in 1967 when 5 inches (*c*.12 cm) of rain fell in 2 hours (Fig. 6.3). Likewise the long abandoned surface drainage system at Malham Cove (Fig. 1.4, p.4) is known occasionally to take surface water when heavy rain occurs, especially when this coincides with surface thaw conditions and frozen ground.

Connections of fluvial systems underground bear no necessary relationship to the surface fluvial system. They may be dendritic (Williams, 1971) where hydraulic gradients are not steep and geological conditions are uniform, as maps of explored cave systems may show (Fig. 6.4). However, geological controls often reorganise drainage basins so that water may drain to output points beneath surface divides (Fig. 6.5a). In addition, water in conduits travelling in a relatively impermeable rock mass may travel independently, crossing without mixing or only mixing at high flows (Fig. 6.5b). Flow nets are only well integrated in highly fissured limestones.

Fig. 6.3 The flooding of Cheddar Gorge, Mendip Hills, UK on 11 July 1967 (from Hanwell and Newson, 1970). (Photo: M. D. Newson.)

6.3 Closed depressions and polygonal karst

On an impermeable bedrock, the hydraulic gradient established in hill-slope soils is downslope, that is water draining under gravity is deflected from a vertical pathway at a vector determined by slope angle as it drains to surface streams. In an unsaturated pervious limestone bedrock, the hydraulic gradient is dominantly vertical. The gravitational potential is dominant over any topographical one and percolation tends to be downwards. This can be deflected sideways if joints are not opened, or if no nearby cave system exists at depth to carry percolation drainage away or if the bedrock is saturated. Then the drainage direction will be at a vector defined by the location of outlet points.

In addition to hydraulic gradient the development of opened joints will also depend upon the ability of percolating water to dissolve limestone. This will depend largely on soil acidity and soil carbon dioxide levels (as described in Chapters 3 and 4). Acid percolation water can thus open up fissures in limestone, initially draining downwards and to outlet points which may be a cave system vertically below, or more obliquely, to outlet points. The flow may become integrated, with the preferential opening of a few vertical joints, especially where two major joint systems

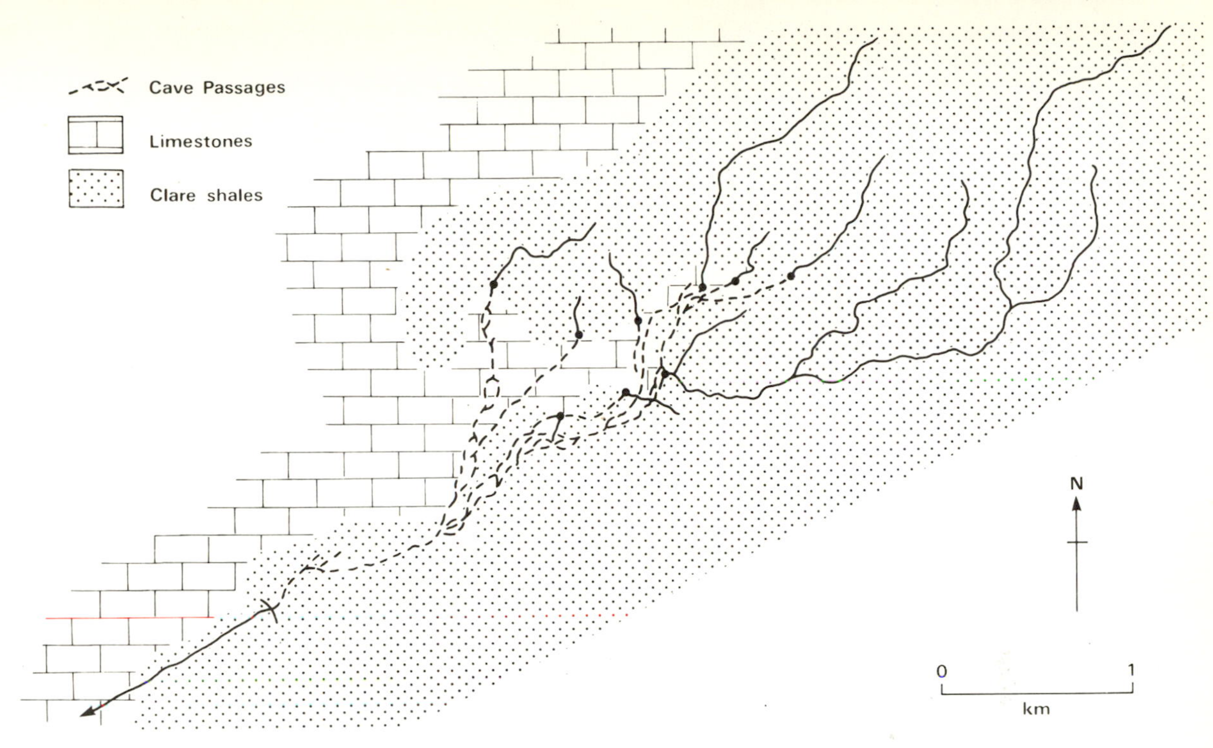

Fig. 6.4 Dendritic cave system, Doolin system, County Clare, Eire in a region of partially shale covered unfolded limestone and gentle (2°) dip to the southwest (modified from Tratman, 1969).

cross. If weathering and erosion products can be carried downwards readily in such opened systems, this will lead to preferential loss of material above the system. The loss will not only be solutional, but also of soil material washed down the opened joints and fissures. Thus, depressions may form on the surface by subsurface abstraction. This is illustrated in Fig. 6.6. Because of the dependence of flow upon the exploitation of fissure and joint systems and because these, in turn, tend to be produced by regional tectonic stresses in regular patterns (as seen in the discussion in sections 2.4 and 4.3, pp. 11 and 63), then the depressions often show regularity of pattern or alignments, frequently in association with structurally guided cave systems below. This is illustrated for the Dan-yr-Ogof and Ogof Ffynnon Ddu systems in South Wales (Fig. 6.7).

Depressions may be isolated in a plateau area or, if exploitable joints occur frequently enough and development can progress through time, the depressions may become contiguous, with an identifiable divide between each depression. Such a landscape has the appearance of an egg box when viewed from above and can be termed polygonal karst (Fig. 6.8). The geometry of the depressions depends to a certain extent on the amount of water they have to transmit. In high rainfall areas they may tend to coalesce as there will be more rainfall per unit area of ground surface; this greater depth of rainfall means that there will be more water per exploitable joint. Thus, for one given joint frequency depressions will coalesce, as erosion rates will be greater where there is greater discharge per depression. Thus, in drier areas, not all joints are fully exploited and depressions tend to occur with intervening flatter areas. In many tropical areas with high rainfall, such as the 'Cockpit' Country of Jamaica and some areas of New Zealand (rainfall 2370 mm,

Fig. 6.5 (a) Diagrammatic representation of two drainage basins with underground limestone drainage. The original surface drainage lines are now dry valleys and the sources for the risings come from an adjacent basin up-dip rather than from the source topographically above the rising. (b) Independence of two subsurface drainage systems developed in joints and bedding planes at different levels with intervening impermeable rock.

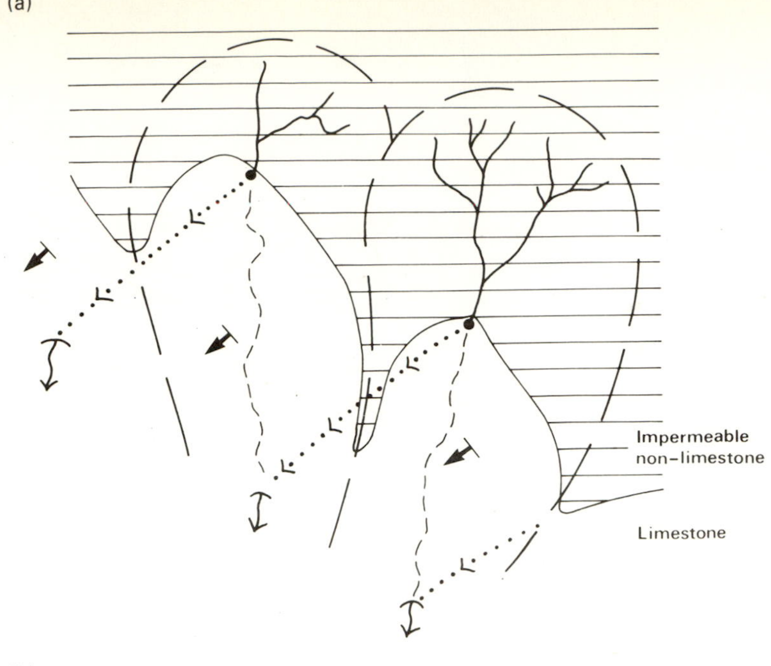

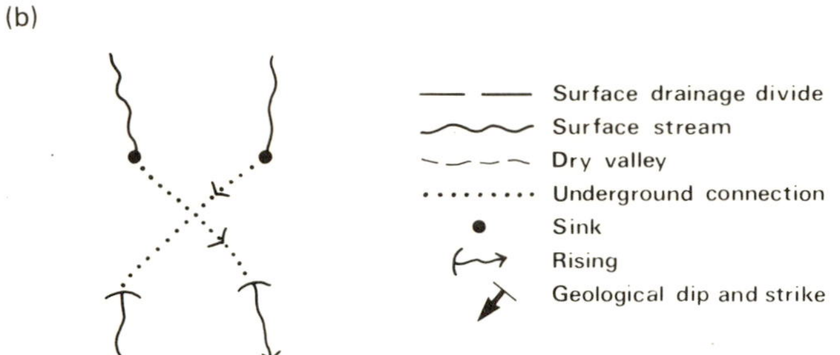

Fig. 6.6 Depression formation over major joints: (a), (b) and (c) show a sequence of solutional joint opening and the lowering of the surface above the joint. The major joint spacing could be 1–2 m or up to 0.5–1 km apart.

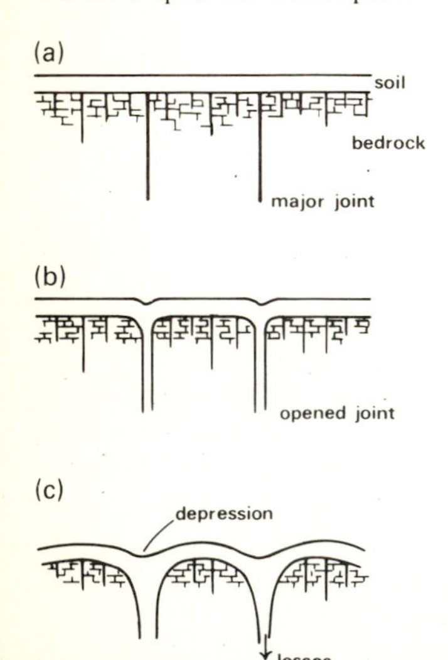

Gunn, 1981a) depressions coalesce, forming polygonal karst. Brook (1977) stresses the significance of joint spacing (Fig. 6.9).

Aside from these outlines above, great variations exist in closed depression morphology and distribution. Gradations also exist in the relative importance of various factors. Thus, while Jennings (1975) distinguishes **solution dolines**, a closed depression where the soil surface is gently lowered from below by bedrock dissolution, and **subsidence dolines**, where the soil itself is actually abstracted below into cave systems, in practice these are difficult to separate, though in subsidence dolines, rapid subsidence may be evident from exposed shear faces in the soil mass. **Collapse dolines** may also occur, more readily identifiable by steep or vertical sides where collapse into a cavity below has occurred. Collapse dolines are less frequent than other dolines, and the distinction between solution and subsidence dolines is a blurred one since some component of each is liable to be present in many cases. Smith *et al*. (1972) support the view that closed depression development proceeds in tandem with the development of an underground drainage network; the network acts as a transport system for the weathering and erosion products derived from the surface and near surface limestone present in the depressions.

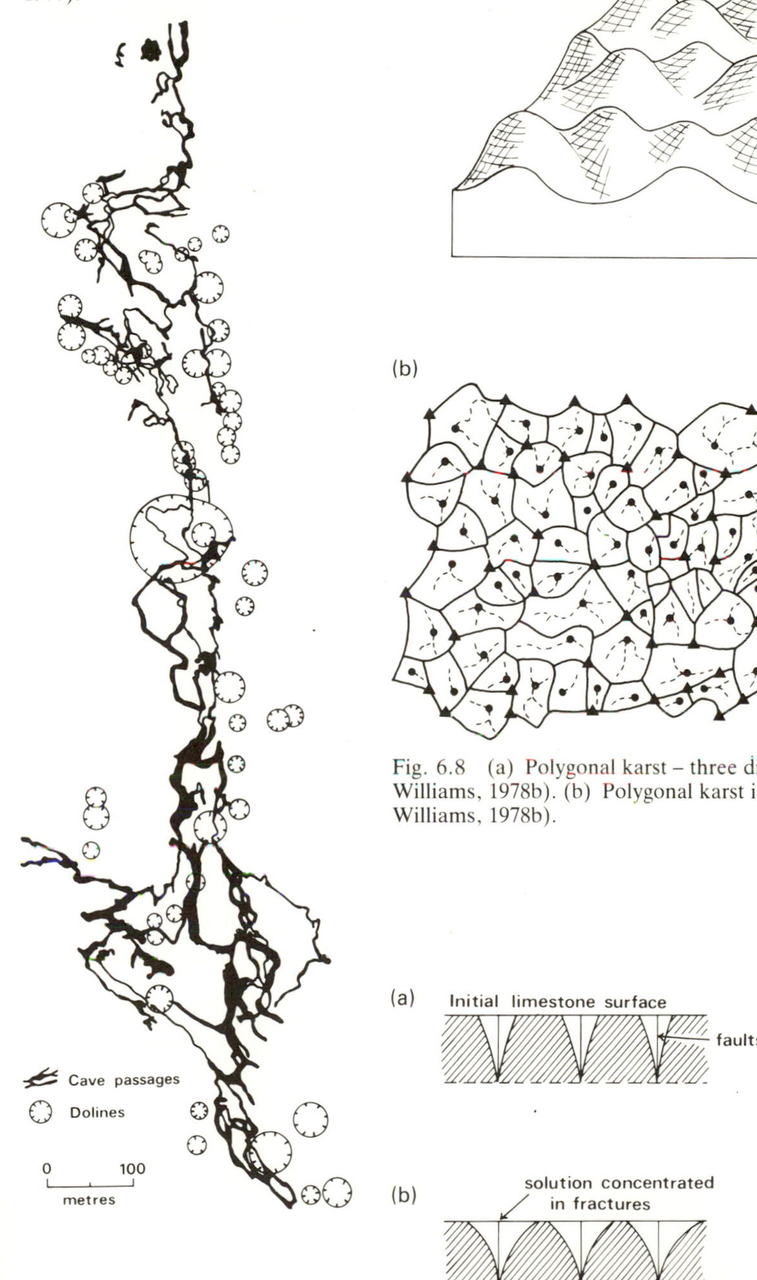

Fig. 6.7 Alignment of surface depressions and cave systems, Dan-yr-Ogof, South Wales, UK (modified from Coase and Judson, 1977).

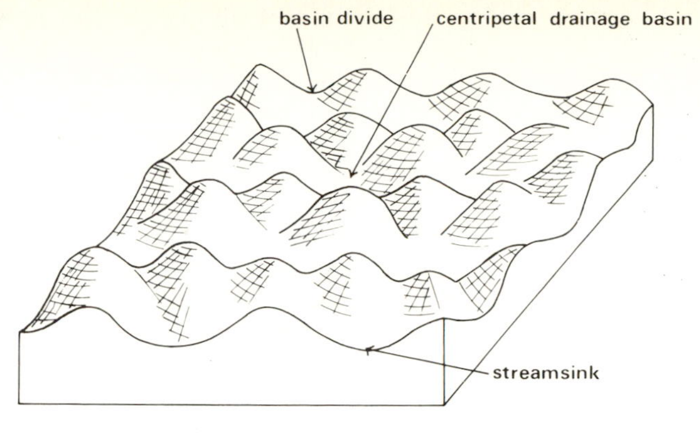

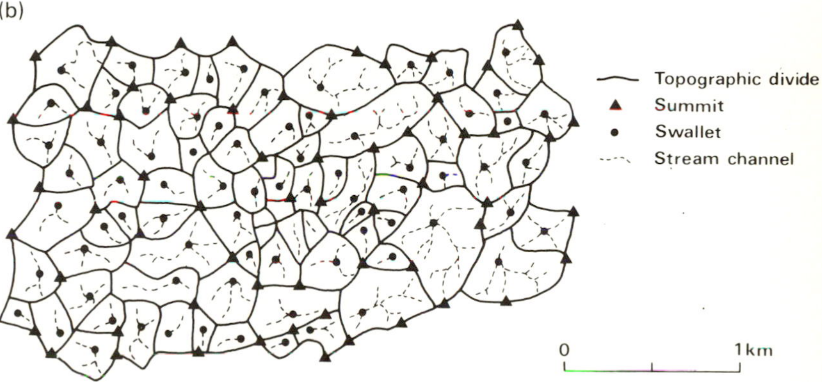

Fig. 6.8 (a) Polygonal karst – three dimensional representation (modified from Williams, 1978b). (b) Polygonal karst in New Guinea seen in plan form (modified from Williams, 1978b).

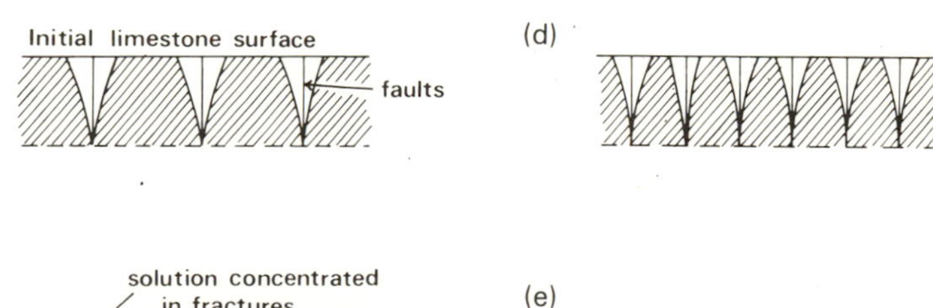

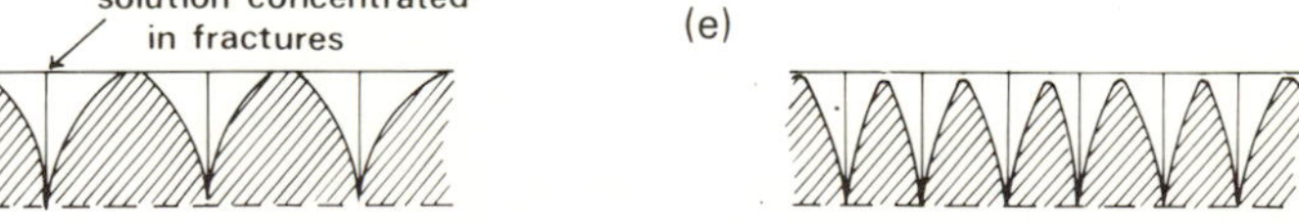

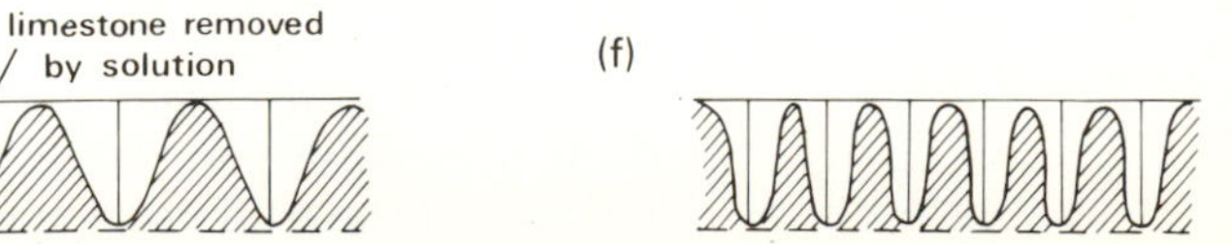

Fig. 6.9 Sequences of depression formation in relation to joint or fault spacing. (a) to (c), sequence with widely spaced joints, (d) to (f) with closely spaced joints (modified from Brook, 1977).

Integration of subsurface flow nets beneath closed depressions is difficult to assess because of the multiplicity of input points and the fact that the alluvial deposits and soils in the base of the depressions make water tracing difficult. However, painstaking work by Gunn (1981a, b) has shown by dye tracing of thirty-four depressions that drainage divide delimitation is possible (Fig. 6.10)

6.4 Interstratal karst

Closed depressions may be formed in thin beds of non-limestone material overlying limestone; Jennings (1975) defines these as subjacent karst dolines, and Bull (1977) shows how they may be related to cave systems below (Fig. 6.11). Karst features developed in limestone between non-limestone strata can be termed interstratal karst. This has been studied by Thomas (1974) in South Wales, where Carboniferous Millstone Grit has collapsed into Carboniferous Limestone (Fig. 6.12). The formation of depressions is not only encouraged by the presence of joints in the limestone and caves beneath, to act as drainage conduits, it is also promoted by the occurrence of very acid soils (pH 3–4) occurring on the gritstone. Acids percolate from the gritstone surface through joints in the grit, leading to very marked erosion at the grit-limestone subsurface contact.

6.5 Towers, plains and poljes

Limestone towers are isolated upstanding masses of limestone surrounded by eroded plains. Poljes are very large closed depressions, surrounded by limestone hills, usually floored with impermeable materials and commonly closely related to regional structure. These features are often regarded as more mature aspects of a karst landscape.

Isolated hills and towers are often regarded as residual features and are frequently surrounded by alluvial deposits. In Puerto Rico, the towers are termed Mogotes and have been studied by Miotke (1973), Day (1978) and Ireland (1979). Erosion appears to be concentrated at the base of the towers, giving steep sides and emphasising lateral retreat (Fig. 6.13). However, not all tower-like residuals are scattered over alluvial plains (Jennings, 1972). Cliff foot caves undoubtedly play a role in the basal erosion of towers, as does recession caused by run-off, leading to pediment formation (Ford, 1979). Towers are essentially residuals left in relation to the increased verticality imposed on karst drainage by the incision of drainage, permitted by the fact that drainage can occur through the rock rather than merely over it. Isolation of rock masses from fluvial erosion is an inevitable consequence of preferential flow along the more exploitable flow lines in the rock.

6.6 A sequence of limestone landforms in a fluvial system

A sequence of limestone landform evolution can be suggested, as outlined in Fig. 6.14. Initially an integrated dendritic system exists on impermeable rock overlying limestone strata (a). Subsequently, fluvial erosion tends to expose local areas of limestone at the surface, leading to the establishment of a hydraulic gradient from the surface to base level through the aquifer (b). Drainage nets become established through exploitable fissures and joints, some of which develop preferentially into cave systems. Erosional loss of cap rock proceeds by subaerial processes (c). Later, run-off water is derived almost wholly from outside the limestone area, with percolation water derived from the limestone mass itself. The retreat of the impermeable cover is an important process as

Fig. 6.10 Water tracing in polygonal karst, Waitomo District, New Zealand (modified from Gunn, 1981a).

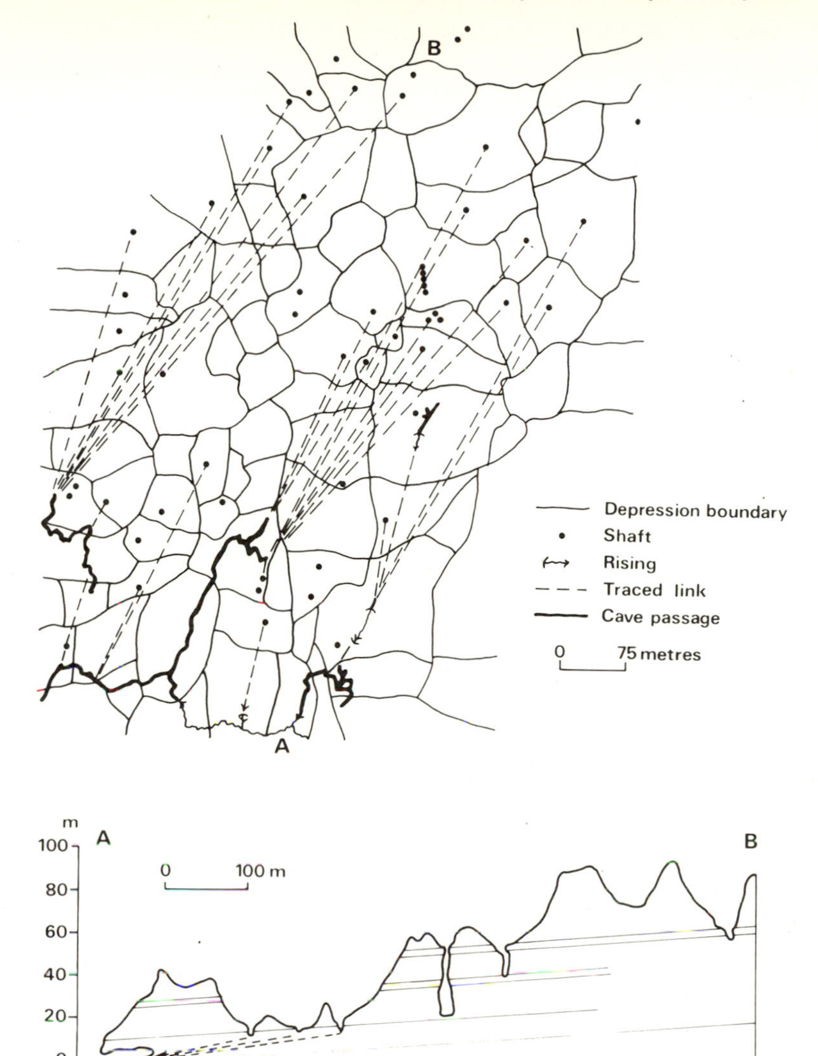

Fig. 6.11 Relationship of surface depressions and cave systems, (a) dissolution with percolation to cave below. (b) dissolution and partial collapse into cave. (c) depression with collapse into cave (modified from Bull, 1977).

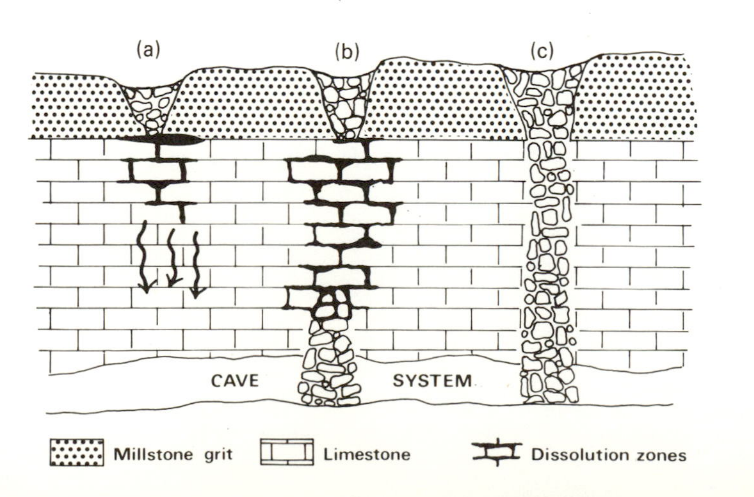

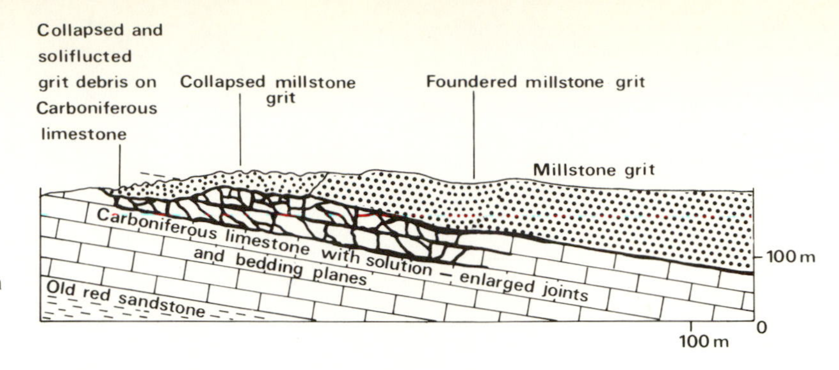

Fig. 6.12 Schematic section with collapsing and subsiding Millstone Grit on the Carboniferous Limestone (modified from Thomas, 1974).

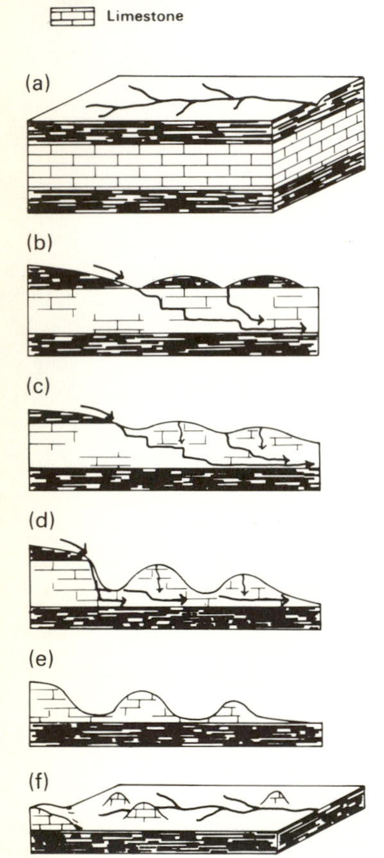

Fig. 6.14 Progression of evolution of karst landscape (a) to (f), with dissection of limestone block between two impermeable rock layers; for explanation see text.

percolation water derived on limestone soils is liable to be alkaline and capable of little opening of joints. Unless an acid drift deposit is present over the limestone, the only source area for acid waters will be the caprock. Thus swallets and initial joint opening is liable to be achieved at the margin of the caprock. In glaciated areas, meltwater may also assist this process, balanced by the glacial erosion losses of near surface bedrock layers with previously opened fissures and joints (d). Dissection of the limestone landscape continues, with the isolation of rock masses by preferential flow zones. The verticality in the landscape depends upon the relative rate of uplift and erosion by the integrated fluvial network (e). As base level is reached, horizontal fluvial erosion occurs and rivers act to erode basal positions of residual hills. Finally a surface fluvial system is established, with isolated limestone hills (f). This progression may be viewed in the critical light of existing evidence and argument. It is likely that rapid downcutting with rapid uplift obviates some of these stages. Similar sequences are proposed for Canadian karst Brook and Ford 1977, Fig. 6.15) and for Chinese karst (Williams, 1978b, Fig. 6.16). The great diversity of combinations of limestones, joint frequencies, rainfall patterns, erosion rates and uplift rates will mean that there will be great variety in detailed landform patterns. However, these variations can be understood with reference to a basic model of landform evolution, if the assumptions of that model are correct. The basic factors influencing limestone landform evolution in a fluvial system have been

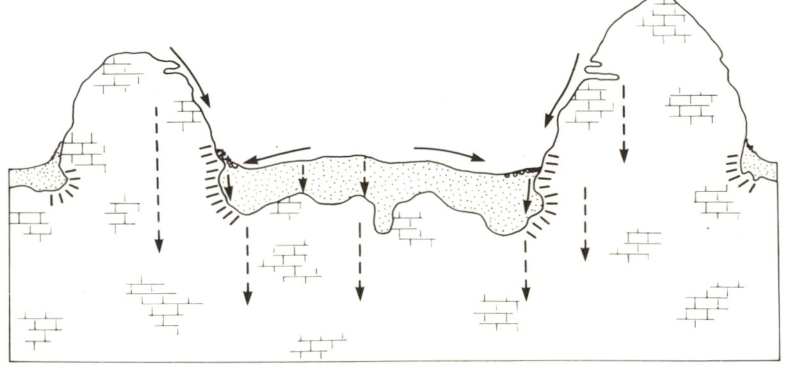

Fig. 6.13 Mogotes in Puerto Rico, showing zones of maximum erosion. Run-off water tends to collect at the edges of the depression leading to a focusing of erosion there (modified from Miotke, 1973).

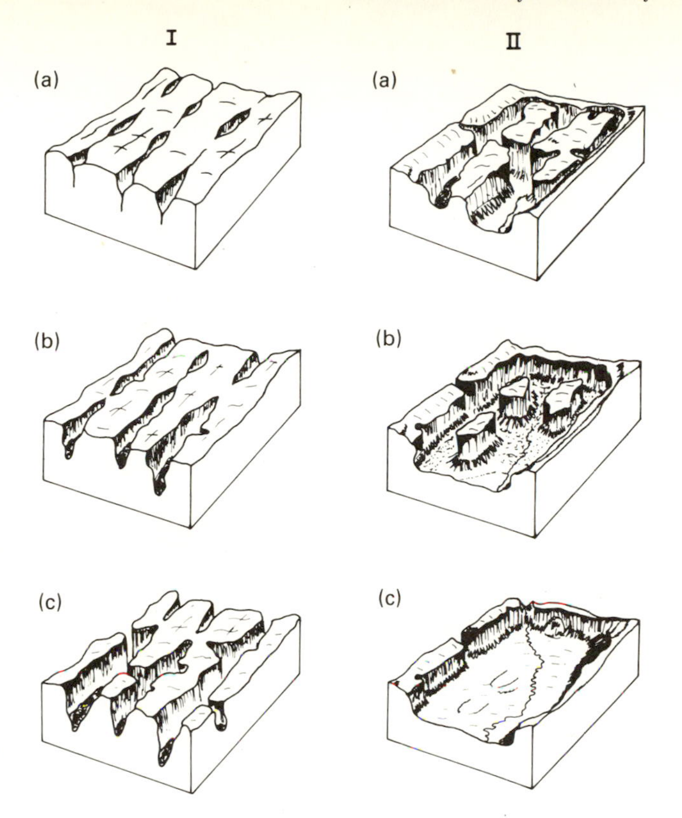

Fig. 6.15 Sequences of karst development (a–c), showing increasing dissection. I, development of fissure system where vertical incision is greater than lateral erosion: II, development of open depressions, with lateral erosion (from Brook and Ford, 1977).

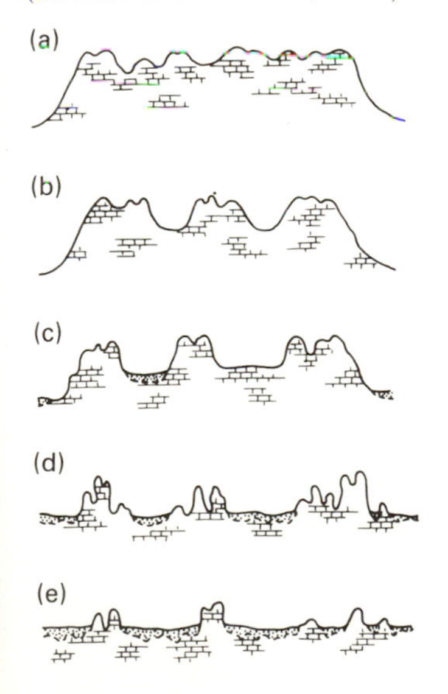

Fig. 6.16 Development sequence for Chinese karst. (a) high erosion surface, polygonal karst; (b) 50% incision, (c) wider valleys present, (d) clusters of towers with wide valleys and plains, alluvial soils, (e) plain with isolated towers (modified from Williams, 1978b).

outlined here; the important features are the surface–underground–surface transitions as the fluvial system works its way through the limestone mass. As with many geomorphological systems, an understanding of differential erosion is important in the explanation of topography. Erosion is dominated by dissolution in the soil and upper bedrock zone; joint and fissure patterns act to focus this action, leading to preferential erosion in closed depressions. Cave conduits act as transport networks for the collected erosion products. Vertical fluvial erosion rates may be slow in comparison to surface and subsurface lowering, unless uplift is rapid. Uplift rates and geological structure are probably the most important factors in influencing the variations in limestone landforms, together with climatic variations, especially rainfall, which act to reduce or increase the effective run-off and erosion rates.

7 The description of limestone landforms

7.1 Introduction

The description and quantification of landform dimensions are often seen as prerequisites for the understanding of the relationships between erosion processes and the evolution of form. However, many studies continue to focus solely on form or on process or on inferences about genesis. Notwithstanding this, if progress is to be made in the study of process–form relationships, the considerations of morphological description should be seen to be just as important as those of process. Thus, having outlined the basic factors which will influence the production of limestone landforms in Chapters 1–6, it will now be appropriate to consider the varieties and differentiation of landforms. Many attempts have been made to attempt to describe and classify limestone landforms and, more recently, attempts have been made to use morphometric analysis to quantify the dimensions of landforms.

7.2 Types of karst

Sweeting (1972) discusses types of karst, distinguishing between **holokarst**, or true karst, where solutional landforms dominate and all or much of the drainage is internal. **Fluviokarst** is more influenced by surface rivers and **glaciokarst** has been modified by glacial erosion. **Tropical**, **arid** and **periglacial** karsts are also recognised, by climatic association. Tropical karsts are often characterised by the presence of towers or cones, but these forms are by no means restricted to tropical areas (Brook and Ford, 1976); evolutionary history and structure are also important determinants of landform.

Because of the uncertainties of the climatic–landform links, as discussed more fully in Chapter 8, such terms as 'tropical karst', or any other climatic terms, are probably best avoided. Classifications and descriptions of karst features which avoid climatic inferences (such as discussed in section 7.3), or those non-genetic treatments which involve morphometry (sections 7.4–7.7), are to be preferred.

7.3 Landform types

The reorganisation of surface fluvial systems and the development of vertical relief, with closed depressions, are the characteristic features of limestone landforms. As such, description, classification and measurements of limestone landforms have focussed upon features such as blind valleys, the extent of vertical development compared to horizontal development and the degree of dissection.

In terms of fluvial features, Jennings (1971) sees a progression from surface fluvial features to the gradual re-routing of water flow underground. Thus, **gorges** and **canyons** represent deep downcutting, without underground flow of the river, as with non-carbonate rocks. Underground development is involved, however, in the formation of **meander caves** and **natural bridges**. Here, subsurface connections may be made between surface fluvial features, either short-circuiting a meander or expanding its dimensions. Bridges and caves are closely related features, in some respects bridges often being merely short lengths of cave, and in some cases representing remnant cave features. Further subsurface development leads to the formation of **semiblind valleys** (Jennings, 1971, p. 109) where the river is diverted underground for much of the time but is able to resume its former surface course during times of high flow. **Blind valleys** occur when the underground course is developed to such an extent that the former surface stream bed downstream of the sink is never deepened by erosion. It is thus left as a high, dry, remnant feature. Eventually completely **dry valleys** form.

A similar sequence has often been recognised for **closed depressions**. These are closed contour features without an influent stream. Simple, conical-shaped depressions, or **dolines**, are thought to coalesce, forming compound features termed **uvalas** (Sweeting, 1972; Jennings, 1971). The author Cvijic suggested that this led to the formation of **poljes**, but as discussed by Sweeting (1972, p. 192) and in this book on p. 105, poljes have a rather different scale and mode of origin, with the role of sediments deposited in the central area also being important.

Dolines have been subdivided according to shape and origin. Sweeting (1972, p. 45) recognises **bowl-shaped**, **funnel-shaped** and **well-shaped** dolines (the latter referring to a deep development rather than to any notion of merit). Sweeting (1972, p. 46), Williams (1969) and Jennings (1971, pp. 120–7) refer to four major types of dolines: (1) **solution dolines**, (2) **collapse dolines**, (3) **alluvial**, **drift** or **subsidence dolines** and (4) **subjacent karst collapse dolines** where non-limestone rock is present above limestone (Fig. 7.1.).

7.4 Doline morphometry

The form of dolines has also been studied by the use of morphometric measurements, as for example, by Jennings (1975). Jennings uses the following measurements of doline size and shape:

L = length (longest diameter)
W = width (greatest diameter perpendicular to L)
D = depth (height difference between the lowest lip of the doline margin and the lowest point within).

Indices of doline development used are

$$\frac{(L + W)}{2} = \text{mean diameter}$$

and depth in relation to mean diameter:

$$\frac{2D}{(L + W)} = \text{cross section ratio.}$$

Fig. 7.1 Classification of dolines (modified from Jennings, 1975).
(a) Solution doline, with bedrock lowering and solute loss.
(b) Collapse dolines, with physical movement of limestone bedrock into solution cavities.
(c) Alluvial, drift or subsidence dolines, with soil subsidence and loss to bedrock cavities.
(d) Subjacent karst collapse dolines where non-limestone rock present above limestone collapses into solutionally opened cavities.

(a)

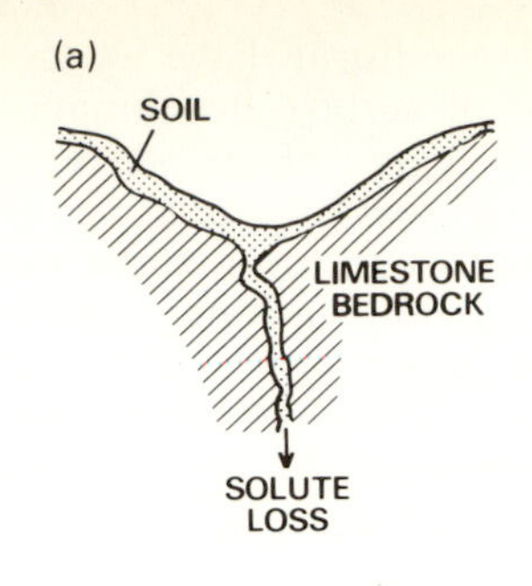

(b)

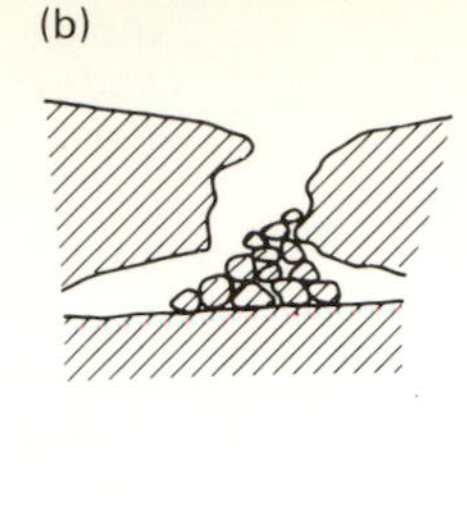

(c)

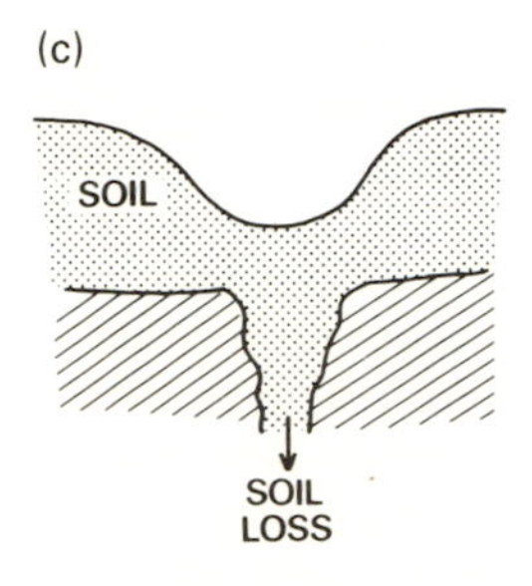

(d)

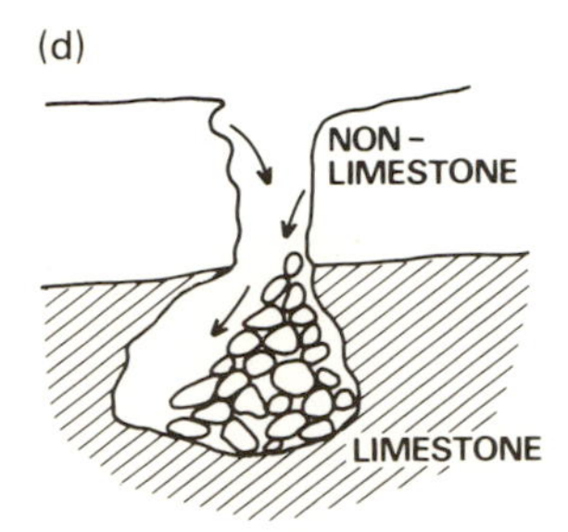

An elongation ratio is also used:

$$\frac{L}{W} = \text{elongation ratio.}$$

This is useful for describing alignment along structural features and consequent departures from an circular shape.

Such measurements are useful in the quantification of karst development. They are of especial value in comparing one area with another. Similarly, karst development can be assessed by the use of pit density:

$$\frac{\text{Number of dolines}}{\text{Area of limestone}} = \text{doline density}$$

or, to take into account the area of dolines:

$$\frac{\text{Area of limestone}}{\text{Area of dolines}} = \text{index of doline area.}$$

Surface complexity can be indicated by:

$$\frac{\text{Total doline area}}{\text{number of dolines}} = \text{mean depression relief}$$

or

$$\frac{\text{\% limestone area covered by dolines}}{\text{number of dolines}} = \text{index of surface complexity.}$$

Such indices are discussed by Williams (1969, 1971) and were used, for example, by Trudgill (1979a) to assess the degree of landform development on a recently emerged limestone surface on Aldabra Atoll, Indian Ocean. The data are shown in Table 7.1 and display a contrast between the older, ridge surface and a newer surface. The older is less complex, and has a higher area occupied by pits, though the total areas occupied by pits is similar in both cases.

Table 7.1 Morphometric parameters in air photograph study quadrats, Aldabra Atoll, Indian Ocean (after Trudgill, 1979c).

Indices	*Older, more karstified area*	*Younger, less karstified area*
Study area, A km^2	1.887	3.108
Number of pits, n	14	98
Density of pits, n/A km^2	7.4192	31.53
Total area of pits, A_p km^2	0.2664	0.2849
Per cent area of pits, A_p/A.100, $\%A_p$	14.1	9.1
Index of pit area, A/A_p*	7.083	10.9091
Index of complexity of pits, $\%A_p/n$†	1.008	0.0929

* The lower the number, the higher the area occupied by pits.
†The lower the number, the more complex and pitted the surface, the complexity of the surface being proportional to the number of pits but inversely proportional to the area of pits.

7.5 Tower morphometry

A morphometric approach has been used on positive relief features as well as negative ones (Fig. 7.2). For example, Day (1978) has made a quantitative study of the limestone hills of northern Puerto Rico. Here, he showed that there was a relationship between tower heights (H) and tower diameter (D), with a mean diameter : height ratio of 3.85. Balazs (1971) has used this index to classify types of tower karst development, with class divisions at ratios of 1.5, 3.0 and 8.0, representing a progression of dissection as the numbers decrease (e.g. if $D = 10$, $H = 1$, $D/H = 10$; $D = 1$, $H = 10$, $D/H = 0.1$). Again, such quantification, together with measurements of the isolation of towers, will give an index of karst development and facilitate comparisons with other areas. This minimises the need for subjective assessments of karst development and of differences between areas and provides a more rigorous basis for comparison.

7.6 Fluvial features

Fluvial features have been studied morphometrically by Williams (1969). Here, a geological map and a map showing fluvial features accurately, including swallets and risings, are necessary. The following indices are measured (in km or m):

A_l = area of limestone outcrop (km^2)
H_l = thickness of limestone outcrop (m)
S_o = swallet stream order (according to number of subdivisions of tributaries)
$\bar{H}_s$ = mean height of swallet (m)
L_s = mean distance to neighbours having same order (km)
ΣK_r = total number of karst risings
$\bar{H}_r$ = mean height of karst risings (m)
ΣL_l = length of every stream flowing across the limestone (km)
ΣS = total number of swallets
$\bar{L}_u$ = mean length of shortest distance of underground flow (km)

The indices which can then be calculated are:

Swallet density: $\dfrac{\Sigma S}{A_l} = D_s$ (No. per km^2)

Rising density: $\dfrac{\Sigma K_r}{A_l} = D_r$ (No. per km^2)

Fig. 7.2 A limestone tower in Jamaica. Tower diameter can be best measured on aerial photographs and tower height either by photogrammetry or on the ground by trigonometry in order to derive the D/H ratio (see text). The tower shown has an approximate ratio of 4–5, indicating only moderate dissection and development of relief; the ratios decrease as dissection increases. (Photo: S. T. Trudgill.)

Swallet : rising ratio: $\frac{\Sigma S}{\Sigma K_r} = R_{sr}$ – indicating the amount of underground branching

Distributive systems will have a low ratio, such as those with a low hydraulic gradient or significant ground water bodies. Integrated systems, more characteristic of surface streams, will have a high ratio where a number of swallets drain to 1 or a few risings.

Vadose index: $\bar{H}_s - \bar{H}_r = V_i$ – a measure of the depth of aeration – an index of vertical karst development

Rising coefficient: $\frac{\sigma H_r}{\bar{H}_r} \times 100 = V_{hr}$

where σ is the standard deviation estimated from the sample. This is an indication of the evolution of the groundwater network as it defines the variation of the altitude of the springs; a high coefficient indicating variability.

Stream density on limestone: $\frac{L_l}{A_l} = D_l$

this is a measure of the length of permanent streams in the area, and thus of the permeability of the limestone and therefore, by inference, of the degree of development of the underground network.

Other indices can be compared from area to area, such as $\bar{L}_u$ and also the relative karst system relief, H_k, which is the height difference between the basin's lowest resurgence point and the highest place in the area studied.

7.7 Indices of karst development

Indices of dissection, and especially of closed depression characteristics, were also discussed in an earlier paper by La Valle (1968). The author used three parameters which are believed to reflect the salient aspects of karst depression morphology: mean depression relief, mean depression area and mean depression flank slope angle.

The first of these, depression relief, describes the 'relative extent of karst degradation in the vertical dimension' or, put more simply, karst dissection. The measure is calculated by taking differences in height between the lowest point in the depression and the highest contour line on the rim; the measurements can be taken per unit area, say 1 km^2, or of morphologically distinct areas of comparable size. This index will increase as karst development increases and it can therefore be expected to increase with age of karstification.

Development in the horizontal plane can be measured by mean karst depression area. This is determined from measurements of the areas within the upper closed contour within a given area. Large values for mean depression area are expected to reflect intense karst erosion while small values indicate less intense karstification.

The mean slope angle of the flank of the solution depression is determined for a given area by averaging measurements of slope inclination from depression bottoms to sets of points randomly selected along the perimeters of the uppermost closed contours. Steeper depression flank slope angles are expected in areas of intense karstification. The indices are illustrated in Fig. 7.3.

In La Valle's study area, south central Kentucky, USA, it was found that all three indices, mean depression relief, area and flank slope angle, had the largest values in areas with denser, purer limestone, high hydraulic gradient and with well-developed cavern systems. A high hy-

Fig. 7.3 A closed depression in the Cockpit country of Jamaica. Indices of description discussed in the test include mean depression relief – from the tops of the skylines to the floor of the depression (centre); mean depression area (from air photographs) and mean slope angle of the flank of the solution depression, as shown by the two skylines to the right and left of the pictures. This area shows moderate karst development, with slope angles of approximately 30°. (Photo: S. T. Trudgill.)

draulic gradient from the highest points in the area to the lowest outflow points encourages flow within the bedrock and the development of caverns and depressions draining into them. The degree of vertical development away from a surface drainage system is perhaps the most characteristic index of a karst system as it is a bedrock drainage system which typifies karst development.

The indices used by Jennings (1975), described above, were able to illustrate the degree of karst development in limestone areas of South Island, New Zealand. Depression depth generally increased with increasing distance to the next nearest depression. Thus depressions were either shallow and closely spaced or deep and widely spaced. This pattern probably relates to joint frequency and also to possible hydrological interaction between depressions during their evolution. The small depressions are not necessarily lost or amalgamated during the formation of the deep ones and the scatter of size data often increases because small depressions also occur between the larger ones. There were also a larger number of deeper shafts in the more highly elevated area studied.

In assessing the role of morphometry in geomorphology, Williams (1971) considers that there is 'an urgent need for an objective system of description and analysis of karst landforms'. This need has arisen because of the uncertainty as to whether apparently similar landforms in different areas are, in fact, similar or not and, if so, what the degree of similarity is. Thus, the need is for an exact quantitative description of morphometry. This can be used as a more precise basis for comparisons of different area, seen by Williams as a *prerequisite* for comparative genetic studies, with the proposal that work on the understanding of the origin of the landforms should include morphometric work.

To date, the promise in this proposal remains still largely unfulfilled, with many parallel but separate works existing on morphometry, process or inferences concerning genesis. Williams, however, does continue in his own paper to show how morphometric and genetic studies may be linked. In particular, he uses a graph of depression order (number of bifurcating channels reaching the swallet in the depression) plotted against percentage frequency of depressions. This was particularly useful because as the depressions grow in size, they increase in order. Thus, as smaller order depressions are eliminated and surviving depressions enlarge, the modal class of depression order will change upwards as the karst becomes more developed. Williams studied two sites in New Guinea: Mt Kaijende, where the modal depression order was between 1 and 2; and the Darai Hills, where the modal depression order was 1. The former can therefore be seen as more evolved than the latter. The Darai Hills example is shown in Fig. 6.8b, p. 92. Such assessments could thus be of use in inferences concerning genesis, age of evolution and differentiation of landforms.

Morphometry can therefore be seen as providing a more objective assessment of karst form than earlier subjective classifications and it is also an assessment which leaves aside prejudgements of genesis, which the older terms such as 'tropical karst' employed. Such assessments should be of great use in further discussions of process–form relationships (Ch. 8), but the links between morphometry, process and landform genesis are, at the present time, not so well developed as they might be. This should be an area of fruitful development in the future.

8 Process and form in terrestrial limestone landforms

8.1 Climate and geomorphology

Considerable effort has been expended on the description of limestone landforms which exist in different climatic regions. Implicit in this description has often been the assumption that climatic variation is a cause of landform variation. This is not necessarily the case, however, and two factors complicate the relationship between climate and landform. First, the limestones which occur in different climates often show contrasts in themselves and are frequently of different ages and have contrasting structures, consolidation and other features. Second, climatic changes in the Pleistocene and Holocene mean that landforms visible today are not necessarily adjusted to the climates occurring at the present day but are, to some extent, fossil features. Thus in a review of karst and climate Sweeting (1980) concluded that although the influence of climate on landforms is vitally important, the earlier work on climate and karst was naïve: 'climate and karst are inextricably related, but in a far more subtle way than was once thought,' she concludes.

The premises for the assumption that climate has an important effect on limestone landform variation are based on two proposals concerning erosion processes. First, increased rainfall will lead to increased run-off and, hence, increased potential for solution products to be removed. Second, temperature will affect biogenic carbon dioxide production, although this will also be controlled by moisture availability. Carbon dioxide production in soils will be at its maximum where warm, moist conditions prevail; here organic matter breakdown will be maximised. Evidence has been collected to show that there is a substantial amount of support for these proposals, but not without qualifications. In addition, in order to assess the importance of these process variations to landform production, the intervening variables of structure and historical conditions of stability, uplift and environmental change have to be considered. Thus, it is clear the *processes* of erosion may vary markedly with climate but this should not be confused with the statement that *different landforms therefore necessarily form* in specific climates, landforms owe their form as much to the nature of geological structure and to history as they do to variations in process. Thus, the search for 'tropical' forms, 'temperate' or 'polar' landforms has been a somewhat unfruitful one, despite evident contrasts in erosion processes in these areas. For example, karst towers are commonly to be found in the tropics, these have therefore often been identified as a tropical form. However, similar towers can also be found in non-tropical areas where structural and uplift conditions conducive to tower formation also exist (Brook and Ford, 1976). Climatic geomorphology is thus an over-simplification; the relationships between climate and process are undoubtedly strong but landforms owe their characteristics to the interaction of process and structure over time. The topic can be explored in greater depth by the consideration of further aspects of the process–form relationships.

8.2 Processes and climate

Erosion processes undoubtedly vary with climate. For example, solution processes are less important than mechanical processes in cold regions. This is because, first, there is less biological activity in cold climates and consequently lower levels of soil carbon dioxide and organic acids and second, freeze–thaw action is prevalent. Obviously, also, solution processes are not dominant in arid regions. These aspects have already been illustrated in Fig. 3.34 (p. 51) which illustrates that solution loads of run-off waters draining arctic and alpine areas often have a lower range than those draining temperate or tropical areas.

Attempts have been made to relate karst denudation rates to rainfall amount, as for example by Engh (1980). This approach has been criticised by Gunn (1980) on logical grounds since evapotranspiration is ignored: Gunn cites the earlier work of Smith and Atkinson (1976) who correlate erosion rate with run-off rather than with rainfall (see Fig. 3.24, p. 46). Gunn also criticises earlier work which attempts to relate karst denudation to climate by Corbel (1957) and Lang (1977) on the basis of an inadequate sampling pattern: the annual solute load should be estimated from a solute rating curve applied to a flow duration curve or directly to the discharge record. Inputs in rainfall must be subtracted and the proportion of non-carbonate rock in a drainage basin must also be taken into account (e.g. Williams & Dowling, 1979). This task has yet to be widely undertaken on a world-wide comparative basis. Other work aimed at quantifying the effect of climate on solution chemistry includes that of Harmon *et al.* (1972). Here statistical manipulation yielded little clear information because of the large numbers of variables involved compared to the small number of samples – despite a very extensive sampling network. In a review of spatial and temporal variations in karst solution rates, Ford and Drake (1982), consider that the run-off variation is a most important variable but they also stress the role of temperature in influencing the levels of carbon dioxide in the soil (assuming adequate moisture for soil organism respiration). They refer to the equation:

$$\log pCO_2^* = -2 + 0.04T \tag{8.1}$$

where pCO_2^* is the potential partial pressure of CO_2 (atm) and T is the mean annual air temperature (°C).

It can be concluded from the examination of the evidence for the variations of process with climate that there is a pattern. This pattern can be both deduced logically and supported by observation. First, the availability of carbon dioxide will control the solute concentrations of run-off water – and CO_2 availability will vary with climate (Miotke, 1974). Second, the amount of run-off will vary markedly with climate in respect of the precipitation–evaporation balance. Thus, warm, moist conditions are conducive to vegetation growth, organic matter decay, carbon dioxide production and high run-off; decreasing temperature will reduce the effects of solutional erosion, but increasing run-off can increase the solutional effects, even if carbon dioxide production remains constant. These statements, are however, far removed from an explanation of landform variation. First, they do not take account of non-solutional processes and, second, they do not take into account the structural and historical dimensions.

8.3 Process and landform

Several isolated studies of karst processes and karst landforms have been undertaken in different parts of the world. Ford and Drake (1982) question the general applicability of such studies and Jennings (1972), in the context of tropical humid karst, admirably sums up the difficulties: 'the risk in extrapolating from the phenomena of particular regions to generalities conceived to have universal application seems especially great'. It is evident that, for example (a) tower karsts are not restricted to the tropics and (b) there are a great variety of landforms to be found in the tropics, with an emphasis on variety and complexity, rather than uniformity of type. Similarly, but in a different case, it is clear from the discourse by Sweeting (1972) that the word 'polje' has been applied outside its original slav meaning of a flat cultivated area, firstly to indicate a karst landform and secondly to all such large, wide karst plains surrounded by limestone hills. On subsequently finding that not all areas labelled 'polje' are similar, subdivisions of types of poljes and terms such as 'semi-polje' have been used. Nomenclature is, however, perhaps of less importance than is a fundamental understanding of process–form relationships, though problems will obviously arise if single terms are applied to a number of landforms and *vice versa*.

Almost by very definition, landforms must either stand up in the landscape or become lowered by differential erosion and deposition (as discussed on p. 1). Thus, it is important to understand what causes differential erosion or deposition. When viewed in terms of vertical development, it is often structural features which have been exploited which are important, as with the joints shown on p. 90. When viewed in terms of horizontal erosion, structural features may again may be important, such as bedding planes; equally important may be differential erosion which occurs in terms of water levels. Such differential erosion is thought to be important in the evolution of poljes. Surface lateral planation is thought to occur by Roglic (1951; quoted from Sweeting, 1972) as shown in Figure 8.1. Here, alluvium deposition is an important feature, with lateral planation occurring at the margin of the alluvial deposit. Roglic's scheme shows an impermeable rock at the centre of the polje, but Sweeting (1972) also describes the occurrence of flat, limestone floors in the centre of poljes. In tropical humid karst, Jennings (1972) also refers to planed limestone surfaces between upstanding lime-

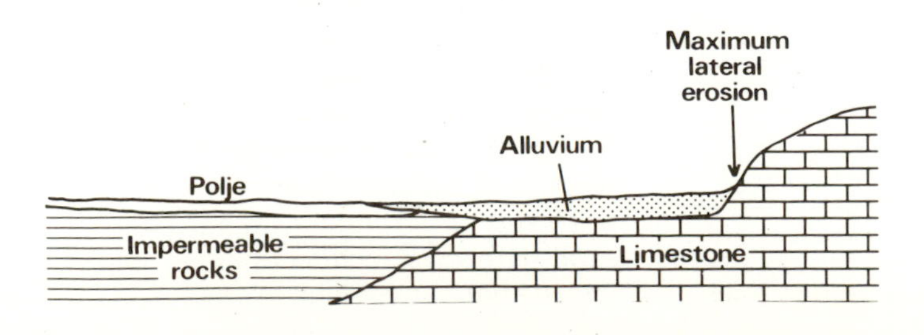

Fig. 8.1 Evolution of a polje. Lateral erosion occurs at the edge of the polje next to a cover of alluvium (modified from Sweeting, 1972).

stone massifs. The consensus of opinion would appear to be that two features of poljes are important: surface lateral planation at the height of watertable levels in sediments and the accumulation of the sediments themselves. Cvijic (1894; quoted from Sweeting, 1972) suggested that the polje was the end product of karst denudation, produced by the amalgamation of smaller dolines. Sweeting (1972) however suggests that this is incorrect because poljes can be seen to be currently breaking up into smaller features. Sweeting places emphasis on the historical dimension, with erosion occurring both during the Tertiary and Quaternary eras, with alternating periods of erosion and sedimentation. As with much of historical geomorphology, elegant though many arguments may be, there is often a lack of concrete proof or refutation of any particular hypothesis because, unless a dated sequence is available, the evidence is often fragmentary. However, it is clear that it is profitable to think of interpreting landforms in terms of the *focus* of erosion – be it a structurally controlled or a process focus. In the case of poljes, the focus of erosion is a lateral one, where sediments and water contact the limestone massif; poljes – or flat plains cutting into limestone hills – thus evolve by lateral planation.

Similarly, McDonald (1976), in a study of hillslope base depressions in the tower karst topography of Belize, suggests that, on the observation of the action of present-day processes, lateral planation is again the important process. The bases of the towers appear to be undercut by mechanical erosion during rainstorms and/or the action of corrosion in swamps and lakes. Thus, again, the landform can be understood in terms of the focus of erosion, as suggested by Figure 8.2. Run-off is at first centripetal and then drains around the base of the tower, thus facilitating erosion at this location.

In softer, younger limestones, upstanding areas may be reinforced by the reprecipitation of carbonate material within the surface of the porous rock. This helps to reinforce and exaggerate relief by making the more upstanding landscape elements more resistant. The reprecipitation is often termed 'case hardening' and its geomorphological role has been discussed, for example, by Ireland (1979), Day (1981) and Monroe (1966). The phenomena appears to be common in the central American and Caribbean area where Tertiary and Cretaceous limestones exist as well as younger Pleistocene limestones. Jennings (1972), however, points out that while Monroe (1966) has made a convincing case for the importance of case-hardening of the outer surfaces of limestone hills in Puerto Rico, many limestones, particularly the older limestones, need no such reinforcement to enable them to stand up in the landscape. Jennings also discusses whether solutional activity or mechanical erosion is important in lateral planation and isolation of residual hills. As McDonald (1976) stresses, their importance will vary with local conditions and it is difficult to generalise on this point. In all cases, it is clear however that lateral planation is important, presumably preceded by uplift, downcutting or lowering base level in order to give the vertical development of the landscape prior to lateral cutting.

Fig. 8.2 Focus of erosion in running water at base of karst towers (modified from McDonald, 1976).

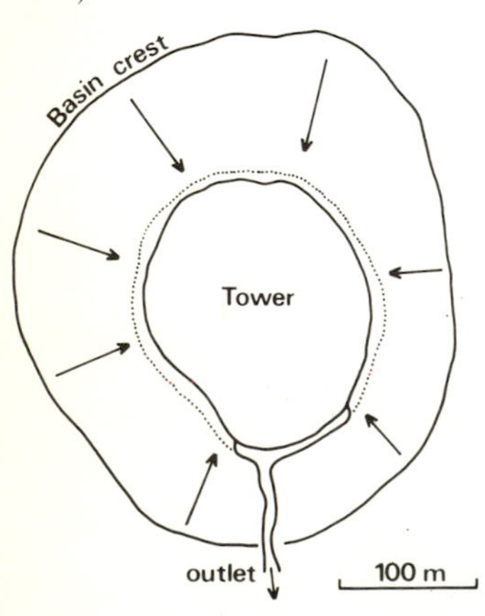

Differential erosion is again implicated in the development of closed depressions or dolines. Here, however, it is not the case that the erosion agent varies in its distribution but that it is the fact that the rock structures vary which is important. Dolines develop, as discussed in Chapter 6, by the preferential flow through a rock mass as encouraged by structural pathways such as opening joints and fissures. Thus, it can be predicted that rock which is fractured into large blocks should give rise to similar sized surface features; closely fissured rock should give rise to a more dissected landscape. The study of joint frequency is aided by the observation of joint spacing in cave sections – and by the spacing of caves themselves. Thus, the study of Wadge and Draper (1977a,b) in Jamaica, for example, is of great use in interpreting the geometry of the landforms above the cave. Not all joints present are equally exploitable, and many are infilled with material that may be less soluble or more soluble than the surrounding rock. However, joint frequency can be used as a first step in modelling the scale of landform variation which can be expected. In addition, the porosity and the permeability of the rock will be important factors. The more permeable the rock, the more dissipated the rainfall and drainage will be over the landscape; the less permeable the rock the more focused the erosion will be and the more dissected the landscape will become. This can be readily seen in northern Jamaica, where the porous Montpelier limestone gives rise to low, rounded landforms, reminiscent of the Chalk landscape in England, while the less porous White Limestone gives rise to a more dissected landscape, with greater focusing of erosional activity down opened joints.

Like all generalisations, however, the foregoing must not be applied simplistically. Dissection also proceeds and develops during the stages of erosion, relief becoming maximal at half-way stages in development. For example, as is well known from the writings of Jennings (1976) and Williams (1978b) (and see p. 95), Chinese karst workers recognise a sequence of development of karst. First, a high level surface is cut into by fluvial action. Second, a 'peak forest' develops, with deep dissection; third, isolated peaks occur, separated by plains. Thus, dissection may not simply be interpreted in terms of structure. Structure provides the *potential* for dissection: the *stage of evolution* also determines the degree of dissection. In addition, climatic change can lead to degradation of sharpening of features. Thus, Ford (1978) talks of a 'Tower karst in decay' when describing an area of tower karst at Chillagoe in northeast Australia. Much of the tower characteristics of the area, with basally planed pediments, are thought to date from an earlier pluvial phase in Australia's history, with little modification at the present day. Similarly, Ford (1983) and Smart and Ford (1983) describe caves cut into by glacial action, glaciation providing a temporal interruption of karstification. In the case of Chillagoe, dissection has become less active with decreasing rainfall; in the case of the North American examples described by Ford and his co-workers, karstic dissection has been substantially modified by glacial action.

Vertical dissection and lateral planation can also be modified by process variation caused by soil covers. Both the presence or absence of a soil cover and the type of cover, if present, are involved. For example, there is discussion as to whether the soil cover present in some karst depressions is merely collecting in an area of low relief or whether its presence actually enhances or causes lateral corrosion. Undoubtedly both cases exist, but the nature of the sediment is probably important. Porous sands and gravels permit the percolation of water to the limestone and encourage a widespread, diffuse erosion at the soil–bedrock interface beneath them. Less permeable clays enhance run-off at the surface and encourage a more focused erosion at the lateral edge of the clay at the limestone contact surface (see Fig. 6.13, p. 94). A carbonate-rich sediment will also discourage the erosion of the rock beneath the sediment while an acid sediment or soil will have the reverse effect. This is generally true for all soils present on limestone surfaces, though an impermeable acid soil will, of course, protect the limestone beneath it if water cannot flow though the material and bring acid water to, and solution products from, the soil–limestone interface (Trudgill, 1977a). Carbonate-rich soils will provide calcium to run-off waters, while there is little in the way of surface lowering of rock – a point emphasised by Beckinsale (1972). Nevertheless, losses from carbonate soils should be regarded as a landforming process as the form of the soil-covered surface will show a lowering over time. The important factor, then, will be the distribution of soils – and of soils of differing permeability, acidity and carbon dioxide levels in the landscape. This is of especial importance in glaciated regions where soils are developed on drift materials (including glacial drift and loess) which may have a character unlike that of any limestone soils which tend to form *in situ*. Thus, acid glacial drift may be deposited over limestone, giving enhanced local erosion. In general, the presence of an acid, permeable soil over limestone will lead to greater dissection relative to an area which is not soil covered and a soil-free area will have a greater erosion rate than that covered by a calcareous soil (as discussed in pp. 63–70).

Landform variation can thus be understood in terms of dissection occurring over time. Dissection will occur in relation to differential erosion. Erosion may be spatially differentiated either because the erosion process may have an uneven distribution – such as a patchy soil cover or localised water-level – or it may be differential because of structural weaknesses. Combinations of the two may, of course, occur. In addition, with changing conditions of uplift and climate, relict and fossil features may occur, especially if conditions change from ones of greater to lesser erosional potential (e.g. higher to lower rainfall).

8.4 The time dimension

Landforms only show a steady progression in their stage of development if their external conditions remain constant; more commonly, however, factors such climatic conditions may alter. Rates of uplift may also be episodic rather than uniform. This means that the term 'stage' applied to a position in a sequence of landform development will not simply mean that there is a stage in *one* simple sequence of development but that there are stages in *several* sequences. It would also involve the notion of adjustment to varying conditions. In addition, other factors may also alter the course of landform development in that the state of landform development may have a feedback effect on the operation of the processes of erosion. For example, the removal of an overlying, impermeable stratum may change the erosion regime from one focused on a subjacent karst percolation system to a more diffuse, subaerial regime. Stages in sequences could therefore be seen as involving sequences which are (a) simple, for example, downcutting under stable climate and crustal conditions; (b) slowly changing, for example, decreased or increased uplift; (c) major changes, with consequent subsequent adjustments, for example, glaciation/deglaciation; and (d) modified by feedback effects.

The problems associated with the time dimension in karst study are reviewed in a useful article by Jennings (1982) on the 'principles and problems in reconstructing karst history'. The article focuses on Australian examples but the principles discussed have wide application. Jennings proposes that karst, like any other terrain, may have three historical components: active, relict and exhumed.

Active landforms are those which have been formed by the processes still at work on them at the present day. **Relict** (or inherited) landforms are those which are a response to former conditions that hold sway no longer. **Exhumed** (or resurrected) landforms are those which once formed part of a former landscape but which were buried by accumulation of younger sediments or volcanic materials: subsequent erosional removal of these sediments leads to the reappearance of the landform at the surface. Exhumation can be readily recognised where it is partial, with an old landform partially mantled in sediments, but where the cover has been completely stripped it is less easy to recognise.

Active erosional features often cut into relict features. Jennings gives the example of the Cooleman Plain, New South Wales, Australia (Fig. 8.3). The relict portion of the landscape consists of a Tertiary erosion surface in the form of a plain at about 1270 m above mean sea-level. There is a widespread palaeosol and other deposits dating from the Tertiary era to be found on this plain. Part of the evidence for the date of the plain comes from the existence of basalt lavas dated at Miocene age over part of the area. The active portion of the landscape consists of the dissection effected by the cutting of gorges into this surface. This down-cutting has a Pleistocene date, continuing to the present day, though probably more slowly now than in rather wetter Pleistocene times.

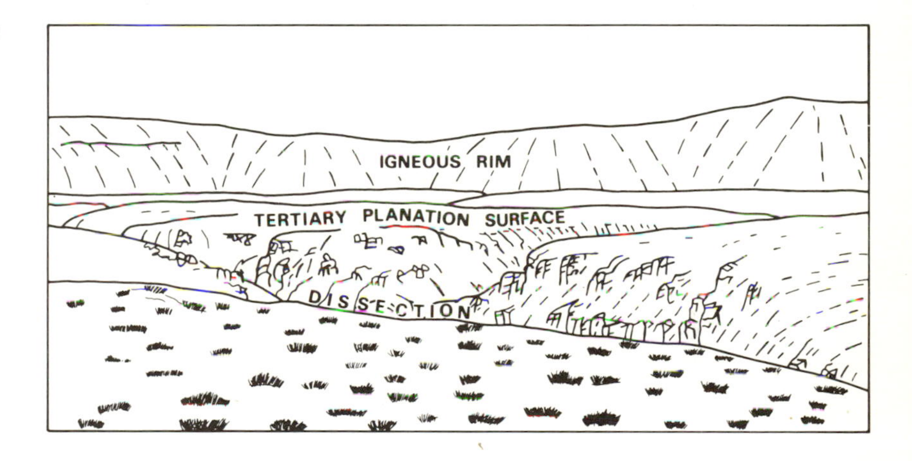

Fig. 8.3 Sketch of the Cooleman Plain area, New South Wales, Australia showing older land surfaces and younger incisions (modified from Jennings, 1982).

The active–relict division is not always a clear one but Jennings suggests that active landforms depend on present climate and vegetation, present sea-level position and the present internal conditions of the earth controlling volcanic activity and tectonic activity. These modern processes operate upon relief which is a product of past earth movements, sea-levels and bioclimates and have not yet been brought into adjustment. In considering the compound history of landforms, it is clear that it is more common for a landform to have both active and relict components rather than simply active components. Jennings' paper thus highlights an important way of thinking: limestone landforms, as any other landforms, have complex histories compounded of many elements. Often the relict elements may be more dominant than the active components.

Several other authors make similar points when studying a wide variety of areas. For example, Pfeffer (1981) describes relict tropical karst forms where the existence of Tertiary deposits in pockets between ero-

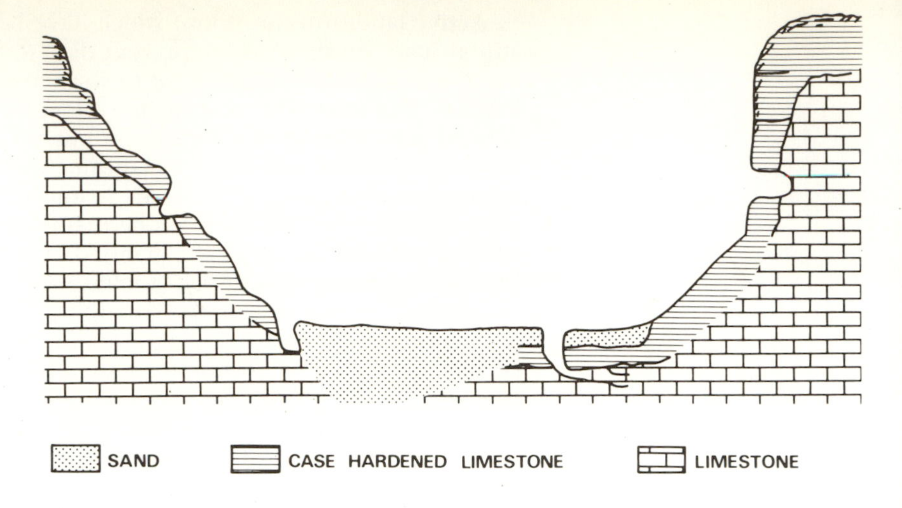

Fig. 8.4 Case hardening in Puerto Rico. Note occurrence of case hardening under cover sands (modified from Ireland, 1979).

sional features is an indication of the antiquity of some of the features. Similarly, Ireland (1979) describes the landforms in Puerto Rico, indicating that much of the relief was initiated before the deposition of sands. His argument is that rock surfaces become case-hardened during the subaerial processes of weathering. That is, the rock near the surface becomes hardened by the redeposition of calcite between the rock grains. This case hardening is present not only on the sides of the mogotes, but also under the cover sands. The cover sands are not dated but the progression of case hardening is seen by the increase in thickness away from the sands (Fig. 8.4).

In a study of the geomorphology of limestone outcrops in the northwest Sahara, Smith (1978) also differentiates between active and relict aspects of landforms. Here, some of the major land-forming processes relate to a previous, wetter climate: drying out of the climate to the present day has left relict solutional features. As is often the case in this situation, the relict landforms are on a larger scale than the currently forming minor features. Thus, cliff foot recesses occur which are thought to relate to seepage erosion – solutional activity by seepage water under a moister climate than at present. Within these relict solutional features,

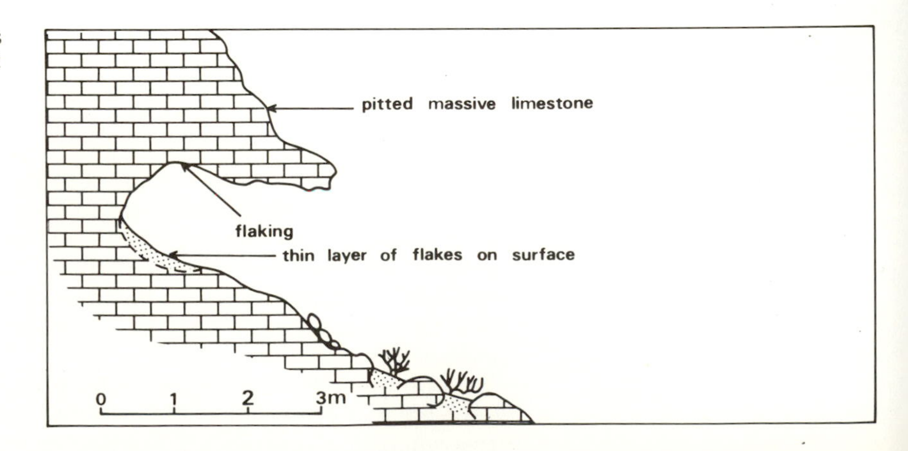

Fig. 8.5 Relict solutional features in the northwest Sahara. Cliff foot recesses formed under former wetter climates now show only minor solutional activity and modification by mechanical and biological processes (modified from Smith, 1978).

which are recessed into the rock by up to 3 m, minor sculpturing is now active on the rock surfaces (Fig. 8.5). The processes now active include flaking, granular disintegration and scaling. Solutional processes operate only occasionally and on a limited scale. Flaking consists of small flakes of rock being dislodged during moisture absorption; in some cases the presence of salt crystals was observed and also implicated in the flaking processes. The crystals were more common in locations nearer the sea and their growth, being behind rock particles, acted to dislodge the flakes from the rock wall. There were also signs of dislodgement by faunal activity, mainly because of the construction of nests behind flakes by spiders, hornets and other insects. Flaking seems best developed where moisture penetration of surface layers is followed by fairly rapid drying out. In more crystalline limestones, granular disintegration may occur. These processes lead to the removal of material from the inside of the roof of recesses, shown in Fig. 8.5, and to the accumulation of the thin layer of flakes on the floor of the recess. Scaling occurs on cliff faces where fragments break off parallel to the surface, often in conjunction with perpendicular cracking. Accumulation of detrital material behind the scales could assist with the dislodgement of the scales. In all the active processes operating today, the scale of the landform produced is a few centimetres in dimensions. This is much smaller than the larger scale relict features of dimensions which are measured in metres.

A similar relict landscape, with present-day micro-scale sculpturing superimposed on the larger scale solutional features is reported by Waltham and Ede (1973) and Trudgill (1976c) in the Kuh-E-Parau ranges of Iran. Here, large-scale, water-formed solutional features were again formed in wetter times during the Pleistocene. The features include the large-scale landforms such as sink holes and dolines. These remain as essentially inactive relict features now that the climate is drier than formerly. Active processes now include frost action and mechanical disintegration of small-scale solutional features. Thus relict features of solutional erosion such as rills and solution pockets are now being degraded by fractures and pitting. Thus the larger scale features are relict and substantially unmodified, but the smaller scale features are becoming adapted to current processes.

These case studies illustrate well the principle that it is the micro-scale features which are the most active and respond most rapidly to climatic change, while the larger scale features tend to be more relict or compound features (as also discussed on pp. 2 and 3). Thus Jennings' division of relict and active features can often be seen to have scale factors involved.

In the examples discussed above, the time dimension has involved climatic changes in terms of wetness and dryness. In other areas, changes from subaerial and fluvial regimes to glacial regimes have been involved. The effects of glaciation on limestone regions has been studied most frequently in temperate lands. Here, subaerial and fluvial erosion regimes in Tertiary times gave way to glacial regimes in Pleistocene times, with several ice advances and retreats (with corresponding wetter and drier periods further south from the glacial limits, as discussed above). In terms of time scales in the glaciated areas, the present-day return to subaerial and fluvial regimes has lasted for 10,000 years at the most and in highland and northern areas deglaciation was later, leaving only 6000 to 8000 years for adjustment to present climatic conditions, while other areas continued to be glaciated. The relative time scales available for

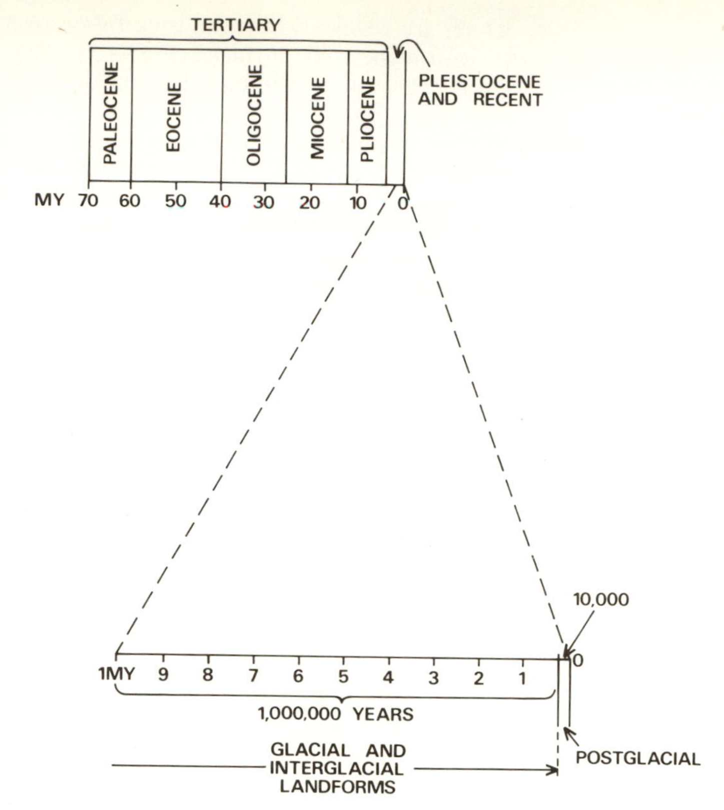

Fig. 8.6 Time scales for landform development. Note, on this scale drawing, the relatively little time available for postglacial modifications of glacial landforms. Notice also the relative time scales of the Tertiary and Pleistocene. Most landscapes visible today were initiated in the Tertiary, modified glacially and then show minor post glacial features. MY = million years. (Dates after Holmes, 1965.)

landforming and adjustment to any one particular regime are shown in Fig. 8.6.

In addition, the rates of erosion should be considered. Glacial downcutting has often been more effective in planation (e.g. stripping limestone pavements) and deepening valleys than fluvial and subaerial regimes have – thus it is important not only to consider the duration of processes but also their intensity.

Rates for the effectiveness of erosion are not widely known, but such rates as are available indicate that cave stream beds may erode at rates around 0.4–0.5 mm a^{-1} (High and Hanna, 1970), with higher rates if abrasional processes are effective. Subaerial rates are often lower, around 0.005–0.05 mm a^{-1} for bare areas, though this may be higher under an acid soil cover or lower under a calcareous one (Trudgill, 1976d, e; Trudgill *et al*; 1981). These rates compare with downcutting of 70–100 m in a few tens of thousands of years by glacial erosion as deduced by speleothem dating (Atkinson *et al.*, 1978). Taking duration and intensity effects together, it becomes evident that the areas under discussion are liable to be dominated by glacial landforms and, with up to 10,000 years fluvial and subaerial modification, at rates of around 0.5 or 0.005 mm a^{-1} respectively, features of the order of 5 m or 5 cm in dimension are likely to have been formed post-glacially, superimposed on a glacial landscape, with glacial erosional features of up to 100 m in scale.

The effects of glaciation on limestone geomorphology and hydrological processes has been extensively studied by Ford (1971b, 1983). Glacial deepening of valleys is one of the more obvious effects upon limestone hydrological systems, leading to steepening of hydrological gradients and the evolution of deeper cave conduit systems, with the abandonment of upper systems, when deglaciation occurs. Karst features such as dolines and sinkholes may become infilled with glacial drift. Carbonate-rich glacial drift may also act to preserve limestone bedrock from solutional erosion if the drift forms a cover over the bedrock. Ford also points out, however, that karst groundwater circulation may be maintained and even initiated beneath the soles of temperate glaciers. Percolation flow systems may be inhibited by freezing but meltwater systems may keep karst conduit systems operational, especially seasonally with considerable flow during the summer months.

Ford considers that, in Canada at least, the destructive processes of glaciation dominate so that in glaciated areas karst systems are by no means as well developed as they are in similar geological areas outside the glacial limits. The destructive processes involve the removal of surface features, such as those discussed in Chapter 4 and the removal of weathered bedrock and subcutaneous (shallow subsurface) water flow routes. Deep dissection disrupts cave systems, often leaving stranded relics of pre-glacial and interglacial cave passage on valley sides (Fig. 8.7).

Clearly, landforms seen at the present day in formerly glaciated areas will have differing degrees of glacial and karst characteristics. Thus, Ford (1979), focusing on the alpine glaciated environments of the Canadian Rockies, makes the useful distinction between the following types of landforms:

1. Postglacial landforms – formed entirely by karst processes after glaciation.
2. Subglacial landforms – formed under glaciers and now relict.
3. Karstiglacial landforms – essentially glacial landforms modified by karst processes.
4. Glaciokarstic – preglacial karst features modified by subsequent glacial erosion or deposition.
5. Mixed forms – karstiglacial + glaciokarstic components combined as a result of repeated glacial and interglacial events.

Fig. 8.7 Relict cave passage seen in glacially eroded rock wall, Morecambe Bay, northwest UK. (Photo: S. T. Trudgill.)

6. Preglacial karst landforms – preserved despite glacial actions.

Postglacial karst landforms cover many of the small-scale features already discussed above, such as solution pits, solutional grooves and channels of dimensions of up to a few metres. In addition, some larger features such as dolines and sinkholes have developed postglacially by local dissolution down joints and other bedrock weaknesses. The essential characteristics of these features is that their geometry has no glacial morphological component.

Going back in time, glacial landforms and modifications include the subglacial forms. There are two categories of subglacial landforms: (1) subglacial calcite precipitates and pressure solution rills formed at the ice–rock interface; (2) subglacial solution sinkholes. These are in the form of shafts which took glacial meltwater drainage. They are often difficult to distinguish from other sinkholes but those unrelated to present-day hydrology may be readily identified, such as those which occur on hill tops and those with no obvious present-day input.

Karstiglacial landforms are defined by Ford as features of glacial origin which have been adapted to karstic drainage, that is by underground drainage through soluble bedrock. This is especially the case for cirques and other glacial hollows which become drained by subterranean outlets. The sinkholes have developed postglacially.

Glaciokarstic landforms are preglacial karst features modified by glacial erosion or deposition (including interglacial forms modified by later glaciations). As with the discussions of scale and landform response to changing conditions outlined above, it is the small-scale karst features which are most readily obliterated by glacial action. Thus, sinkholes which have been ridden over and modified by glacial action can be seen as glaciokarstic but much of the original karst form can still be seen. With smaller scale karren features, many have been erased by glacial action, especially in higher mountain areas where scour will have been most effective. Some karren features do appear, however, to have survived in lowland areas where glacial scour was less effective.

Some landforms show evidence of repeated glacial action together with subglacial, interglacial and postglacial karst action. Such features are liable to be large-scale features where compound effects are noticeable, thus the term mixed forms may be applied to whole drainage basins, within which individual components may be recognised as belonging to one or other of the categories mentioned above.

In the Canadian Rockies, where Ford's work is focused, the main preglacial landforms still existing are fragments of cave passages, isolated at high elevations by glacial dissection; some of these fragments cannot be distinguished from interglacial fragments unless speleothem dating techniques have been used.

It can be seen from Ford's work that much of the landforms in the upland glaciated areas, such as the Canadian Rockies he studied, show a complex history of erasure and superimposition. The concept of relict landforms, as discussed above in the context of Jennings's work, can be seen to be important but it is useful to consider several stages of relict landforms. Essentially, preglacial karst forms have been obliterated, modified or preserved during glaciation. The consequent glacial forms have then been modified or preserved during subsequent karstification. Glacial modification has been effective to the extent that the larger scale landforms are still often glacial, while small-scale landforms are now karstic. Intermediate features such as sinkholes and dolines may be postglacial or preserved subglacial, interglacial or preglacial (Fig. 8.8).

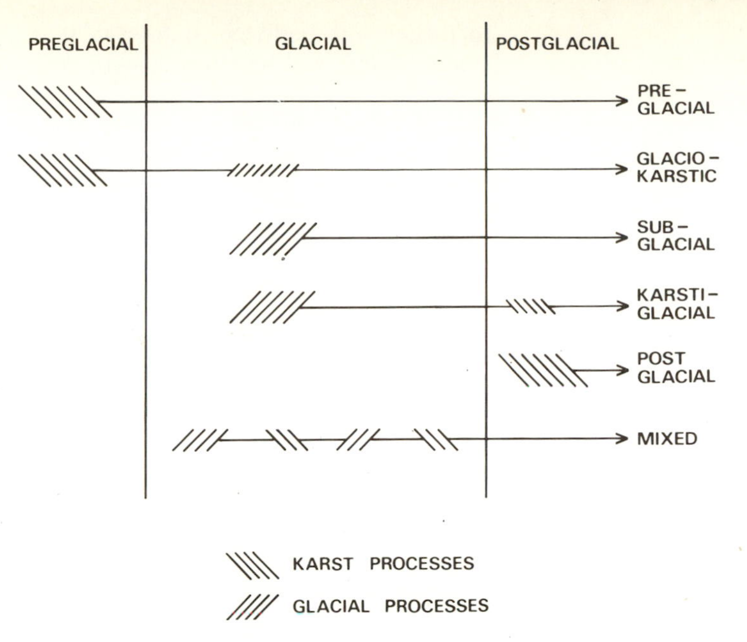

Fig. 8.8 Illustration of terminology of preglacial, glacial and postglacial landforms using the terminology of Ford (1979).

Speleothem dating (see p. 80) can assist in the dating of landscape evolution. In the Canadian Rockies discussed above, glacial effects are mostly clearly marked. However, in other areas this is not necessarily the case. For example, in the Derbyshire limestone district of the UK, an area which was not glaciated in the last glaciation, Ford *et al.* (1983) have provided evidence for cave formation, filling, abandonment and degradation. This sequence is related to progressive valley downcutting. Progressive lowering of the watertable has permitted cave development at progressively lower altitudes, though abandonment of older levels is by no means complete. Thus, the presently visible caves and surface features possess a spectrum of ages from greater than 350,000 years BP to the present day. Given that many older levels still take water in wet weather and thus are still being actively modified on occasions, it becomes clear that the distinction between 'relict' and 'active' may not necessarily be an easy one to make. It can be suggested that the division between active and relict forms may be one of *degree* rather than of clear division. Thus, older, high level caves may be seen of around 350,000 years old or more and which are now more or less completely relict, while younger caves are still wholly active. In between there will be a spectrum of activity, with some degree of running water present. Jennings (1982) distinguished between active (stream) and dry caves and also discusses the distinction between live caves with a stream, or with no stream but with drips of percolation water present, and a dead cave without either. However, the notion of frequency is also involved with conduits acting occasionally in flood but being substantially dry for much of the time.

Postglacial modification of landforms is often substantially influenced by the presence or absence of a soil cover, and the nature of any cover present, as discussed in Chapter 4. Erosion rates and landform production are most active under well-drained, acid soils where acid soil percolation water can modify the form of the soil–bedrock interface. Here, bedrock erosion is generally greater than under subaerial conditions and this, in turn, is greater than under calcareous soils which tend to protect the bedrock. It follows that if soil loss occurs then the erosional regime will be altered, with decreasing rates of erosion if an acid soil is lost and increasing rates if a calcareous soil is lost.

The extent and causes of soil loss in limestone areas has been the subject of much debate. One viewpoint is that it is often noticeable that limestone areas are free of soil, especially when compared with other bedrock types. One inference is that this may have been consequent upon deforestation by early man, especially in mesolithic and iron age times. The present lack of soil could be related to soil loss down open fissures or to the fact that only thin organic soils existed prior to deforestation – these would easily be lost by desiccation, deflation and oxidation after the removal of a protective tree canopy. A possible parallel present-day situation can be seen in parts of the New Zealand karst (Fig. 8.9). Here, where native limestone woodland is preserved, the rock is covered with a carpet of moss and thin organic soil. Bedrock dissolution thus proceeds under a layer of 10–20 cm acid organic matter, giving rise to etched forms including pinnacles and small pockets. These forms are, however, only visible after deforestation. Then, limestone outcrops in pinnacle form amid present-day pasture. The observer viewing these forms might reasonably conclude that the pinnacles had a subaerial origin without the knowledge that deforestation had occurred. Similarly, various apparently subaerial forms seen today in bare areas

Fig. 8.9 Limestone surfaces under forest cover, North Island, New Zealand. Note patchy cover of moss, rather than any cover of mineral soil. For full discussion, see text. (Photo: S. T. Trudgill.)

Fig. 8.10 Dissected limestone forms revealed by the subsidence and retreat of peat soils, Scar Close, North Yorkshire, UK. (See also Fig. 4.16, p. 62) (Photo: S. T. Trudgill.)

may have originated under soil. Such forms can only be identified by comparative work on subsoil forms which are visible when the forms are excavated from under contrasting soil covers. Trudgill (1976d, e) presented some such evidence for UK sites (principally North Yorkshire and the Burren, in West Eire), as discussed in Chapter 4, p. 63 (Fig. 4.16) and as shown in Fig. 8.10. In general dissection increased with soil acidity. Arcuate or cuspate forms were found under acid mineral or organic soils, with sharp-edged runnels under wet peaty soils. More subdued relief – closer in form to the original glacially planed surface – was found under less acid soils. Smooth, glacially striated forms could be observed under the highly calcareous drift soils. By inference, it may be argued that the occurrence of such forms under present subaerial conditions provides evidence for the loss of soil. If the soil loss is recent, rough surface textures, formed under acid soils, or glacial striations, from under calcareous soils, may still be visible. In addition supporting evidence may be provided by lichenometry (Trudgill *et al.*, 1979), with larger, older lichens growing on rock surfaces away from recent soil retreat and smaller, younger lichen colonies occurring nearer the site of the most recent soil retreat (Fig. 4.26, p. 69, and Fig. 8.11).

If wholesale soil loss has occurred, it can be argued that evidence should be present for the wholesale deposition of the eroded soil, for example in joints, dolines, caves or lake basins. This, however, would not be the case if the thin organic soils or moss covers were all that had been lost, as mentioned above. Indeed, the evidence for such deposition is fragmentary. Pigott and Pigott (1959) were unable to find evidence for deposition of soil material in Malham Tarn, north Yorkshire, but this does not mean to say that soil erosion could not have occurred at the soil surface, with subterranean deposition. Further evidence comes from the archaeological work of Bell and Limbrey (1982), working in the Burren district of County Clare, Eire. Here, they describe the occurrence of mineral soils under archaeological sites, while such soils appear to be absent from either side of the site (Fig. 8.12). The inference is that the sites were constructed when extensive soil covers existed and that they have now been lost from around the sites whilst patches have remained preserved under the site. The difficulty with this argument is, however, that sites could have been built on isolated patches of soil in

Fig. 8.11 Sizes of lichen colonies, illustrating progression of soil retreat. Large white colonies (*Lecanora* sp.) seen in the centre of photograph, indicate longer emergence than peripherally where lichen colonies are smaller or absent. (Photo: S. T. Trudgill.)

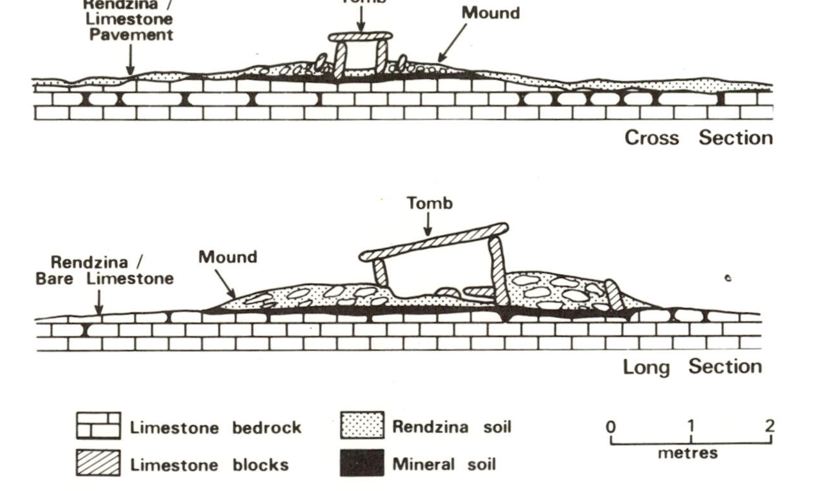

Fig. 8.12 Evidence for the existence of former soil covers in the Burren district of County Clare, Eire. Note mineral soil beneath archaeological structures (modified from Bell and Limbrey, 1982).

the first place, the soil-covered area having been selected preferentially because of ease of construction. The authors do, however, provide further evidence of mineral soil preserved in grikes and also in dolines and other depressions (Fig. 8.13). They also point out that the chemistry of cave deposits can frequently show a change in composition to a more iron and manganese rich nature. The argument is that these elements have been mobilised during soil acidification prior to soil loss. Such deposits have been dated in one instance to 1000–1500 years BP (by Ura-

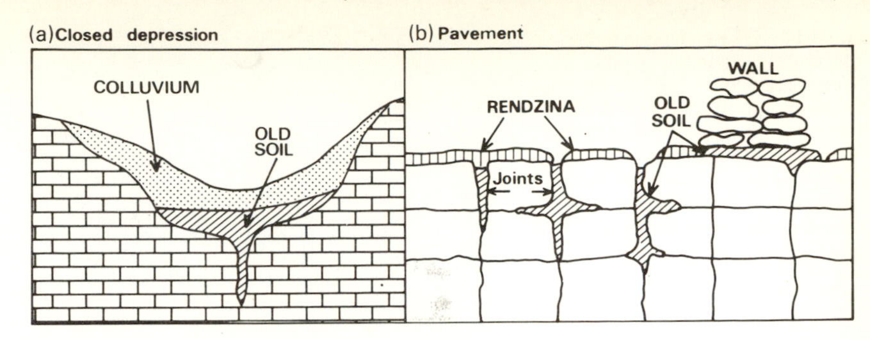

Fig. 8.13 Evidence presented for soil losses from limestone pavements by showing the accumulations of soil material in grikes, dolines and other depressions (modified from Bell and Limbrey, 1982).

nium–Thorium dating). This is a similar date to a ^{14}C date of 1500 years BP provided for a peat deposit present over limestone, inferentially formed after loss of mineral soil. Drew (1983) extends the discussion to talk of accelerated soil erosion in the Burren district, the accelerated loss being due to man. Drew again points out that archaeological sites such as Megalithic tombs, and also ancient walls, are placed over reddish mineral soil material. Elsewhere, only thin organic rendzina soils are formed, the reddish material again occurring – inferentially by deposition – in grikes and closed depressions. The inference is that there has been widespread loss of soil mineral soils following deforestation. This inference does not take into account, however, the observation of undisturbed woodland soils on limestone, in the form of organic soils under moss, elsewhere in the world and which would be relatively easy to erode. Certainly in the Burren there are areas which can be observed to have solutional forms associated with a subsoil environment but which are now in a subaerial environment. Equally, there are flat areas which show no signs of such forms and which, by inference, need not have had a soil cover and, at the most need only to have had a light humus and moss cover, probably under readily cleared light scrub.

Evidence for soil loss in the form of cave sediments is doubtful. Bull (1976, 1981) records the occurrence of fine grained sediments in cave deposits but there are few observations of sedimentation later than the late glacial times (Bull, personal communication). There is no evidence for widespread soil deposition in caves during postglacial time.

A further consideration is that soil disappearance from the surface can take place with only local movements to the subcutaneous zone. This is difficult to observe except in quarry sections (Fig. 8.14) Here there is evidence of soil at some 1–2 m depth, preserved in deeper fissures without having been washed into cave systems; this is paralleled by observations of soil and peat washed into the shallowest caves only a few metres below the surface. Here, soils and peats may be observed on cave walls. Such shallow caves are not widely distributed however, and it is likely that much of the soil remains in the top few metres of rock, neither visible from the surface nor visible from below in caves. In addition, soil can disappear from the surface without any physical washing of the material. It can be suggested that because of subsoil dissolution of the bedrock, the soil may be gradually lowered into the rock mass as it becomes solutionally dissected. Here, the loss is not one of soil, but of calcium carbonate in solution, derived from the bedrock. The sequence envisaged is shown in Fig. 8.15. Subsoil dissolution opens up joints and other weaknesses, into which the soil is lowered. The upstanding portions of the bedrock appear as they break through the surface of the soil during the soil lowering process. This process can be argued to have occurred

Fig. 8.14 Quarry face in soil-covered limestone pavement. Note the evidence for soil washing down the quarry face but note also that this could be accelerated or caused by the removal of rock during quarrying. Soil subsidence into solutionally eroded limestone is a more likely explanation of soil loss from the surface (see Fig. 8.15). (Photo: S. T. Trudgill.)

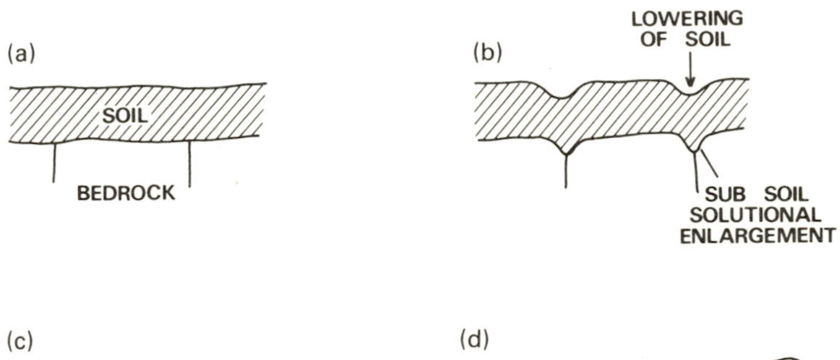

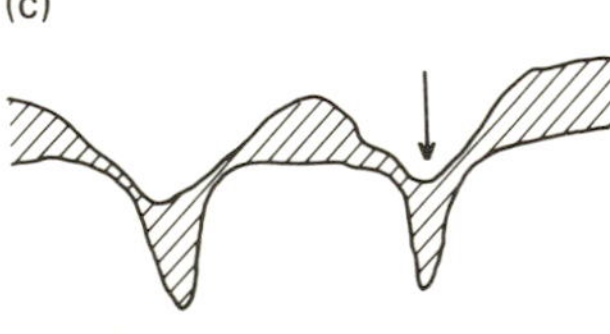

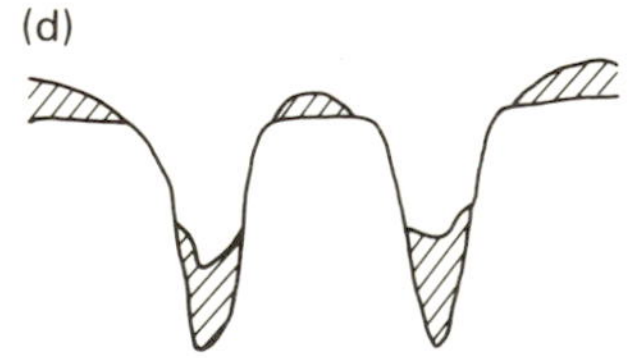

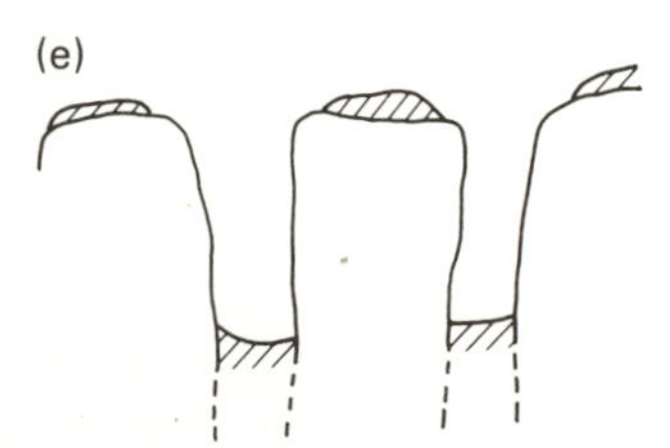

Fig. 8.15 Sequence of soil subsidence: soil is gradually lowered during solutional loss of subsoil limestone.

where eroded limestone appears in patches, surrounded by acid drift soils. The soil has not been eroded in the usual sense of the word but has simply been fragmented as it has been lowered through a solutionally opened bedrock.

Undoubtedly, the processes envisaged and discussed above can have occurred in a variety of different combinations and different histories will apply to different places. As yet, the evidence is fragmentary. However, it is clear that limestone surfaces are prone to denudation of their soil covers because of the opening of joints which will provide pathways for subsidence or erosional loss; equally some areas will have differed in the nature of their soil covers and thinner, organic soils will have been easier to lose than the thicker mineral soils. In addition, from morphological arguments, other presently bare areas probably have never had a complete soil cover. Certainly, there is evidence for the postglacial alteration of erosion environment, with consequent effects on morphology. The discussion above thus emphasises some points. First, the dangers of correlating landforms which have evolved over time with present-day erosional environment – even with smaller scale landforms. Second, the need for caution in an approach to landform explanation in terms of changing environmental conditions: apparently active forms may be partially relict and also active landforms may be subject to rapidly changing conditions in the last few thousand years.

It can be concluded that while the classifications of geomorphological activity proposed by limestone geomorphologists are conceptually sound, they may be difficult to apply strictly: landforms are liable to show a whole spectrum of inheritance and modification from being completely fossil to being completely modern, with every shade of intermediate gradation. Nevertheless, the 'active' and 'relict' classifications proposed provide useful subdivisions for dealing with this spectrum and these are important starting points for landform study. Moreover, they are important conceptually for emphasising the complex history of landform evolution through time.

8.5 Modelling process–form relationships in limestone geomorphology

There have been a limited number of attempts at modelling process–form relationships in limestone geomorphology. Many attempts have simply involved placing an observed spectrum of landforms into a logical sequence of inferred evolution. Thus, for example, Jakucs (1977) shows a sequence of karst development in European folded mountain development. Other sequences envisaged for tower karst in China and fluviokarst are presented elsewhere in the book (pp. 95 and 94). Carson and Kirkby (1972) have attempted to model the evolution of limestone slopes from first principles (Fig. 8.16). The authors assume that evapotranspiration on a hillslope is limited by soil moisture availability. Since soil moisture tends to increase downslope, evapotranspiration will tend to increase downslope. They then argue that the solutional removal will be proportional to rainfall multiplied by p, where p is the proportion of oxide present, modified downslope by evaporative loss of rainfall. Assuming p to remain constant downslope, losses in run-off will decrease downslope so that the upper part of the slope will become lowered to a greater extent than the base of the slope (as effective run-off will increase upslope, more being lost by evaporation downslope). Thus the slope will evolve by decline, as if there was a pivot point at the base of the slope, as shown in Fig. 8.16.

Fig. 8.16 Evolution of limestone slopes, with lowering of upper part of slope greater than lower ('slope decline') (modified from Carson and Kirkby, 1972).

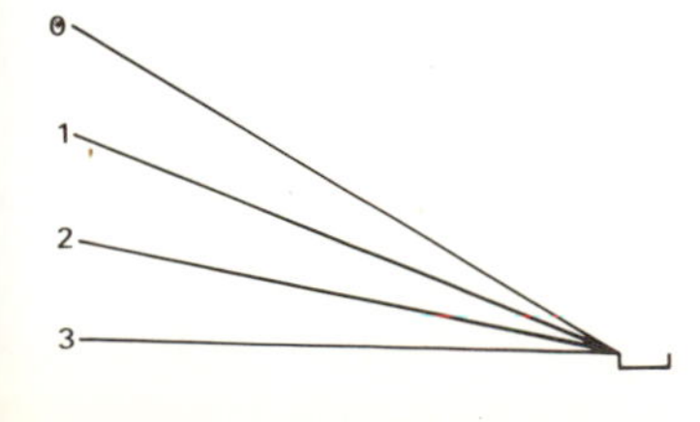

A more descriptive approach is adopted by Clark and Small (1982)

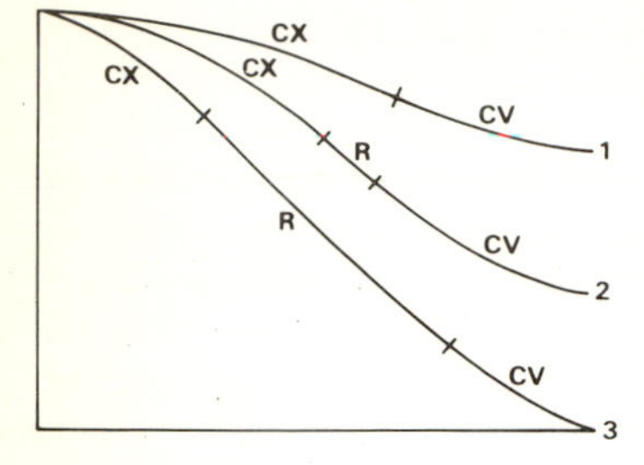

Fig. 8.17 Slope development on chalk, with the gradual expansion of the central rectilinear portion (modified from Clark and Small, 1982).

which takes into account lithological variation and variable stream downcutting. Without stream downcutting, lateral retreat is more important. From observations of actual slope development down valleys on the UK Chalk, the authors also suggest that early convex–concave slopes (Fig. 8.17) develop a central rectilinear portion at around 33° slope. This rectilinear portion gradually expands until only the upper and lower slopes are convex and concave respectively.

Hillslope evolution is modelled by Smith *et al.* (1972) in the context of the evolution of cockpits (closed depressions) in Jamaica. The authors assume an initial gently undulating surface, derived from an uplifted sea floor. As soil develops, it will collect in the lower-lying areas. They then assume that erosion will increase proportionally with soil depth as soil carbon dioxide content tends to increase with soil depth (this assumes a freely draining soil, as discussed in Chapters 3 and 4). The erosion rate, k, is envisaged as being proportional to soil thickness, T, as shown in Fig. 8.18. Using the notation b for the base of the slope and c for the crest, as $T^b > T^c$, then $k^b > k^c$ and, as erosion proceeds, the slope evolves by steepening, as shown. Slope steepening promotes the mechanical loss of soil so that, at about 30°, all soil cover is lost, giving uniform subaerial rates of erosion on the bare areas. Soil accumulation at the base of the slope promotes further erosion in these areas, plus the probable development of a shaft which can act as a conduit for further soil loss.

With all models, the validity of the model rests on the validity of the assumptions; this kind of approach is relatively undeveloped in limestone geomorphology, but it does help to understand landforms from first principles. It can be suggested that this kind of modelling is a profitable exercise and that the endeavour is one that could be usefully developed further.

The role of a soil cover is an important one. In many situations, and in contrast to some of the discussion above, acid soils tend to occupy

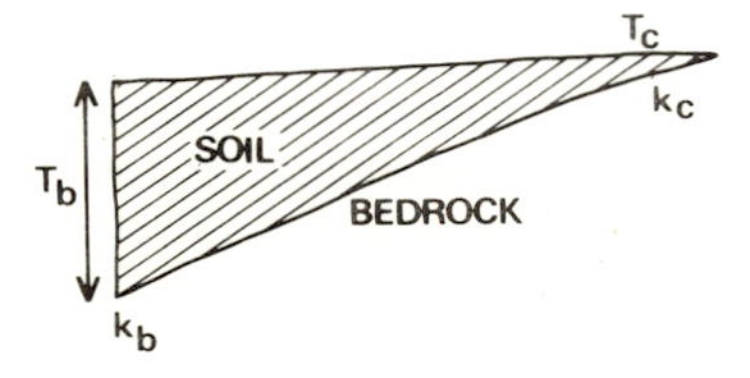

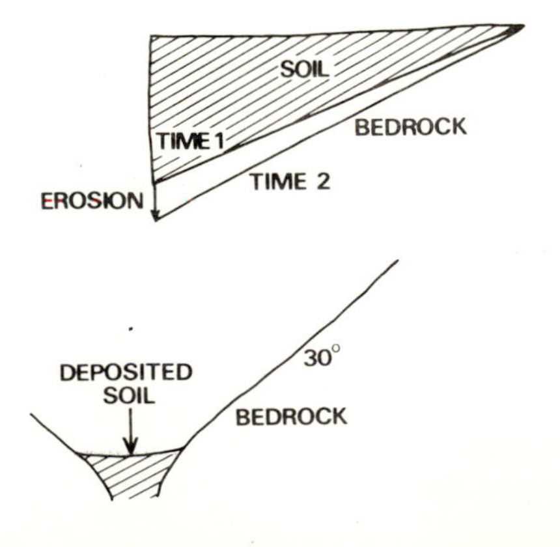

Fig. 8.18 Evolution of a tropical cockpit landform, Jamaica, West Indies. For explanation, see text (modified from Smith *et al.*, 1972).

upslope positions, while downslope positions are covered by calcareous soils. The latter will be derived mechanically by the downslope transport of limestone fragments, be it by slope wash or by solifluction processes (Fig. 8.19(a)). Where vertical permeability of the bedrock is limited, this situation will be reinforced by lateral throughflow and subcutaneous flow (such as described by Williams, 1983). Such flows will bring solutions rich in calcium carbonate to the slope foot site (Fig. 8.19(b)). With vertical permeability in the bedrock, lateral transfer will diminish and each slope segment can be considered independently (Fig. 8.19(c)). In either case of (b) superimposed on (a) or (c) on (a), slope decline – with higher rates of erosion on the upper slope – will lead to the preferential lowering of the slope crest (Fig. 8.19(d)) unless mechanical removal has led to the loss of soils (as envisaged by Smith *et al.* for Jamaican cockpits). Here, the areas of thinnest soils will tend to have the lowest rates of chemical denudation as there is less opportunity for acidity to develop in the upper part of the soil profile (Fig. 8.19 (e,)(f)). This model is likely to provide a useful starting point for understanding slope evolution where variations in soil chemistry exist downslope.

8.6 The role of present-day process studies in the understanding of process–form relationships

Crucial to the understanding of process–form relationships is the assessment of the spatial distribution of processes – and of the rates of operation of the processes. Thus, in Fig. 8.19, below, a study of processes and rates of slope erosion in relation to soil characteristics provides an initial starting point for modelling the evolution of form. Many process studies have not, however, considered the spatial distribution of processes directly and, even less, how they might relate to the evolution of form.

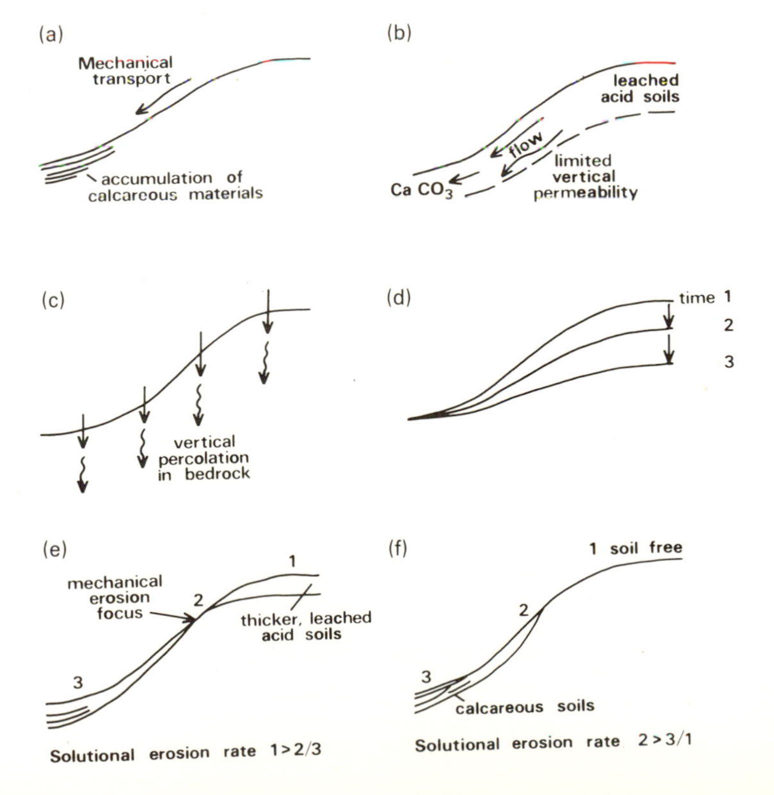

Fig. 8.19 Models of slope evolution on limestone. For explanation, see text.

Many process studies have focused on the topics of overall rates of erosion (e.g. Williams & Dowling, 1979) but, as Beckinsale (1972) has observed, there is little knowledge of where the source areas for solutional erosional loss are. Other studies have focused on fluvial processes, such as on magnitude and frequency of erosional events (as by Gunn, 1981c, 1982).

It is often the *unequal* distribution of rates which gives rise to differential erosion and hence to the production of erosional landforms. Thus, for example, in terms of process variations, McDonald (1976) reports on the focusing of drainage – and hence erosion – around the base of karst towers in Belize, a focus leading to greater differentiation of relief. The review by Ford and Drake (1982) tackles the question of spatial and temporal variations in karst solution rates in a substantive manner. They recognise intra-regional variability within areas of karst where processes and rates differ, giving rise to landform variation, for example within drainage basins. They also recognise inter-regional variability, where in different combinations of lithology and climate, contrasting landforms may be produced.

Studies of hydrochemical responses to rainfall events and of equilibrium solute levels provide information on variations within and between areas, but in order to fully understand landforms it is necessary firstly to see landforms in the historical context and secondly in terms of the interaction of process and lithology – often via the study of erosion rates. In conclusion, however, the crucial factor is the study of the distribution of erosion rates – whether in relation to the nature of lithology or process – and this both at the present day and, as far as possible, in past erosional regimes.

9 Coastal erosion processes

9.1 Introduction

Limestones exposed on coasts are often eroded into the shape of a notch if the rock surface is near vertical (Fig. 9.1) or, if the rock surface is near horizontal, pinnacles and potholes can often be seen (Fig. 9.2). Both these features are most well developed in the mid-intertidal zone and it may be deduced that whatever processes are responsible for these forms are focused in this zone.

There are a wide range of processes which should be considered as capable of contributing to the evolution of these features, including wave action, wetting and drying, spray impact, abrasion by sand and pebbles, salt weathering and hydration as well as the processes often emphasised in limestone terrains: dissolution and biological action. Some of these are clearly at their most powerful in the mid-intertidal: for example, wetting and drying cycles will decrease in number both upshore and downshore in relation to tidal cycles, upper shores being wetted and lower shores being exposed to drying only at extreme high spring tides. However, other processes, for example abrasion, may be at a maximum

Fig. 9.1 Intertidal notch in limestone, Aldabra Atoll, Indian Ocean. Picture taken at half-tide. (Photo: S. T. Trudgill.)

Fig. 9.2 Intertidal pinnacles on Carboniferous Limestone, County Clare, Eire, mid-intertidal at half-tide. Pinnacles with darker growths of the mussel (*Mytilus edulis*) are seen, with intervening potholes. (Photo: S. T. Trudgill.)

in the lower shore, leading to planation at that level. Thus the relative importance of each process, its distribution in the intertidal zone and the relationships between processes and morphology have to be closely examined. In addition, sea-levels have not been constant in the past and the roles and erosion at higher and lower levels than at present have also to be taken into account in explaining present-day morphology.

Considerable effort has gone into the study of the chemical and biological processes of intertidal limestone erosion but this has often been to the neglect of the processes common to all costs, especially the physical processes. However, investigation has often revealed that some of the most powerful forces shaping intertidal limestone landforms are, in fact, biological. Both direct biological actions, such as grazing and boring by molluscs, and indirect actions, such as the alteration of carbonate chemical equilibria, may be involved. Indeed, where dissolution processes appear to be important this is usually in association with biological processes. Naturally, on very exposed shores the relative importance of physical actions increases and this may be at the expense of biological processes.

Solution processes will be discussed in section 9.2 and biological processes in section 9.3. The relative importance of these, and other, processes will be discussed in Chapter 10 together with other factors influencing the relationship between process and form such as exposure, sea-level changes and lithological variation.

There is an extensive literature on the carbonate chemistry of seawater, but much of it is concerned either with the precipitation of carbonates from seawater and the formation of limestones by inorganic and organic processes or with open ocean conditions. Thus, much of this geological or oceanographic literature may not always be relevant to the topic of the solutional erosion of intertidally exposed limestones. However, it will be useful to first outline the chemistry of seawater and to summarise the theoretical points concerning the solubility of calcium carbonate in it. Attention will then focus on solution processes in inshore waters which are relevant to intertidal limestone erosion.

It will also be necessary to discuss the details of the processes involved in biological erosion. The biology of grazing and boring organisms has to be studied closely in order to establish the relationships between the distribution of eroded morphology and the distribution of erosive organisms: some organisms merely lodge in cavities and others actually produce them. Grazing is carried out by a wide number of intertidal molluscs and some fish; algae is merely removed from the rock surface by some, but others carry away a little carbonate substrate during their feeding process. Grazing organisms will be discussed in detail in sections 9.3.2–3. Boring is effected by a wide range of organisms including algae, sponges, bivalve molluscs and echinoderms. These will be discussed in sections 9.3.4–7. The wider issues involving process–form relationships are then detailed in Chapter 10 in the light of the discussions of solution and biological processes in sections 9.2 and 9.3.

9.2 Solution processes

9.2.1 Introduction

Considerable effort has been expended on the question of whether or not limestones can be dissolved in seawater. The crux of the problem is as follows: chemical analyses usually show that surface seawaters in many parts of the globe are saturated with respect to calcium carbonate; if this is so, then it is difficult to envisage how calcium carbonate can move into solution in seawater from limestones (conversely, of course, it helps to facilitate explanations of the formation of carbonate rocks by processes involving chemical precipitation). However, in intertidal situations delicate, sculptured rock surface occur which, by analogy with freshwater situations, would appear to owe their origin to solution processes. Either these surface forms are produced by processes other than solution (for example, salt weathering or biological action) and the analogy with freshwater solution forms is both misleading and fortuitous, or there are solution processes which facilitate the dissolution of limestones in seawater but which are not revealed by routine standard chemical analyses. Research has shown that both these possibilities may be true. The standard chemical oceanographic analyses of seawater often avoid inshore conditions where local variations may occur, but it is precisely these local conditions which are of interest in the dissolution of limestones in inshore waters. When inshore waters and intertidal rock pools have been investigated they have revealed that their chemistry can be substantially different to open ocean chemistry. The existence of respiring organisms in pools or other situations of closed or restricted water circulation can substantially increase the amount of carbon dioxide in solution, and therefore the potential for the dissolution of limestones. This will be especially the case at night when respiration output of carbon dioxide will not be balanced by photosynthetic uptake of carbon dioxide. In addition, theoretical carbonate equilibria calculations may not be valid when substantial amounts of organic matter are present in the seawater. The organic matter may make calcium carbonate more soluble than would be apparent from theoretical calculations based on aqueous ionised species since calcium can be taken up in large quantities in organic chelates. Calcium carbonate may also be isolated from aqueous interaction by organic coatings surrounding small particles of carbonate. A further consideration is that limestones are not pure calcium carbonate or pure calcite: the mineralogical forms of aragonite and high magnesium calcite may occur and these are more soluble than the calcite form. In addition, trace elements or substantial amounts of other elements and compounds may be present in the limestones which may act to raise or lower the solubility of the limestones. Thus, many mechanisms may exist whereby limestones may go into solution. Indeed, both carbonate precipitation and dissolution may occur in seawater, according to local conditions, and both these processes are often strongly influenced by organic processes.

Section 9.2.2 outlines the main points concerning seawater chemistry for background information; then the theoretical aspects of carbonate equilibria are outlined in section 9.2.3. The roles of organic complexes, the methods of assessing saturation status, the role of mineralogy and diurnal variations in saturation status are discussed in turn. Thus, the theories concerning the dissolution of limestones in seawater are elaborated to include natural variation in both seawater properties and rock type.

9.2.2 Seawater chemistry

Chemical elements have been concentrated in seawater by the action of the hydrological cycle – evaporation of water vapour from the sea surface having left behind elements of terrestrial orgin – and the result is a solution of high ionic strength. The ratios of chemical elements are practically constant for oceanic waters and the average concentrations for the major constituents in seawater are given in Table 9.1. The dominance of sodium and chloride are evident; but note also that there is over five times as much magnesium as calcium and that, as well as the dominance of the anions by chloride, there is more sulphate than hydrogen carbonate. This is in marked contrast to most of the freshwaters hitherto discussed where calcium normally dominates over magnesium (except in dolomitic limestone areas) and where hydrogen carbonate is normally the dominant anion. Also, in contrast to freshwater, the oceanic pH is on average pH 8.2 and seawater is strongly buffered.

Table 9.1 The major constitutents of seawater
(a) Modified from Sillen (1961), p. 549

Cations	*Concentration* ($mol\ l^{-1}$)	*Anions*	*Concentration* ($mol\ l^{-1}$)
Na^+	0.47015	Cl^-	0.54830
Mg^{2+}	0.05357	SO_4^{2-}	0.02824
Ca^{2+}	0.01024	HCO_3^-	0.00234
K^+	0.00996	Br^-	0.00083
Sr^{3+}	0.00015	F^-	0.00007

(b) Modified from Hem (1970), p. 11

Cations	*Concentration* ($mg\ l^{-1}$)	*Anions*	*Concentration* ($mg\ l^{-1}$)
Na^+	10,500	Cl^-	19,000
Mg^{2+}	1,350	SO_4^{2-}	2,700
Ca^{2+}	400	HCO_3^-	142
K^+	380	Br^-	65
Sr^{3+}	8	F^-	1.3

The sulphate anion may be combined with calcium or magnesium in an ion pair and the relative proportions of free calcium and magnesium and ion pairs with sulphate, hydrogen carbonate are shown in Table 9.2(a and b). In addition, the relative proportions of the hydrogen carbonate ion to the carbonate ion and $CO_{2\ aq}$ is shown in Table 9.3.

pH buffering is achieved by the system shown in Table 9.3 and also, to some extent, by $B(OH)_3^-$ and $B(OH)_4^-$.

These are all general statements which can be made concerning the chemistry of sea water but local conditions may alter this picture. It should be emphasised that the average values given may show large devi-

Table 9.2 Ion pairs in seawater
(a) Modified from Garrels and Christ (1965)

Ion	Free ion	$CaSO_4$	$CaHCO_3$	$CaCO_3$
Ca^{2+}	91%	8%	1%	0.2%
		$MgSO_4$	$MgHCO_3$	$MgCO_3$
Mg^{2+}	87%	11%	1%	0.3%

(b) Modified from revision by Kester and Pytkowicz (1969)

	Calculated	Measured
Free Ca^{2+} ion	88.5% (± 0.5%)	86.3% (± 0.3%)
Free Mg^{2+} ion	89.0% (± 0.3%)	88.1% (± 0.3%)

Table 9.3 Aqueous carbon dioxide system (modified from Raymont, 1963)

CO_2 + H_2O = H_2CO_3 = H^+ + HCO_3^- = H^+ + CO_3^{2-}		
1 ml	40 ml	5 ml

ations in places with high biological activity. Dilution by rainwater may also occur or concentration by evaporation in enclosed areas, but the ratios of the ions remains constant unless a selective removal occurs, as is the case with some biological systems.

9.2.3 Carbonate equilibria

In seawater, carbon dioxide is dissolved and dissociates to yield the ions shown in Table 9.3. If solid phase calcium carbonate is in contact with seawater and carbon dioxide is introduced into the seawater the dissolution and dissociation of the gas will increase the solubility of the solid phase carbonate, with the formation of HCO_3^- ions from the H^+ derived from the carbon dioxide input and the CO_3 from the $CaCO_3$. Ca^{2+} and HCO_3^- are thus dominant ions in seawater in the carbonate system; 88% of all calcium is present as the Ca^{2+} ion and 64% of all hydrogen carbonate is present as the HCO_3^- ion (Pytkowicz and Kester, 1971; data for 25 °C, 19.375% chlorinity, 1 atm. and pH 8.0). Other calcium ion species present are: $CaSO_4^0$ (11%), $CaHCO_3^+$ (0.6%) and $CaCO_3^0$ (3%) and other carbonate species present are: $MgHCO_3^+$ (16%), $NaHCO_3^0$ (8%), $CaHCO_3^+$ (3%), CO_3^{2-} (0.8%), $MgCO_3^0$ (6%), $NaCO_3^-$ (1%) and $CaCO_3^0$ (0.5%). In practice, in the study of carbonate equilibria reference is only made to the total calcium, whatever ion species are actually present. The hydrogen carbonate system may be described by referring to the total CO_2 in the system (i.e. the equivalent of CO_2 present in $CO_2 + H_2CO_3 + HCO_3^- + CO_3^{2-}$) or just to the dominant ion species, HCO_3^- and CO_3^{2-}, together with H^+ derived by dissolution and dissociation of CO_2 in water.

The study of carbonate equilibria thus relates the carbon dioxide–water system and the calcium carbonate–water system. It has relevance to the dissolution and precipitation of carbonates in seawater and also to the relationships between atmospheric gaseous composition and the oceans as a sink or possible source of atmospheric carbon dioxide. There is a large literature on the atmospheric gaseous carbon–oceanic dissolved carbon dioxide system in the light of current interest in the global effects of carbon dioxide from the burning of fossil fuels; a useful review of global carbon dioxide values in the atmosphere is given by Bolin and Keeling (1963).

To be able to calculate the tendency of seawater to dissolve calcium carbonate, the saturation status of the water can be studied using the Ion Activity Product (IAP) and comparing this with the thermodynamic solubility product, K_S. The IAP for calcium carbonate is simply the product of the activities of calcium and carbonate in water (Berner, 1965):

$$IAP = a_{Ca^{2+}} \cdot a_{CO_3^{2-}}$$

where a = activity.

Activity is a dimensionless parameter related to concentration by an activity coefficient:

$$a = ym$$

where y = activity coefficient and m = total molality (including free ions and ion pairs). Thus for calcium:

$$a_{Ca^{2+}} = y_{Ca^{2+}} \cdot m_{Ca^{2+}}$$

and for carbonate:

$$a_{CO_3^{2+}} = y_{CO_3^{2-}} \cdot m_{CO_3^{2-}}$$

The values for $m_{Ca^{2+}}$ and $m_{CO_3^{2-}}$ can be determined analytically and Berner (1965; 1971, p. 58) calculates the activity coefficients as $yT_{Ca^{2+}} = 0.20$ and $yT_{CO_3^{2-}} = 0.021$. However, the IAP of calcium carbonate alone will not fully define the system and the reaction $HCO_3^- = H^+ + CO_3^{2-}$ must also be defined. Concentrations of HCO_3^- and CO_3^{2-} are given by carbonate alkalinity: Ac

$$A_c = mT_{HCO_3^-} + 2mT_{CO_3^{-2}}$$

This is derived from titration alkalinity, which is the number of equivalents of hydrogen ion per kilogram of solution necessary to convert all anions of weak acids to their respective acids. In practice this is effected by titration of seawater with hydrochloric acid to pH 4.5. when all hydrogen carbonate ions are converted to H_2CO_3 (Fig. 9.3), aH^+ can be measured using a pH electrode. Berner (1971, p. 58) defines the IAP for the whole system as:

$$IAP = \frac{mT_{Ca^{2+}} \cdot A_c \cdot yT_{Ca^{2+}}}{\dfrac{2}{yT_{CO_3^{2-}}} + \dfrac{a_{H^+}}{yT_{HCO_3^-} \cdot K_{II}}}$$

$$\text{where } K_{II} = \frac{a_{H^+} \cdot a_{CO_3^{2-}}}{a_{HCO_3^-}}$$

Values for the activities of calcium and carbonate are given above and that for yT_{HCO3-} is given as 0.55; the value for K_{II} is given as $10^{-10.33}$. Thus the IAP for the system may be worked out using values for $mT_{Ca^{2+}}$, A_c and a_{H+} derived from calcium analyses, titration alkalinity and pH measurements of sea-water using the formula:

$$\frac{mT_{Ca^{2+}} \cdot A_c \cdot 0.20}{\dfrac{2}{0.021} + \dfrac{a_{H^+}}{0.55 \cdot 10^{-10.33}}}$$

For warm shallow seawater at $P = 1$ atm and $T = 25$ °C values are given by Berner as $mT_{Ca^{2+}} = 0.0103$; $A_c = 2.25 \times 10^{-3}$ and pH = 8.15. Substituting these values into the equation above gives a value of IAP = 12.5×10^{-9}.

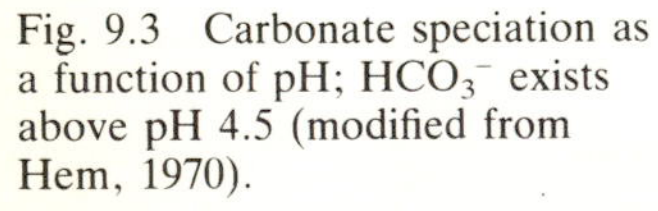

Fig. 9.3 Carbonate speciation as a function of pH; HCO_3^- exists above pH 4.5 (modified from Hem, 1970).

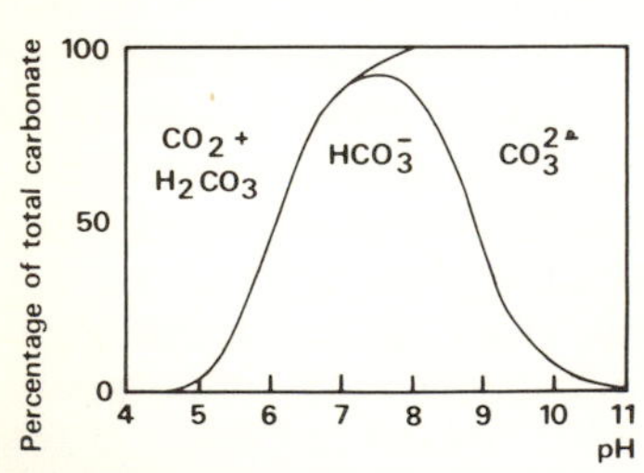

The value of K_s is derived from activities of the ions in question at equilibrium. K_s for calcite is 4.0×10^{-9} (Berner, 1971, p. 58). If IAP is greater than K_s then water is supersaturated; if it is equal then the water is saturated; if IAP is less than K_s then the water is undersaturated. The expression IAP/K_s therefore gives a saturation ratio where unity is saturation; figures less than this indicate undersaturation and figures greater than this indicate supersaturation. Thus, in the example of warm shallow seawater above where IAP = 12.5×10^{-9} and given that K_s for calcite = 4.0×10^{-9} then IAP/K = 3.1, indicating a supersaturated state.

The chemical dissolution of calcium carbonate in seawater is discussed by many authors. Useful reviews can be found in the papers by Cloud (1965); Revelle and Fairbridge (1957) and Pytkowicz (1969) but a more up-to-date article is one by Morse *et al.* (1980). These papers will provide greater insight into the details of the processes involved. All emphasise the importance of carbon dioxide in the system and while stressing that oceanic waters are supersaturated with respect to calcite, local variations in biological production of carbon dioxide could alter this picture. The paper by Li, Takahashi and Broecker (1969) provides a survey of the Pacific and Atlantic oceans and gives extensive data for IAP/K. Their data are summarised on Table 9.4. It is clear that in all cases that the values for IAP/K are substantially greater than unity, indicating supersaturation for surface oceanic waters.

Between about 4000 m to 5000 m deep in the oceans, calcium carbonate deposits are extremely rare; this depth is referred to as the lysocline and below this the rate of dissolution far exceeds the rate of precipitation in relation to values of IAP/K of less than unity. At these

Table 9.4 Values for carbonate saturation (IAP/K) for surface ocean waters (modified from Li *et al.*, 1969)
(IAP/K > 1 = Saturation, < 1 = undersaturation).

Location	*IAP/K*
Northwest Atlantic	6.5
	5.2
	5.4
Caribbean Sea	5.7
	6.1
Northwest Pacific	5.2
Eastern Equatorial Pacific	3.5
Central Equatorial Pacific	5.0
	5.4
Western Equatorial Pacific	5.7
	5.9
Northwest Pacific	5.2
North Pacific	2.3

depths organically derived carbon dioxide is present in high concentrations at high pressures and low temperatures.

The study of the factors involved in carbonate equilibria demonstrate that, first, several ion species may be involved in reactions involving calcium and carbonate ions; this emphasises the importance of analysis of total molality to include calcium complexed in all forms of ion pairs, and, second, the importance of consideration of H^+, HCO_3^{2-} and CO_3^- ions. Consideration of the activity of these species in relation to calcium concentrations can be used to calculate Ion Activity Products. Comparison of these with equilibrium constants reveals that surface oceanic waters are generally supersaturated with respect to calcite. However, since in the carbon dioxide–calcium carbonate system the amount of calcium is theoretically dependent upon the amount of carbon dioxide flux into the system it may be readily envisaged that biological respiration may locally increase the solubility of calcium carbonate. Moreover, organic matter in seawater and variations in carbonate mineralogy may substantially alter this picture.

9.2.4 Organic compounds

Organic compounds may affect carbonate equilibria in two ways. First, calcium carbonate solubility may differ from the predictions based on theoretical calculations involving the CO_2 – H_2CO_3 – H^+ + HCO_3^- – H^+ + CO_3^{2-} system. This is because calcium can be taken up as a chelate within organic structures or complexed with organic matter. Thus the capacity for calcium carbonate to move into seawater from the solid phase may not be totally dependent on the potential supplied by the carbon dioxide system. Secondly, organic compounds may interact with carbonates to produce organo-carbonate compounds; the solubility of these will not be predicted by inorganic equilibrium calculations and the presence of organic coatings round carbonate particles can inhibit free inorganic equilibriation between calcium carbonate and seawater.

Both dissolution and precipitation processes can be affected by these considerations, but their effects are liable to be limited to areas of high biological activity and restricted circulation, rather than the open oceans. For example, dissolved and particulate organic matter in the form of humic acids are to be found in tropical mangrove swamps where the leaf litter of these marine trees falls into the seawater at their roots (Fig. 9.4). In coral reefs large amounts of mucus are secreted by reef organisms and these organic compounds of considerable chelatory properties may be found in suspension in the water. Dissolved organic compounds present in seawater range in molecular weight from less than 700 to greater than 100,000 (Rashid, 1971) and the lower ranges appear to be the most efficient in complexing metal ions. Values of 1.1–1.2 mg were obtained for dissolved organic carbon by Chave and Suess (1970) for seawater analyses around Hawaii. In laboratory experiments stearic acid and purified egg albumin were shown to interact with calcite and also to inhibit calcite precipitation. In the case of stearic acid it was con-

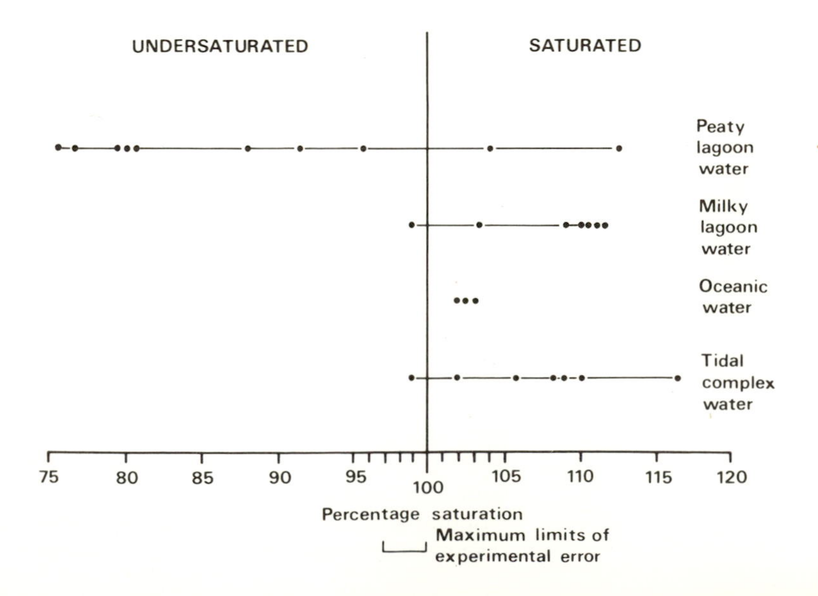

Fig. 9.4 Carbonate saturation status of marine waters, Aldabra Atoll, Indian Ocean. Method: double titration (see p. 17). Note that only peaty lagoon waters, in contact with mangrove swamps, are substantially undersaturated with respect to calcite (from Trudgill, 1976a).

cluded that a calcium stearate layer may be formed around calcium carbonate particles. In natural conditions several organic compounds exist which may behave in a similar way, including fatty acids, proteins and lipids.

Organic acids, as well as having chelatory and complexing properties, can decompose to produce H^+, by direct dissociation, and also CO_2 by the oxidation of organic carbon. Thus, the solubility of calcium carbonate in seawater may be directly enhanced if decomposing organic matter is present.

It can readily be appreciated that saturation status calculations of open ocean waters may be an unreliable guide to saturation status in micro-environments because of, first, the probably greater concentrations of carbon dioxide and higher acidity produced by decomposing organic matter and, second, the theoretical calculations and analyses of seawater invariably do not allow for chelatory uptake of calcium or inhibition of equilibriation by organic compounds.

9.2.5 Measurement of saturation status in the field

It is clear that local variations in seawater chemistry and biological conditions demand that saturation status should be measured at sites of interest and at a local scale. Moreover, some empirical method which would allow for the effects of organic compounds and other effects which are difficult to specify would clearly be useful. Two, related, empirical field techniques exist which involve *in situ* field measurements. Both involve the examination in changes of carbonate equilibria after the artificial addition of powdered calcium carbonate. In the first, pH change is measured, in the second, calcium uptake or precipitation is measured. Given that CO_3^{2-} ions produced by calcium carbonate dissolution combine with H^+ ions in solution to produce HCO_3^- ions, then if carbonate dissolution occurs when powdered calcium carbonate is artificially added to a sample of seawater the amount of free H^+ will fall and pH will rise; conversely, if calcium carbonate precipitation occurs, then pH will fall (Table 9.5). Similarly, a rise or fall in the amount of calcium in solution can be measured before and after the artificial seedings of a seawater sample with powdered calcium carbonate. The pH change method is described by Weyl (1961) and Ben-Yaakov and Kaplan (1969), while the calcium analysis double titration method was used by Trudgill (1976a) (Fig. 9.4). These methods give an empirical indication of carbonate saturation status of seawater but both have been criticised in

Table 9.5 pH change method of assessing saturation, Aldabra Atoll, Indian Ocean

pH	*pH upon addition of calcite*
Open sea	
8.2	8.2
8.2	8.2
8.2	8.2
Lagoon	
8.5	8.3
9.0	8.9
8.4	8.4
8.6	8.6
8.9	9.2*
8.2	8.3*

* Peat-rich mangrove water.

terms of a lack of replication of natural conditions. Reaction rates of small water samples with powders may be relatively fast compared to natural limestone dissolution processes but the final stages of equilibriation may take several hours or even days; if, however, samples are kept for long periods of time de-gassing of carbon dioxide may occur or biological processes may substantially affect carbonate equilibria in the reaction vessel. In addition, crystal growth during precipitation may not replicate natural conditions as many nuclei are artificially provided. Thus, these measurements are usually taken over the order of a few hours to one day equilibriation time and can only be taken as a field guide to saturation status; values close to saturation should be carefully interpreted and those within about 2% of saturation cannot be differentiated from a saturated state.

All methods of assessing carbonate saturation status rely on the chemical analyses of seawater: the IAP/K calculations rest upon analyses of pH or calcium in solution. These parameters are so closely dependent upon the gaseous phase (CO_2) that care has to be taken in sample treatment: rapid analysis is essential and careful sample handling in tightly stoppered thick-wall bottles has to be involved to prevent de-gassing if any sample transport occurs. In addition, if substantial amounts of calcium are involved in organic complexes then results may depend upon analytical methods: complexiometric titrations for calcium in seawater (Culkin and Cox, 1966) may not totally extract all organically involved calcium. Spectrophotometric analyses may give a better result for total calcium but with more sophisticated less mobile equipment more sample transport may be involved. These factors become important when marginal saturation status occurs and saturation values depart from unity by amounts which are within analytical error. It is in these cases, and many others, where a further technique is useful, that is the observation of the dissolution of sample blocks, or natural surfaces, by the monitoring of weight loss or by microscopal examination. Weight loss of sample blocks was used by Peterson (1966) to study deep-sea dissolution of calcium carbonate (Fig. 9.5). Alexandersson (1976) has clearly illustrated the existence of etch patterns by the use of electron microscopy to study limestones in contact with seawater in the Baltic (Fig. 9.6). These types of approaches probably provide a more definite answer as to whether or not limestones can be dissolved in seawater at any one particular place.

In summary, there are several methods available for the assessment of the saturation status of seawater – the comparison of ion species activity in natural water samples with activity at equilibrium; the monitoring of pH and Ca changes during artificial seeding and the direct observation of solutional weight loss or etch patterns. All these methods suggest that seawater is mostly saturated with respect to calcium carbonate at the surface and that undersaturation in surface waters only occurs where a high degree of biological activity occurs, namely respiration and the production of organic acids. However, all these approaches suffer from one major deficiency so far described; they involve considerations of pure calcium carbonate in the calcite form and limestones may often substantially depart from this composition and form.

9.2.6 Dissolution and mineralogy

Limestones are not pure calcium carbonate, as already discussed in Chapter 2. There are many mineral variations which may be present and substantial impurities may also be present. Mineralogically, four main forms may be exposed to seawater: low magnesium calcite, high mag-

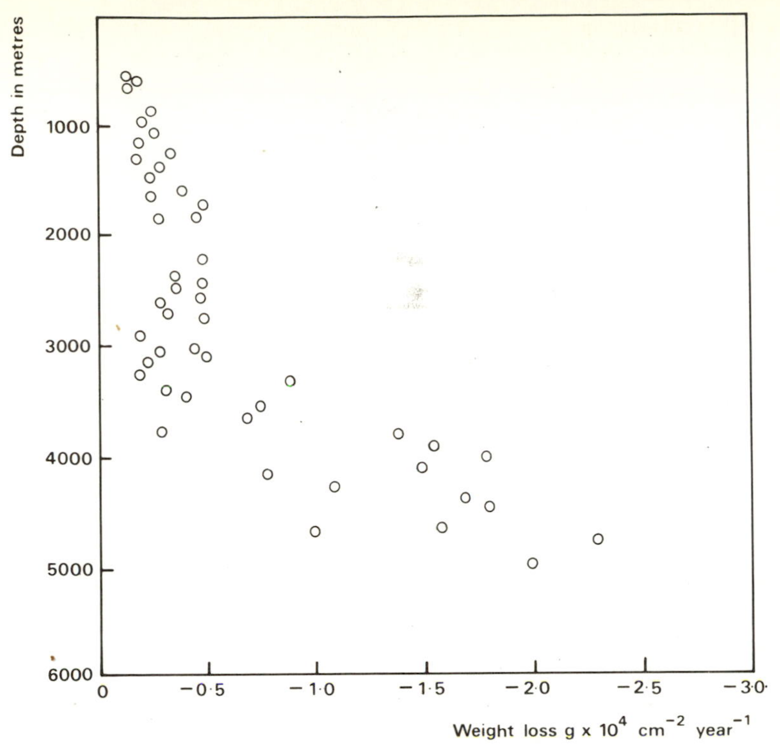

Fig. 9.5 Plot of dissolution of calcite spheres with depth in the oceans. Note the increase in solutional weight loss at around 4000 m (modified from Peterson, 1966).

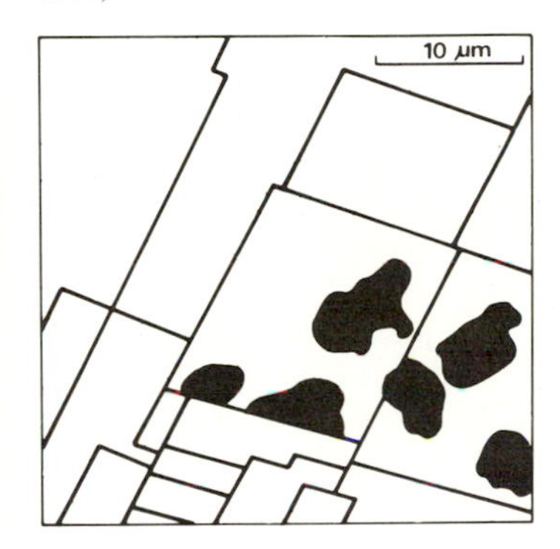

Fig. 9.6 Drawing of electron microscope photograph with calcite crystals appearing as rhomboid outlines; solid areas are cavities made by boring filamentous algae. Note 10 μm scale bar (drawn from a photograph from Alexandersson, 1976).

nesium calcite, aragonite and dolomite. In terms of chemical composition, many compounds and elements may be present, including phosphates, iron compounds, lead and zinc, many of which inhibit calcium carbonate solubility.

Calcite is calcium carbonate in the hexagonal crystal system, rhombohedral class and aragonite is calcium carbonate in the orthorhombic crystal system. If calcite contains less than 4% magnesium carbonate it is referred to as low magnesium calcite (Wolf *et al.*, 1967, p. 87). Aragonite is found in many tropical carbonate deposits, but usually inverts to calcite during diagenesis. However, it forms an important constituent of many Pleistocene and Recent limestone outcrops in tropical situations. Aragonite is more soluble than calcite. The K_s value for aragonite is 6.3×10^{-9} (Berner, 1971, p. 58). Since the K_s value for calcite is 4.0×10^{-9} it is possible for a value for IAP to lie between these two and for water therefore to be theoretically unable to dissolve calcite but able to dissolve aragonite.

Using the method of assessing saturation status by titration for calcium before and after artificial seeding with calcite powder, Trudgill (1976a) found that inshore waters around Aldabra Atoll, Indian Ocean, were saturated with respect to this mineralogical form. However, using powdered limestones instead of pure calcium carbonate it was found that the limestones could in fact go into solution in the seawater and rises in calcium were recorded in some cases. In addition, magnesium also went into solution from the rock powders (Table 9.6). Pure calcium carbonate powders are therefore an unreliable guide to the dissolution of limestones.

Recent work by Morse, Mucci and Millero (1980) has suggested that actual solubilities of calcite and aragonite are substantially different from

Table 9.6 Per cent saturation of seawater samples using double titration method and comparing results of adding pure calcium carbonate with powdered rock samples, (modified from Trudgill, 1976a)

Powder added	*Seawater samples* *1*	*2*	*3*	*4*
'Analar' Calcite	111	102	101	100
Aragonitic Limestone	100	98	100	97
Algal limestone	100	92	100	109
Calcarenite	108	93	102	87
Phosphatic limestone	–	–	98	–

< 100 = undersaturation
– = no data

those predicted by thermodynamic calculations. Dissolving minerals may undergo the formation of a surface layer during the dissolution process. A magnesium-calcite surface containing between 2 and 6 mol % magnesium may form, the magnesium being derived from seawater. Measured solubilities were about 20% less than calculated values.

Thus, for many reasons, the apparent saturation status of seawater with respect to calcite calcium carbonate may be a poor predictor of limestone dissolution: indeed, limestone may be dissolved even when water is apparently saturated with respect to calcite.

9.2.7 Diurnal variation in saturation status

During daytime plants utilise carbon dioxide in photosynthesis. During the night this process is not operative but the respiration of marine organisms continues. Thus, Emery (1946), and many other workers, have been able to demonstrate that at night in seawater, in pools, carbon dioxide levels rise, pH falls and the potential for calcium carbonate dissolution rises. That the process is due to organism respiration was demonstrated by control experiments where organisms were excluded (Table 9.7). Revelle and Emery (1957) were able to demonstrate that rock pools were undersaturated with respect to calcium carbonate at night. Thus a mechanism exists for the dissolution of calcium carbonate in enclosed waters, even though other data demonstrate that the oceans are saturated with respect to calcite. Studies in open waters by Schmalz and Swanson (1969) and by Trudgill (1976a) (Fig. 9.7) indicate that well-mixed open waters show small fluctuations in diurnal saturation ranges but that inshore waters show marked fluctuations, including the occurrence of undersaturation at night. In waters away from rock pools containing gastropods and other similar organisms, it is presumably planktonic respiration which leads to the observed pattern.

Table 9.7 Laboratory tests on the effect of gastropod molluscs on seawater pH (modified from Emery, 1946)

(Initial pH of all samples, pH 8.38)		*pH Values*					
Material	*Treatment*	*Day 1* 0900	1600	*Day 2* 0900	1600	*Day 3* 0900	1600
1. Water, gastropods, rock chips	Dark night and day	7.90	7.70	7.75	7.70	7.70	7.68
2. Water only	Dark night and day	8.25	8.05	8.10	8.00	8.00	8.00
3. Water, gastropods, rock chips	Dark at night, sun in day	7.90	8.15	7.70	8.18	7.70	8.10
4. Water only	Dark at night, sun in day	8.28	8.10	8.10	8.00	8.03	8.03

9.2.8 Summary

The study of the theory of carbonate equilibria is of fundamental importance to, but not a complete predictor of, the dissolution of limestones in seawater. Calculation and measurement show that oceanic

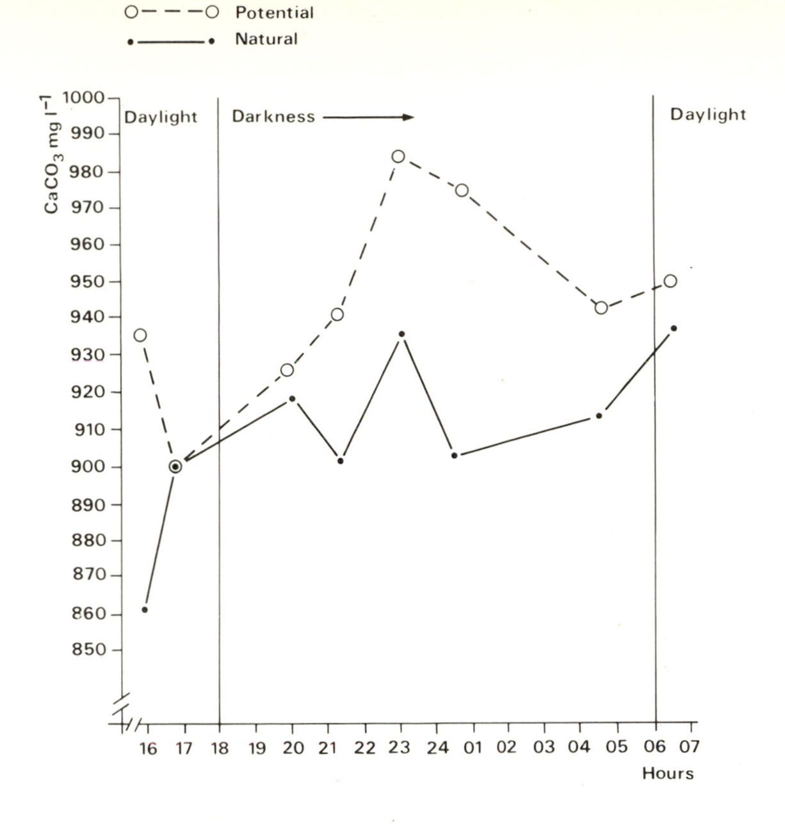

Fig. 9.7 Nocturnal variations in seawater chemistry, inshore waters, Aldabra Atoll, Indian Ocean. Ca^{2+} expressed as $CaCO_3$ (see Appendix). Solid line is natural level, pecked line is potential assessed by double titration method (see Ch. 2 and Trudgill, 1983a). The maximum difference between natural and potential is during the hours of darkness when respiratory production of CO_2 occurs (from Trudgill, 1976a).

waters are invariably saturated with respect to calcium carbonate at the surface, but, however, this may not be the case at night nor where high biological activity releases carbon dioxide and organic acids into restricted circulation conditions, nor where organic compounds may act to chelate calcium. In addition, limestones containing mineral phases other than calcite, and impurities, may be more soluble or less soluble than calcite. These effects have been shown to exist by research in different parts of the world but it is not known how important they may be on a systematic global basis; certainly limestone dissolution is possible in seawater under certain conditions and mostly in relation to biological activity.

9.3 Biological processes

9.3.1 Introduction

The biological processes involved in the intertidal erosion of limestones can be both direct and indirect. The indirect actions involve the way in which the activity of organisms may affect local chemical environments (this has been discussed in section 9.2). The direct actions involve the rasping away of rock surfaces during grazing activity and boring into rock surfaces. Together these direct processes constitute the process of **bioerosion**. An additional, and important, consideration is that some organisms may encrust a rock surface so as to actually protect it from erosion by other processes.

Many marine molluscs and other organisms graze algae which are growing on rock surfaces but some also remove rock during this process; this is discussed in sections 9.3.2–3. Other organisms, including algae, sponges, molluscs and echinoderms, bore directly into the rock, mostly for protection. This not only removes rock from beneath the surface but it also weakens the rock in relation to the physical action of the sea. Boring is discussed in sections 9.3.4–7.

The processes involved in bioerosion frequently constitute the major processes of erosion of limestone coasts and are therefore worthy of detailed study, both in terms of the biological mechanisms involved and also of the effects of bioerosion upon intertidal limestone morphology. Indeed, any cause–effect inferences about the relationships between organisms and morphology must be based on a sound knowledge of the biology of the processes involved. This is because correlation of the presence of eroded holes and surfaces with the presence of organisms by no means proves cause and effect. Many organisms merely lodge in holes or crevices produced by other organisms or by non-biological processes and they themselves have taken no active part in their production of the holes. For example, a boring sea urchin may produce a large hole 10 cm across and 10–15 cm deep in a rock surface, but on the urchin's death, the hole may become occupied by other organisms, such as crabs, which take little part in rock erosion. The biology of each animal in question has to be studied with detail to establish whether or not it is involved in an erosive activity.

9.3.2 Grazing – mechanisms and strategies

Grazing is effected by species of molluscs, echinoderms, crabs and fish. The process involves the rasping of algae and, in the cases of interest, the scraping of the surface and the removal of up to 1–2 mm of rock. Carbonate substrate is ingested together with algae growing on the surface and a growing within the rock. Algae which actually grow within the rock are discussed in detail in section 9.3.5. Several levels of grazing exist: some grazers have little or no effect on rock erosion while others remove considerable amounts. The algae are digested by the grazing organism and any carbonate also ingested is disposed of in faeces, usually in a finer grained form than orginally.

The grazing activities of marine invertebrates have been well documented in the biological literature for some time but their geomorphological effects have been less widely recognised. A useful summary of the biological literature is to be found in the book edited by Crisp (1964) on grazing in terrestrial and marine environments. Most of the biological interest is focused upon the nutrition of the invertebrates and on the effects which invertebrates have upon algal growth. For example, the effects of limpet grazing on algae have been studied by Southward (1964) while Lawrence (1975) and Paine and Vadas (1969a) have studied the effects of sea urchin grazing upon algae. Paine and Vadas (1969b) have studied the calorific value of marine algae in relation to food preference of invertebrates, concluding that the availability of algae plays a significant role in feeding habits. The geomorphological role of grazers is confined to those which ingest carbonate rock as well as algae and this role has been recognised by McLean (1967; 1974) working on gastropods and chitons in Barbados; by Craig *et al.* (1969) working on *Siphonaria pectinata* (a patelliform – limpet like – mollusc); and by Healy (1968) and Trudgill (1976a) who both recognised the importance of several groups of grazing animals.

Fig. 9.8 Radula of the gastropod *Siphonaria pectinata*, face view (left), *c.* 3 mm long, showing rasping surface and side view (detail omitted) (modified from Craig *et al.*, 1969).

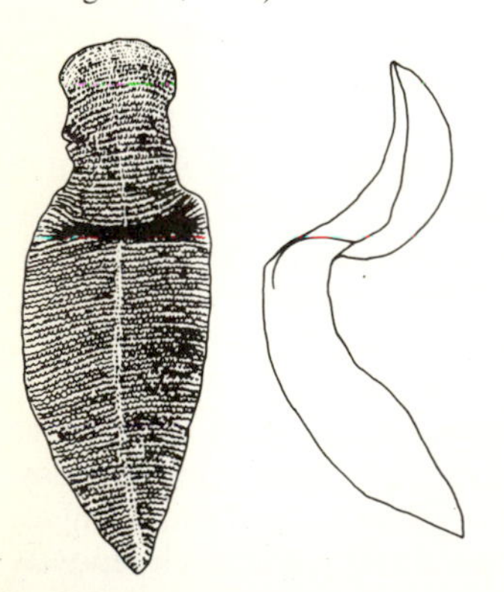

The grazing by molluscs is effected by the use of the rasping *radula*. The radula of *Siphonaria pectinata* is shown in Fig. 9.8. Different groups of molluscs have radulae of differing size and rasping capacity. It is probable that each group may exploit different food sources: the smaller gastropod radulae may be used to rasp thin films of algae or small growths

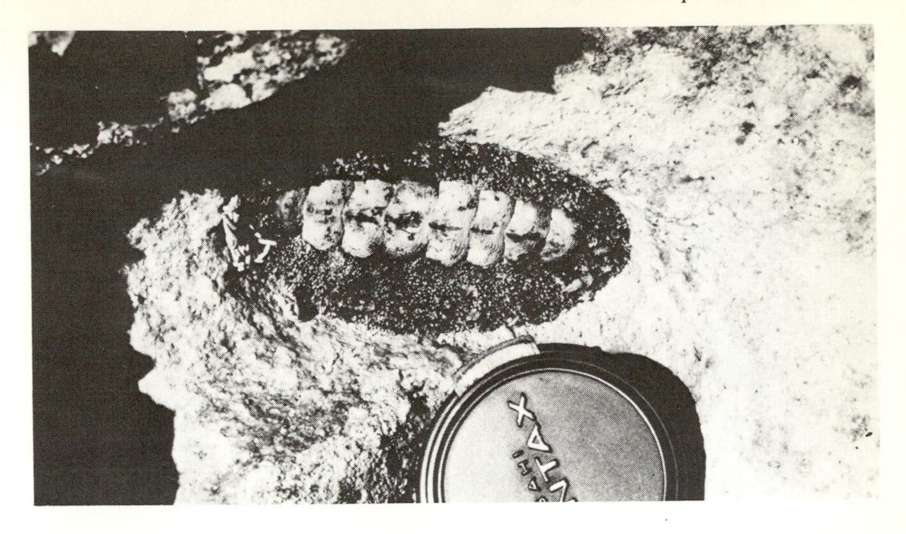

Fig. 9.9 A chiton, *Acanthozostera* (= *Acanthopleura*), on beach rock, Heron Island, Great Barrier Reef, Australia. Lens cap 55 mm across. (Photo: S. T. Trudgill.)

of surface living or **epilithic** algae, thus causing little geomorphological effect. The larger radulae of the chiton (Fig. 9.9) is employed to rasp algae which are present in the rock (which are described in section 9.3.5). These latter algae may be **endoliths**, that is true boring algae present *within* rock grains or **chasmoliths**, present in the spaces *between* grains.

The radula of the chiton *Acanthopleura* was shown by Lowenstam (1962) to be comprised of magnetite which has a hardness of 6 on the Moh scale as compared to 4 for the calcite which it is eroding. Figure 9.9 shows the chiton *Acanthopleura brevispinosa* and Fig. 9.10 demonstrates the erosive marks made by the passage of one feeding chiton. The

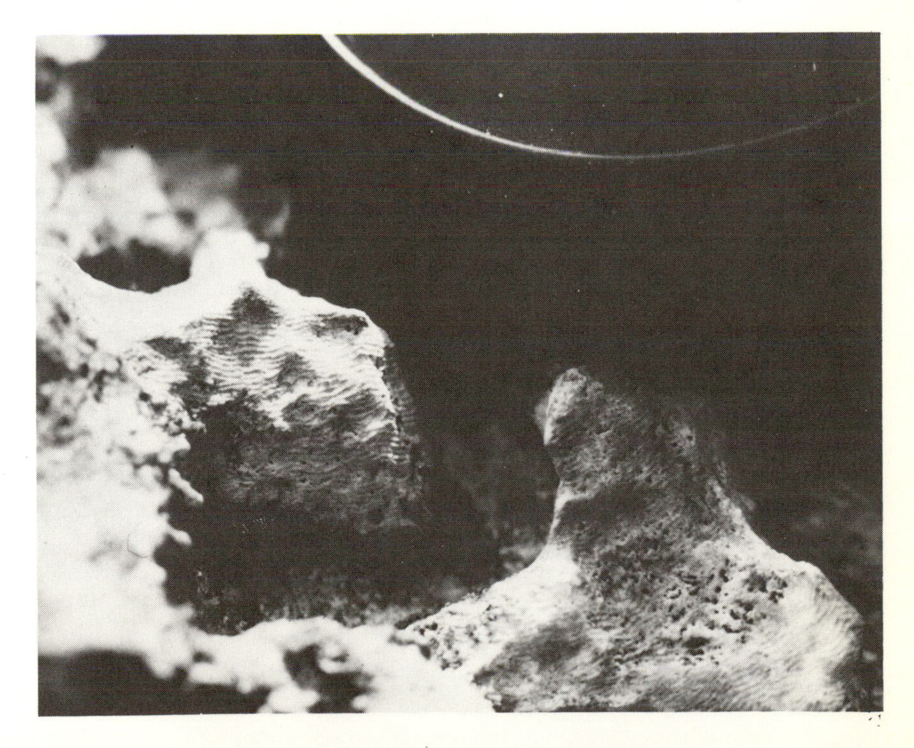

Fig. 9.10 Rasping marks made by chiton *Acanthozostera* on intertidally exposed beach rock, Heron Island, Great Barrier Reef. Lens cap (top right) 55 mm diameter, with horizontal marks in centre of picture. (Photo: S. T. Trudgill.)

grooves illustrated are up to 2 mm long and up to 0.5 mm wide and deep. The radula grows progressively outwards as it is worn during grazing.

Gastropods, patelliform molluscs and chitons all possess a radula and are known to graze and erode rock; erosion by fish has been little studied for carbonate rocks (as distinct from algal grazing which has been relatively well studied) but many fish, especially parrot fish, ingest coral fragments. Fish grazing marks can be identified on tropical carbonate rocks and Fig. 9.11 shows parallel grooves in beachrock on Heron Island, Great Barrier Reef, but the effect is thought to be limited to about 1–2 mm in the rock. This is probably also the depth limit of the superficial scraping performed by *Grapsus* spp. of crabs, also in the tropics.

9.3.3 Molluscan grazing – rates and effects

The rates of molluscan grazing have been largely studied by faecal collection, either in the laboratory or in the field and much of the quantitative work was pioneered by McLean (1967; 1974). In the laboratory, the experimental set up he used is shown in Fig. 9.12, with a set of gastropods grazing on algal colonised rock. A 'rain' of faecal pellets was collected on sample dishes, retrieved, weighed and analysed. Erosion rates were calculated and were 0.2–2.4 g a^{-1} for 6 gastropods (Table 9.8). For a single species, the erosion rate increased with the size of the organism (as measured by shell height) but between species there was no rigid relationship between shell size and erosion rate: some

Fig. 9.11 Fish grazing marks (most probably Parrot fish) on intertidally exposed beach rock, Heron Island, Great Barrier Reef, Australia. The marks penetrate 1–2 mm into the substrate. Lens cap 55 mm diameter. (Photo: S. T. Trudgill.)

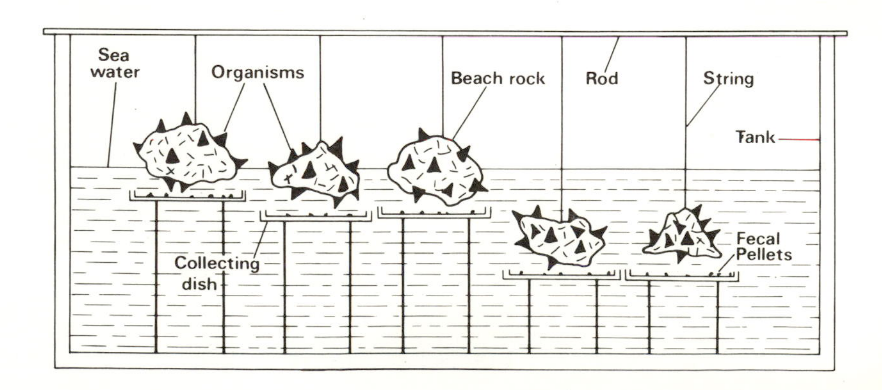

Fig. 9.12 Apparatus used by McLean to collect faecal peelets from grazing organisms. Samples of beach rock were suspended from a rod in an aquarium tank, with collection dishes below (modified from McLean, 1967).

Table 9.8 Gastropod erosion rates, Barbados beach rock, West Indies (modified from McLean, 1967).

Gastropod	*Mean shell height (cm)*	*Erosion rate, g* a^{-1}
Littorina ziczac	1.1	0.6
Littorina meleagris	0.5	0.2
Nerita tuberculata	0.9	1.2
Nerita tesselata	1.2	0.7
Nerita versicolor	1.7	1.4
Cittarium pica	1.4	2.4

species were more efficient at erosion than others of similar size. Depth of algal penetration influenced the rate of erosion, suggesting that algae not only act as a food source but also that they may have a pre-softening effect on the rock. Where algal penetration was up to 1 mm deep *Littorina ziczac* and *Nerita versicolor* were four to five times as efficient at producing faeces than when algal penetration was only 0.1–0.5 mm deep, though in the latter case the carbonate content was higher and so the procedure of searching for algae when algae are not abundant may also cause considerable erosion.

Faecal collections have been also made in the field by McLean (1974) from *Acanthozostera gemmata* (now called *Acanthopleura gemmata*). On beachrock on Heron Island, Great Barrier Reef, this chiton yielded pellets of average length 4.5 mm and maximum diameter 1.0 mm. On analysis they were shown to contain 60–80% carbonate derived from beachrock and 19–33% organic matter, mainly derived from algae. One specimen was calculated to yield 18.0 cm^3 a^{-1}. When the activities of gastropods and chitons were added together, taking into account erosion rates for single specimens and the number of specimens per square metre, surface lowering rates were calculated at around 1–2 mm a^{-1}.

Several grazers may also erode rock by the production of a 'home scar'. They return to the same place after foraging and progressively excavate a shallow depression in which they tightly fit and pass times of desiccation stress, i.e. at low tide during the day. The best known examples of these are the limpets and the chitons. The occurrence of these species is therefore of dual interest since they can contribute to surface removal by grazing and also by excavation of a home scar. The homing by chitons has been studied by Thorne (1967) and an example of two grazing trails are shown in Fig. 9.13. Trails are longer where algae are scarcer; rocks covered with thick algal mats show relatively short browsing trails. McLean (1974) suggested that erosion rates within the home site exceeded 0.5 mm a^{-1} or the equivalent of about 11 cm^3 a^{-1}.

Fig. 9.13 Grazing trails of *Acanthozostera gemmata*. The chiton returns to the home scar, a shallow excavation, after ingesting algae and carbonate material during grazing (modified from Thorne, 1967).

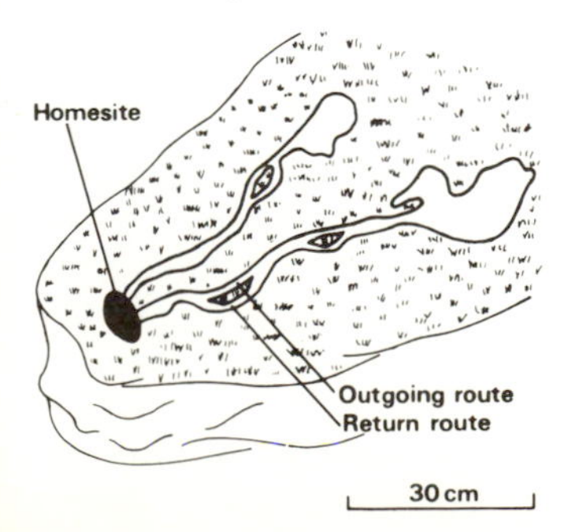

The effects of grazing will depend upon density of grazers and any concentration of grazers in one particular intertidal zone can be expected to lead to more intense erosion in this zone – however, the population must in some way be balanced by rates of algal growth and recolonisation after grazing. This topic has been little researched. Studies of the concentrations of intertidal gastropod distributions are discussed further in section 9.3.3. McLean (1974) gives the term **epirelief** to the small-scale features produced by grazing (such as those shown in Fig. 9.10).

Grazers are thus a potent force of erosion in the intertidal zone but a distinction must be made between a surface-scraping action (with little geomorphological effect) and the deeper gouging action, such as the chitons, with considerable geomorphological effect. Rasping may be up to 0.5 mm at any one time and the overall effects may be up to 1–2 mm a^{-1}. The relative importance of grazers is not widely known, but it will certainly depend upon the population density which is present: numbers will increase where algal growth is encouraged but will decrease in exposed situations and especially where abrasive sand particles are present. Grazing is therefore liable to be important where mechanical processes are less important. It may also have less geomorphological effect where more luxuriant algal growth is present as there will be less need to search in the lithic substrate for algae.

9.3.4 Boring mechanisms

Boring is the direct removal of rock fragments by plants and animals, normally leading to the formation of a tubular passage connected to the surface. A wide variety of organisms bore, producing what McLean (1974) terms **endorelief**. The term **calcibiocavitology** is used by Carriker and Smith (1969) to cover boring or excavation on carbonate substrate. The effects of boring are twofold. First, the borer removes particles from the surface, causing active erosion itself and, second, the rock which is bored is weakened and can be broken off by the mechanical action of waves. In addition, boring algae provide a food source for the grazers discussed in sections 9.3.2–3.

The chief boring organisms are algae, sponges, bivalves, barnacles and echinoderms. These organisms are discussed in sections 9.3.5–7; they are grouped not according to taxonomy but according to the type and scale of excavation. Algae and sponges produce networks of small cavities (0.01–0.5 mm diameter) near the surface of the rock; bivalves and barnacles produce a tubular burrow of high depth to entrance diameter ratio while boring echinoderms produce large spherical pits (of 2–10 cm diameter).

9.3.5 Boring algae and sponges

One of the first stimuli for the study of boring algae came from petrologists interested in the texture of carbonate rocks. Some rocks showed small patches of finer sediment, termed 'micrite envelopes'. It is now known that these are produced by infilling of algal borings. The present-day processes, fossil traces and petrological effects have now been well studied (see, for example, Bathurst, 1966; Flugel, 1977; Kobluk and Risk, 1977(a, b) and Swinchatt, 1969). It has also become now widely recognised that algal boring is a major process of intertidal erosion, (e.g. Purdy and Kornicker, 1968; Schneider, 1976); Swinchatt (1969) suggests that it is limited to about 20 m deep in seawater.

The mechanism is thought to be one where the algal filaments release extracellular chelating or acid fluids from the terminal cell (Fogg, 1973; Golubic, 1973). In addition, Alexandersson (1975) suggests that boring organelles may exist. He studied borings made by *Hormatonema* in limestone where the algal filaments were 10 μm in diameter; small grooves of about 1 μm in width were visible and probably ascribable to the action of thread-like organelles.

The organisms involved are often blue-green algae (cyanophytes) which are able to fix nitrogen and therefore may be of significance in intertidal food chains for this reason. The depth limitation of 20 m observed by Swinchatt (1969) was thought to be in relation to decreased effectiveness of solar radiation with depth, a problem increased by the fact that the organisms are more or less surrounded by carbonate material, and indeed Wilkinson (1974) showed that light could be a critical environmental factor in the ecology of boring algae. Algae penetrate both shells and inert carbonate substrate; the concern in this book, is, however, with calcite in carbonate rocks. In this context, Kobluk and Risk (1977a, b) immersed crystals of Iceland Spar calcite in a shallow

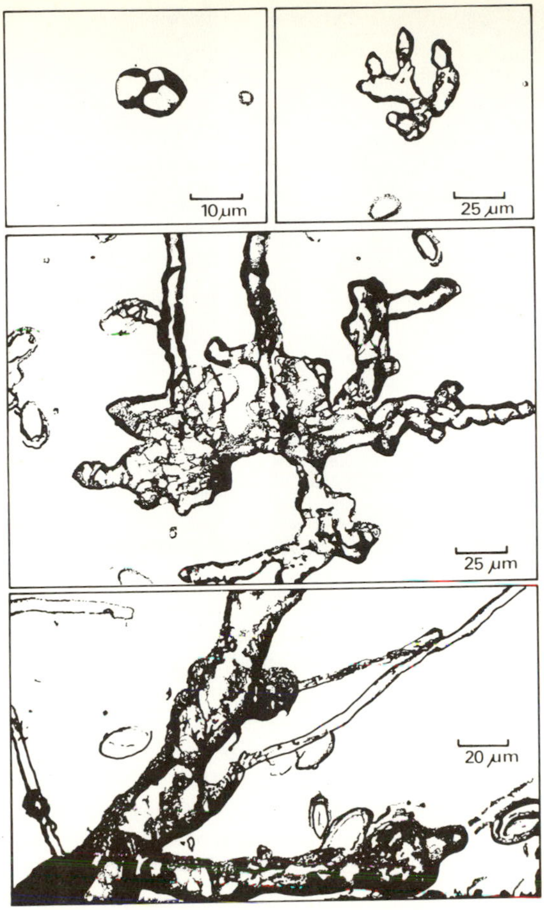

Fig. 9.14 Growth in the boring alga *Ostreobium* in blocks of calcite spar immersed in seawater. Initial, small excavations join into tubular excavations penetrating the block (drawn from photographs from Kobluk and Risk, 1977a).

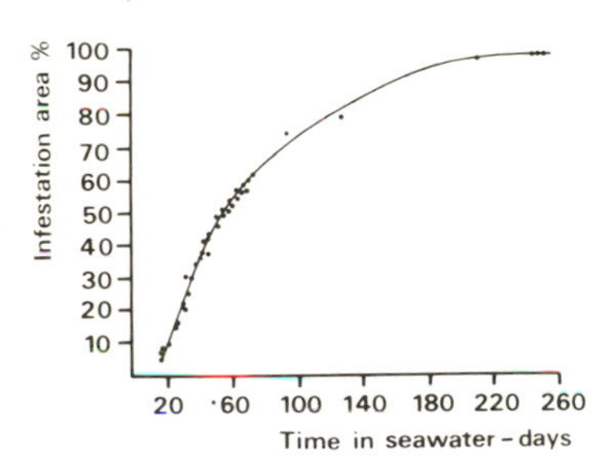

Fig. 9.15 Graphical representation of infestation of calcite spar by *Ostreobium* (modified from Kobluk and Risk, 1977a).

subtidal marine environment at Discovery Bay, North Jamaica. The crystals rapidly became infested with the boring chlorophyte (green alga) *Ostreobium* sp. (Fig. 9.14) with 100% surface infestation after 213 days (Fig. 9.15). After 253 days the crystals were heavily bored to depths of 300–500 μm. The cleavage planes of the crystal exercised a strong control over the orientation of the algal filament growth.

Similarly to the colonisation of experimental calcite blocks, the surface of an intertidally exposed carbonate rock may become so bored so that up to 50% of the surface may be void spaces. Figures 9.16 and 9.17 show the rock surface of intertidally exposed Carboniferous Limestone, County Clare, Eire and scanning electron microscope photographs of that surface at high magnification. The algal produced voids are about 10–20 μm wide.

Purdy and Kornicker (1968) associated the production of small pits in limestones with algal boring by the blue green algae: *Entophysalis*, *Hyella*, *Calothrix* and *Gomphospaeria*, small rock flakes being produced by algal action. Folk *et al.* (1973) used the term **phytokarst** to describe algal-bored fretted limestone surfaces, though they were describing subaerial surfaces. One of the most extensive studies is that by Schneider (1976) working in the Adriatic. He used resin casts to demonstrate the depth and action of algal penetration, removing the algae before resin impregnation and dissolving the surrounding rock by acid digestion after

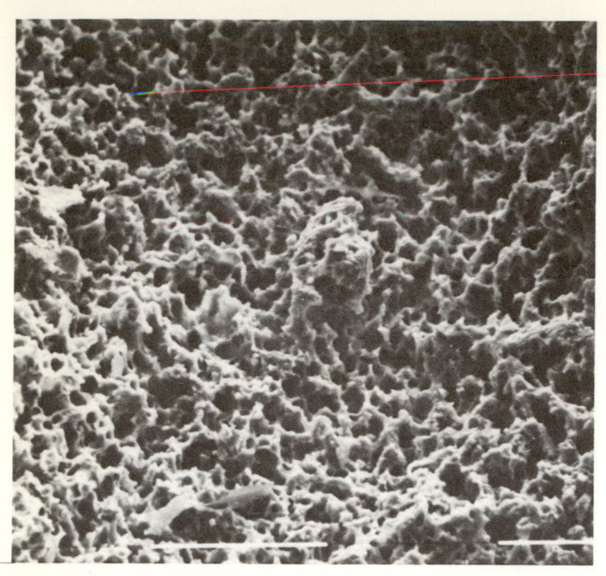

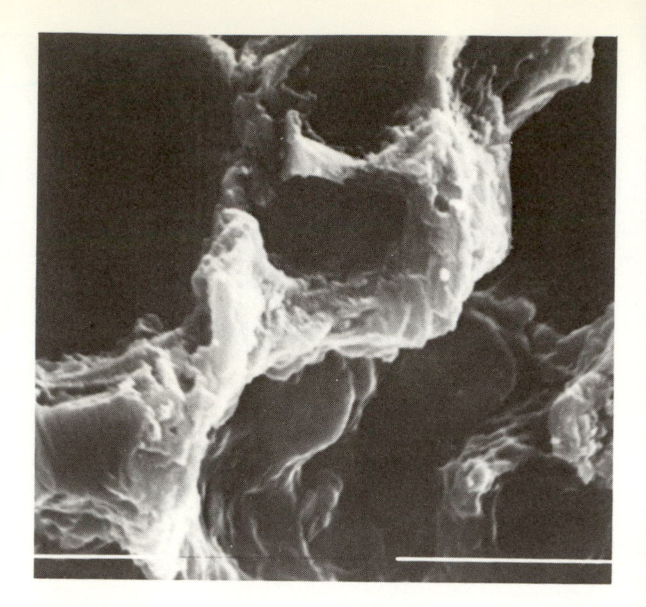

Fig. 9.16 Intertidally exposed surface of Carboniferous Limestone, County Clare, Eire at high magnification, showing that the surface is bored by algae; algal voids are 10–20 μm wide; scale bar = 100 μm. (Photo: S. T. Trudgill.)

Fig. 9.17 Enlargement of portion of Fig. 9.16. Scale bar 10 μm. Note that a substantial proportion of the area is, in fact, void space. (Photo: S. T. Trudgill.)

Fig. 9.18 Drawing of a resin cast of algal filaments penetrating limestone. The resin fills the voids spaces and then sets, after which the limestone is dissolved in acid leaving the resin cast as shown (drawn from a photograph from Schneider, 1976): scale: 10 mm = 100 μm.

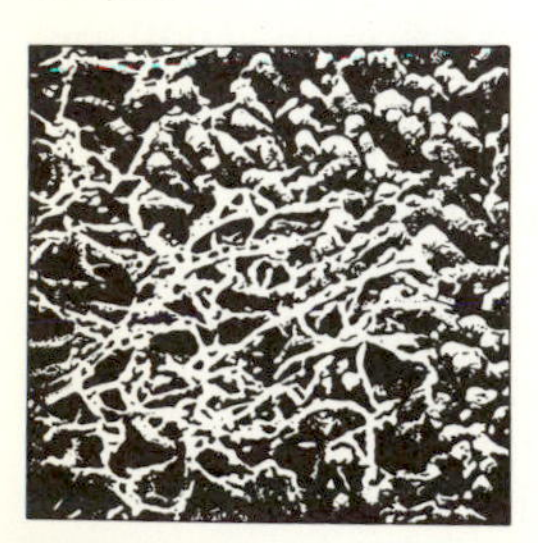

the resin had set (Fig. 9.18). Schneider discusses the problem of the taxonomy of the cyanophytes; the problem is that growth form is used in taxonomic classifications but that it is strongly dependent upon environment. Golubic (1969) also discusses growth form, concluding that it is related to taxonomy and mineralogical factors. Schneider questions the procedure of Drouet and Dayly (1956) who included the majority of forms in one species: *Entophysalis deusta*. He recognises several different forms, including *Hormatonema* (with short, thick threads); *Hyella* (with long threads) and two with longer looped threads: *Kyrtuthrix* and *Mastigoculeus*. These thread-like blue green algae penetrated up to 800–900 μm into the rock. He differentiates these thread like organisms from the non-filamentous algae which may be present up to 1–3 μm below the surface. These occur well below the boring algae and produce a visible green band.

The reasons for the endolithic habit of the algae is suggested as being in relation to two tropisms: hydrotropism and phototropism. The inner parts of a rock are always damp and positive hydrotropism accounts for the direction of boring; algae also bore to greater depths in drier areas. The algae also cease to photosynthesise in intense sunlight; negative phototropism is therefore also implied in the boring habit. Naturally, below a certain depth light penetration will be minimal and thus the boring depth is thought to be an optimum in relation to light intensity.

True endoliths, or **perforants**, are distinguished by Schneider from **cariants** which are epiliths which corrode the surface, giving the surface a carious or fretted appearance. It is these finely fretted forms which may have given rise to the idea of the existence of solutional fretting forms discussed in section 9.1 (Fig. 9.19).

In addition to algae, Schneider discusses boring fungi and lichens. The fungi play a significant role as they bore deeper than the algae. They are not dependent upon light and as heterotrophic organisms they simply require organic substances for nutrition; these are supplied by endolithic algae which can be tapped by those parts of the fungal network in contact with decaying algal matter. Endolithic lichens replace algae higher

Fig. 9.19 Finely fretted surface produced by algal boring, Carboniferous Limestone, County Clare, Eire. Note presence of *Littorina* shells, molluscs which graze on epilithic algae. (Photo: S. T. Trudgill.)

Fig. 9.20 Enlargement of surface produced under intertidal lichen growth; note rounded pit and lichen hyphae. Scale bar 100 μm. (Photo: S. T. Trudgill.)

up the shore profile and the whole surface may be pitted with fungal threads and reproductive bodies. An example of a lichen pit with hyphae is shown in Fig. 9.20 from the upper shore of County Clare, Eire. Similar penetration of calcareous shells by marine fungi is recorded by Kohlmeyer (1969).

There is no doubt that algal penetration of intertidally exposed rocks is widespread and also understimated as a factor in intertidal limestone erosion. Schneider (1976, p. 58) notes that 'The endolithic microflora (algae, fungi, lichens) has a destructive effect on the substrate throughout the entire littoral system'. The geomorphologist, whose task it is to attempt to produce explanations of landforms, should note that delicate fretted, corroded and pitted forms have a distribution which is correlated with that of these endoliths and that at a microscopic level pits and fretting can be seen to be caused by the erosive action of these organisms. In addition, it should be emphasised that they have a dual action of providing food for grazing organisms which also erode the rock to obtain them.

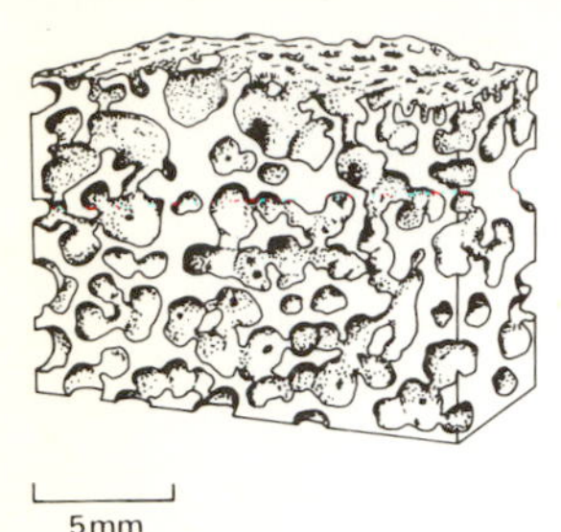

Fig. 9.21 Drawing of a coral block with excavations made by the boring sponge *Cliona lampa* (from Bromley, 1978).

Boring sponges are less able to withstand desiccation than boring algae and thus their distribution is more restricted than that of algae and they are most evident in the lower intertidal and in moist overhangs and other damp places. Like algae and fungi they can penetrate both shells and carbonate rocks. Their excavations often take the form of small spherical chambers but can also be in the form of honeycombed networks (Fig. 9.21); their size is usually 5–10 times larger than the algal borings and at 0.5–1.00 mm in diameter the entrances to sponge borings are often clearly visible to the naked eye on rock surfaces (Fig. 9.22).

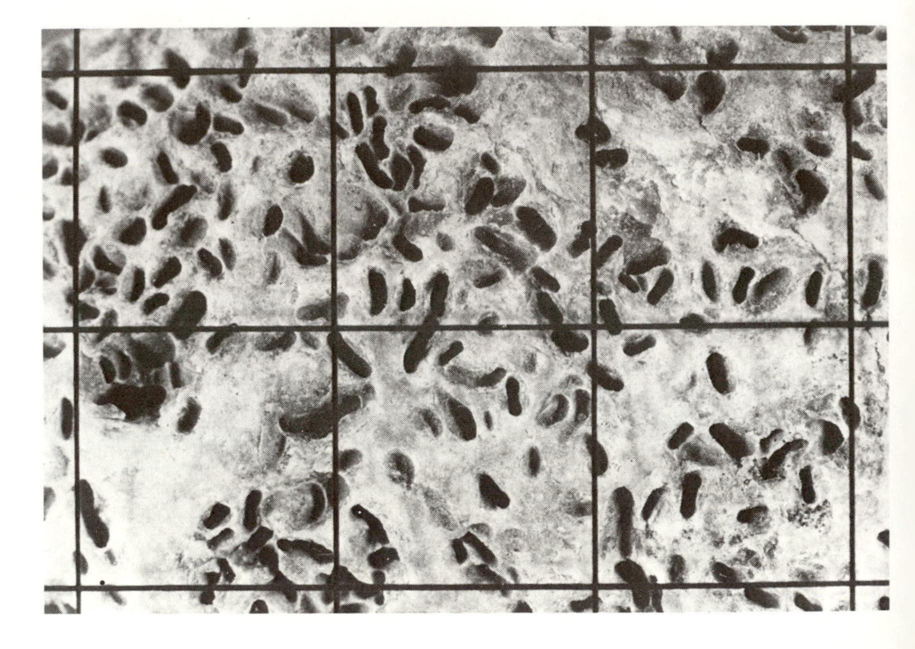

Fig. 9.22 Photograph of the surface of a Carboniferous Limestone sample from the intertidal zone at Penmon, Anglesey, Wales, UK. The surface is bored by the sponge *Cliona* and the excavated slits are 1–2 mm long, grid = 1 cm square. (Photo: J. Owen.)

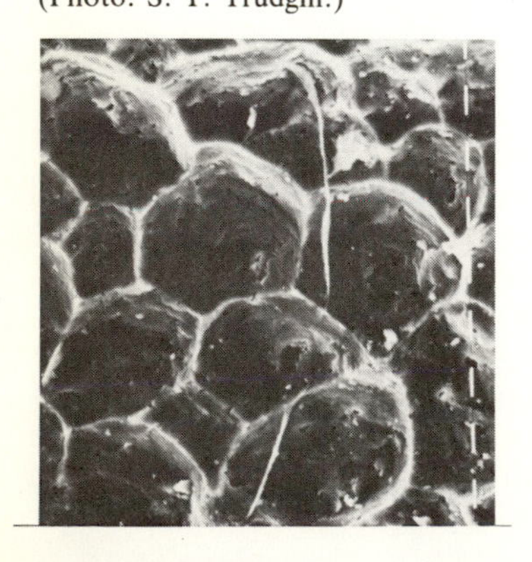

Fig. 9.23 Enlargement of a surface produced by *Cliona*; note small semi-circular pits about 50 μm across; scale bars 10 μm. Samples from the base of a *Cliona* excavation, Carboniferous Limestone, County Clare, Eire. (Photo: S. T. Trudgill.)

Sponges are filter feeders and set up inhalant and exhalant currents, and thus their excavations need to be in contact with the rock surface and water; they appear to bore primarily for support and protection.

Most species of boring sponges belong to the genus *Cliona*, including *C. lampa* and *C. celata*. Pomponi (1977) also recognises *Anthosigmella* and *Spheciospongia* in tropical environments. The boring mechanism was studied by Cobb (1969) for *Cliona* in shell material and in calcite crystals. Extensions of the sponge tissue, termed *etching amoebocytes*, are able to penetrate calcite in a semi-circular cutting. Chips about 60–80 μm wide are discarded by the sponge. They leave corresponding elliptical to circular pits, arranged in reticulate patches. Typical pits are shown in Fig. 9.23. The etching cells and excavated carbonate substrates have been studied in detail by Pomponi (1976, 1977). The pits excavated in the carbonate range in size from 20–90 μm in width. Acid secretions

Fig. 9.24 Colonisation rate of limestone blocks by boring sponges (from Rutzler, 1975).

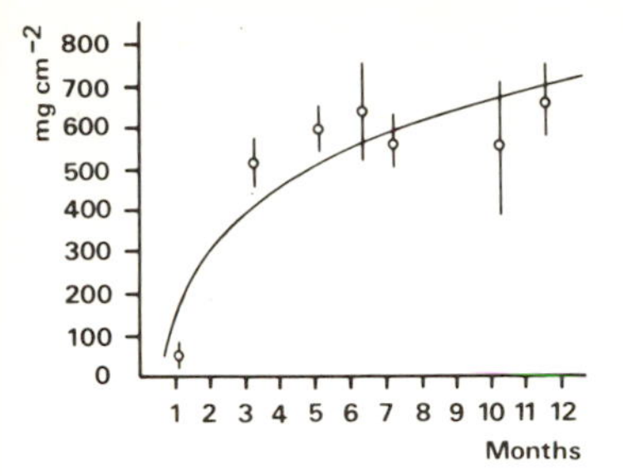

are thought to be derived from cell processes which grow into the substrate.

Coral dwelling sponges have been studied by Pang (1973) and MacGeachy (1977). The sponges were common in dead coral but were found also in living specimens and had a wide vertical range. The volume removed by boring sponges was up to 23% of the total mass, with penetration of up to 2 cm into the coral. Moore and Shedd (1977) estimate that bioerosion rates by sponges approach 7 kg m^2 a^{-1} on reefs at Discovery Bay, Jamaica.

Few studies have quantified the rates of erosion by sponges for carbonate rocks other than for reef carbonates. However, Neumann (1966) observed deep undercut notches in a calcarenite rock in Bermuda where the notch was infested with boring sponges. Experiments with colonisation of initially sponge-free limestone blocks suggested that rates were in the region of 5–7 kg limestone m^2 in 100 days or equivalent to up to 1 cm coastal retreat per year. However, Rutzler (1975) has shown that such measurements are almost certainly too high as rates fall off once the substrate colonisation has passed the initial stages; the rate curve flattens out after 6 months (Fig. 9.24). It is of interest that the notch reported by Neumann is subtidal, thereby confirming that the action of sponges is most marked at these levels; it should be added that other boring and also grazing organisms were involved in the bioerosion he described and that coastal retreat in the area studied was not due to sponges alone.

The mechanism of sponge erosive action upon carbonate substrates has been relatively well researched; their action is undoubted. Their effectiveness in a reef context has been demonstrated and some tentative erosion rates estimated; their action also contributes carbonate grains to reef sediment. Their presence, actions and erosion rates in carbonate rocks as a whole must be significant but have not been widely researched.

9.3.6 Boring bivalves barnacles and other tube-producing borers

Certain species of bivalve molluscs, barnacles, sipunculid worms and polychaete worms produce tubular borings which may penetrate into carbonate rocks up to several centimetres; some also bore into live coral, sandstone, wood, peat and clay. The processes in carbonate material have been well documented especially since the time of Otter (1937) who drew the attention to the processes involved during The Great Barrier Reef Expedition of 1928–9. Several species were noted as being effective but the principal bivalve mollusc was *Lithophaga* and the rock-boring barnacle *Lithotrya* was also present. *Lithophaga* are shown in Fig. 9.25 and, together with the borings, in Fig. 9.26. In the latter figure the calcareous siphon tubes can be clearly seen round the keyhole-shaped tube which allows for the positioning of the inhalant and exhalant siphons. The rock has also clearly been eroded by a few millimetres around the calcareous tube secretions during the lifetime of the molluscs.

There is now a wide biological and geological literature on boring bivalves and useful summaries are provided by Yonge (1961), Ansell and Nair (1969) and Carriker and Smith (1969). They are widely distributed in tropical and temperate regions and bore both into corals and carbonate rocks. Boring bivalves, like many borers, are not confined to limestone and bore widely into clay, peat, wood and sandstone.

The mechanisms involved are mechanical in both carbonate and non-carbonate substrates and there is also evidence for acid secretion by borers in carbonate rocks. Ansell and Nair (1969) compare the methods by

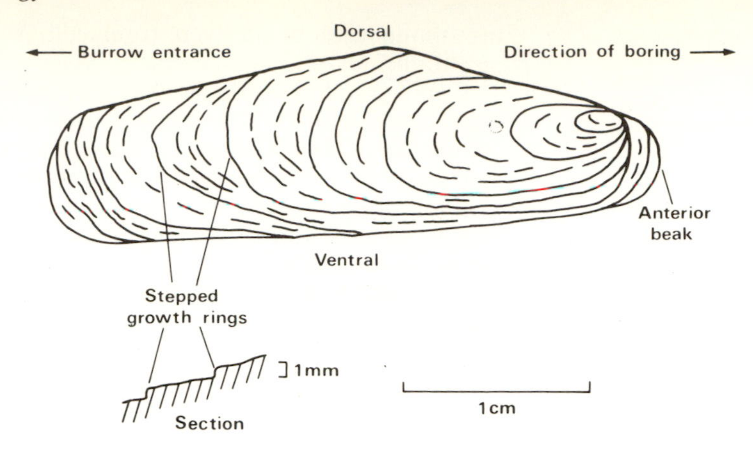

Fig. 9.25 Drawing of *Lithophaga nasuta*.

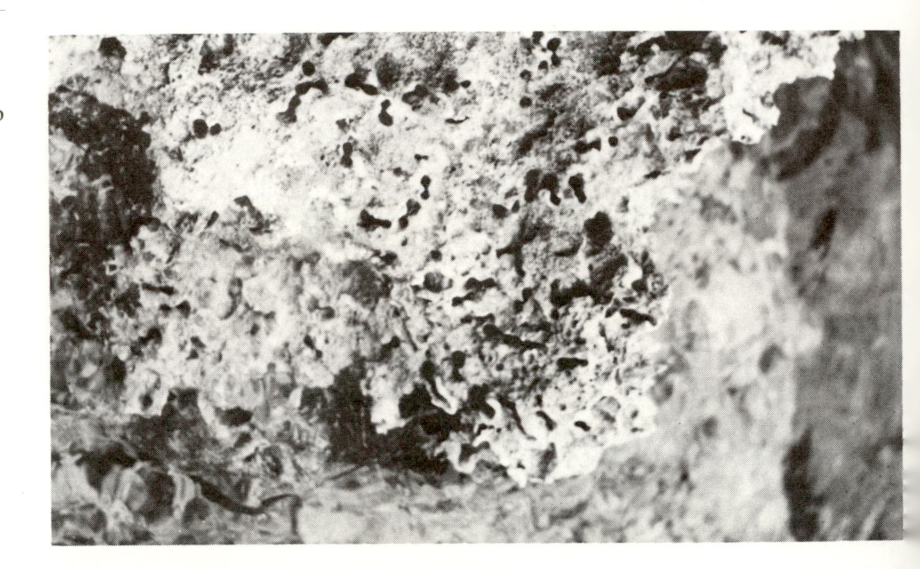

Fig. 9.26 View of entrances to borings made by *Lithophaga* in coral limestones, Aldabra Atoll, Indian Ocean. The voids are up to 1 cm long and the keyhole shape is clearly visible. (Photo: S. T. Trudgill.)

which bivalves bore, concluding that rocking motions and rotational movements may be involved, the impetus coming from muscular contraction. Where chemical action is involved, as in *Lithophaga*, this is often by means of glands at the pallial edge of the molluscan mantle (furthest away from the burrow entrance) and which have an acid reaction. There is a local secretion from these glands which causes dissolution and pre-softening of the rock. The actual removal of the particles is achieved by the mechanical movements of the shell.

Predatory shell-boring molluscs, such as the gastropod *Thais lapillus* (Dog Whelk) have boring organs with which they penetrate prey shells and here the enzyme carbonic anhydrase may be used to hydrate carbon dioxide and produce hydrogen ions (Chétail and Fournié, 1969). Similar enzymes may be involved in rock-boring molluscs. Travis and Gonsalves (1969) suggest that the organic layers present in carbonate shells may act as primary pathways for shell breakdown by boring gastropods, carbonate crystals becoming removed mechanically.

Acid secreting glands are widely found in the mollusca. They are present in many molluscs which do not show a boring habit and they are used

Fig. 9.27 Longitudinal section of 1 cm of valve of *Hiatella arctica*, a boring bivalve, showing the stepped nature of the outer shell surface (above) and internal growth bands. (Photo: J. Owen; slide M. Whyte.)

Fig. 9.28 Photograph of outer surface of 2 cm ridged shell of *Hiatella arctica*, showing well-marked ridges, presumed to be annual. (Photo: J. Owen.)

for defence. Therefore, the production of acids by boring molluscs cannot be taken alone as conclusive proof of a chemical boring mechanism unless the secretion is observed during undisturbed boring activity.

Studies of rates of boring have been undertaken by the observation of the rate of colonisation of carbonate blocks by borers or by the counting of growth rings to determine age and dividing the depth of the hole bored by the age to obtain a rate in depth produced per year. There is some discussion as to whether clear growth bands are readily visible on the surface of bivalve shells, and while the bands are more readily visible in a long section of the shell (Fig. 9.27), their annual nature is not always unequivocal. There is a large literature on periodicity in bivalve shells (for example, Craig and Hallam, 1963; Farrow, 1971; Merrill *et al.*, 1962; Orton, 1928 and Pannella and MacClintock, 1966). At a very high magnification daily growth rings can be visible but there is sufficient evidence from controlled growth experiments that major steps in shell morphology can be equated with annual growth rings (Fig. 9.28) though this has not been investigated in detail for all the species involved. Any work presenting growth rates based on the counting of visible growth rings should be viewed in the light of the qualification that such rings may not be annual. However, in this context, Comfort (1957) has investigated the duration of life in molluscs and he concludes that life spans are of the order of 5 to 10 years. Therefore, age estimates which fall within this span are likely to be plausible estimates in the light of lack of any other concrete evidence.

Boring by *Lithophaga* has been studied by Hodgkin (1962), Kleeman (1973), Trudgill (1976a) and Warme and Marshall (1969). The latter authors estimated boring rates to range from 2 to 10 mm per year while Trudgill counted growth rings of *Lithophaga nasuta* and compared this with shell length, showing a slowing down of growth after 5–7 years; boring appeared to proceed at a mean rate of 9 mm per year (Fig. 9.29). Kleeman (1973) determined values of 4.3–12.9 mm per year for *Lithophaga lithophaga* in different limestones, and he suggested that the slower rates occurred in the harder limestones.

In temperate regions the bivalve *Hiatella arctica* is a frequent borer in low intertidal and subtidal situations in limestone. It may also occur in other rocks such as soft sandstone. Rowland and Hopkins (1971) recorded *Hiatella arctica* in the Pacific from Alaska to Mexico and also in the Arctic and Atlantic Oceans. The presence of a borer in both limestone and non-limestone substrate is often taken as an indication that the borer does not excavate by means of acid secretion alone. Indeed, observations by Hunter (1949) on *Hiatella* could detect no acid secretion glands and he concluded that the boring process was largely mechanical; however, respiration may enrich the fluid within the burrow with carbon dioxide and assist dissolution. Recent work using a scanning electron microscope suggest that this might be the case because etch patterns are clearly detectable within *Hiatella* borings in County Clare, Eire (Fig. 9.30). Hunter (1949) suggested that boring is a self-protection mechanism, occurring most in rocks of smooth outline; if crevices are abundant and afford adequate protection then the animal simply lodges in the crevices and is attached to the rock by a byssus (a network of threads) in a similar fashion to the mussel *Mytilus*.

Rock-boring barnacles have been studied by Ahr and Stanton (1973) and Trudgill (1976a); the former records that the species *Lithotrya* is widely distributed in tropical seas; *Lithotrya valencia* is shown in Fig. 9.31. Using the rings clearly showing in Fig. 9.31 as annual growth

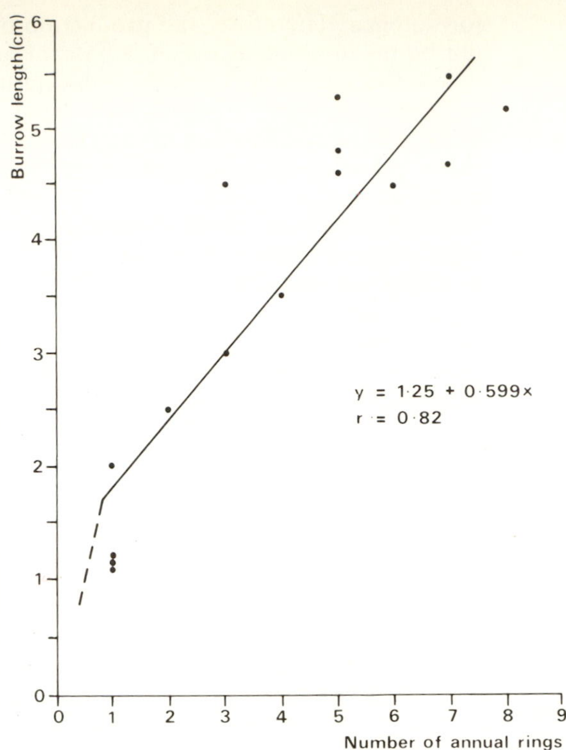

Fig. 9.29 Boring rate of *Lithophaga nasuta*, Aldabra Atoll, Indian Ocean, assuming stepped growth rings to be annual (from Trudgill, 1976a).

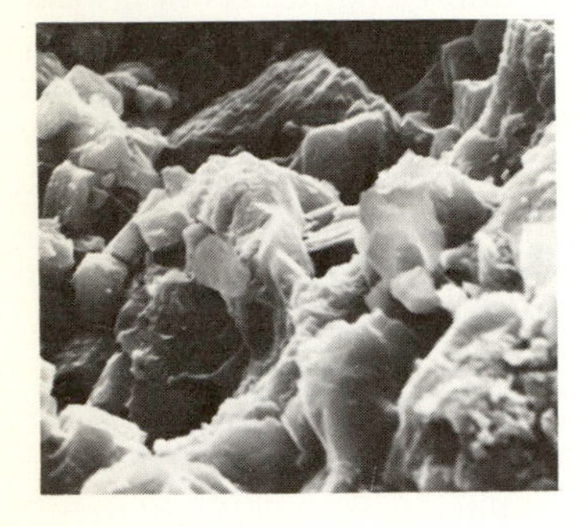

Fig. 9.30 Etched limestone crystals from the base of a boring by *Hiatella arctica*, County Clare, Eire. (Photo: S. T. Trudgill.)

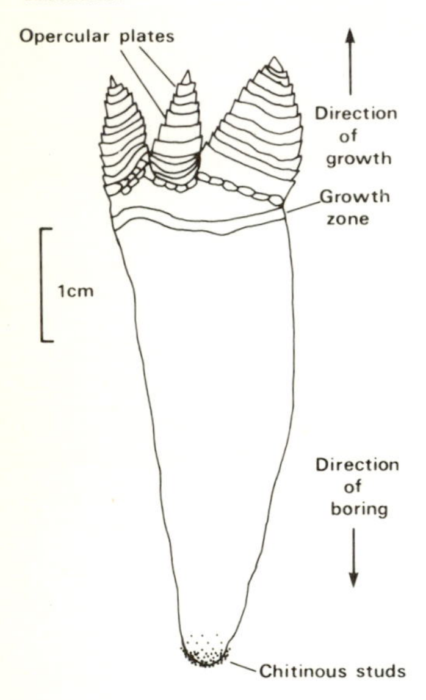

Fig. 9.31 *Lithotyra valencia*, boring barnacle.

rings, Trudgill (1976a) calculated a rate of 0.8 cm per year for the rate of production of boring, on Aldabra Atoll, Indian Ocean.

Other borers include polychaete worms (Blake, 1969) and sipunculid worms (Rice, 1969; Hutchins, 1974) but their occurrence and roles in bioerosion are not widely known.

The role of borers is suggested by Hunter (1949) to be limited by the erosive mechanical action of the sea, such as the fracture of rocks during storms. New borings will continue to be started until the outer layers of the rock are honeycombed, then no new borings will be initiated and further boring will not occur unless the bored surface layer is removed by mechanical action (or by other bioeroders such as sponges). Thus, boring is cyclic, with a sequence of boring–fracture–boring occurring, with a non-active stage between the fully bored stage and the occurrence of fracturing. Thus, rates for boring calculated by various means should be put into this context when considered in terms of geomorphological role. The rates represent a maximum rate of removal. The actual effective rate of removal may be much less, for removal of the bored rock by other means is minimal. However, there is no doubt that the boring process and fracture is effective and honeycombed rock fragments are frequently to be seen cast up on a shore after major storm events. The extent which boring can have is shown by the use of resin casts of bored rocks, where the rock has been dissolved away after the resin has set (Fig. 9.32).

9.3.7 Boring echinoids

Several species of echinoid (sea urchins) bore into limestone, producing hemispherical or cyclindrical pits, or occasionally grooves in the rock (Figs. 9.33 and 9.34). Rock boring by echinoids has been widely known

Fig. 9.32 Resin cast of pebble from the high tide zone, where the pebble had been cast up after being broken off from the low tide zone, presumably in storms; Penmon, Anglesey, Wales, UK. Casts of *Hiatella* occupy much of the volume and they are up to 1 cm across; smaller casts (above and below) are of *Cliona* and algal borings. (Photo: J. Owen.)

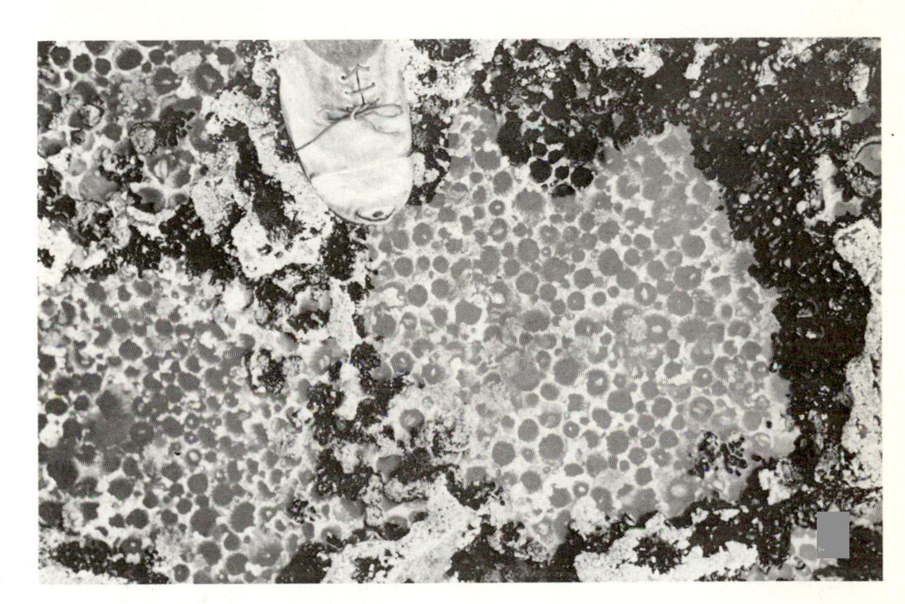

Fig. 9.33 Rock boring sea urchins, *Paracentrotus lividus*, in an intertidal pool, County Clare, Eire. Each individual is in a hemispherical pit which it has excavated. (Photo: S. T. Trudgill.)

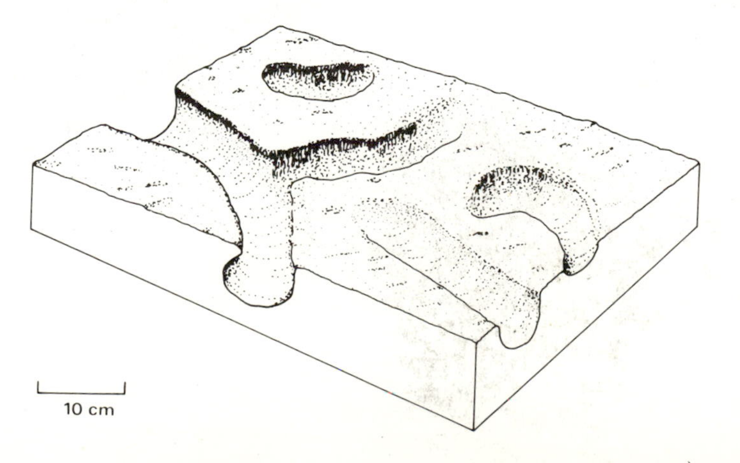

Fig. 9.34 Grooves cut into limestone by the sea urchin *Echinometra lucunter* (from Bromley, 1978).

Fig. 9.35 Scanning electron microscope photograph of the base of a pit excavated by *Paracentrotus lividus*, County Clare, Eire. Scale bar 100 μm. Note lack of relief and smooth, abraded surface (contrast can be made with the etched surface relief shown in Fig. 9.30). (Photo: S. T. Trudgill.)

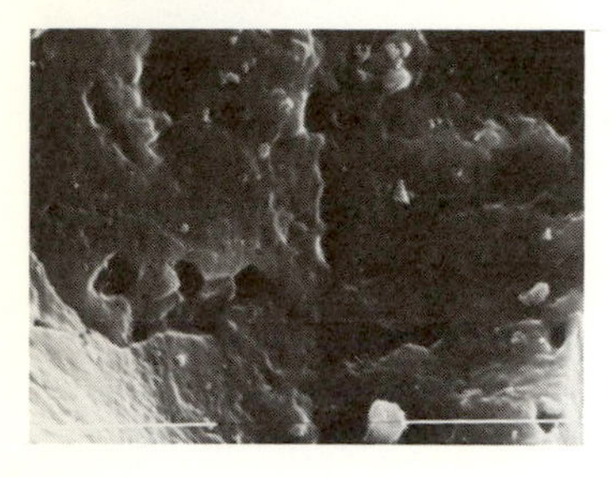

since the review by Otter (1932), who concluded that the deepest boring may occur in the most exposed situations, boring being a response of exposure to wave action and rough intertidal conditions. For example, *Paracentrotus lividus* (Fig. 9.33) seldom bores in the relatively tideless Mediterrranean, whereas it bores from 5–15 cm in exposed Atlantic environments. Boring is thought to be achieved by the action of the basal teeth, assisted by the spines, a supposition confirmed by scanning electron microscope examination of the smooth abraded walls of echinoid pits (Fig. 9.35).

Tight fitting excavation minimises moisture loss and maximises protection against dislodgement (Kowsmann, 1972), but if the organisms are fixed and sedentary, the interesting question arises of how the organisms feed since most other echinoids which are free-living move about, browsing on algae. Many boring species may, in fact, feed on floating algal detritus by passing pieces to the mouth using their tube feet, as observed for *Echinostrephus molaris* by Campbell *et al.* (1973). *Paracentrotus* can be seen to have algal and other debris attached to it (Fig. 9.33) which, while it is thought plays some part in reducing insolation stress, may also be used in part as food. In addition work by several authors (de Burgh, West and Jeal, 1977; West and Jeal, 1973 and West *et al.*, 1977) suggests that dissolved organic matter can be absorbed directly into the tissues from seawater, a suggestion confirmed by the tracing of the uptake of isotopically labelled amino-acids from seawater into *Paracentrotus* tissues.

Fig. 9.36 Inside of a *Paracentrotus* test treated to show growth rings: these are the three faint lines present on each individual plate (especially left centre), chevron shaped and parallel to the zig-zag growth zones running from top to bottom of the test; specimen shown is about 3 cm across, (Photo: J. Owen.)

The growth rates of sea urchins have been widely studied (e.g. Regis, 1972; Allain, 1972) and Jensen (1969) outlines how age determinations may be made by the chemical treatment of the sea urchin test (shell) which reveals annual growth rings (Fig. 9.36). Using this technique Crapp and Willis (1975) counted growth rings and coupled this with measurements of size to produce a growth rate curve (Fig. 9.37). Growth rates fall off during the later stages of growth, and this author (unpublished) used a similar method for ageing and measured the depth of borings to produce a boring rate curve for individuals in exposed and sheltered situations in County Clare, Eire (Fig. 9.38). This work confirmed that boring was minimal on sheltered coasts but that on exposed coasts the rate approached 1 cm per year.

It is clear that large amounts of rock material can be removed by boring echinoids and this must be taken into account when bioerosion is evaluated (e.g. Healy, 1968). Their distribution is widespread in both tropical and temperate seas, with several species being involved, *Paracentrotus* being common in temperate seas and *Echinometra* species being common in tropical seas. In addition the large, dark-coloured, long-spined sea urchin *Diadema* is common in coral reef environments where its action involves the ingestion and comminution of coral reef sediment (Hunter, 1977; Stearn and Scoffin, 1977).

9.3.8 Erosion and protection

It should be stressed that some marine organisms can encrust rock surfaces and thus may protect them from physical erosion (Focke, 1977). Dissolution may proceed under such a cover (especially in relation to respiration) but this will only become apparent if the organism itself is removed.

The organisms which may have a protective role include the following:

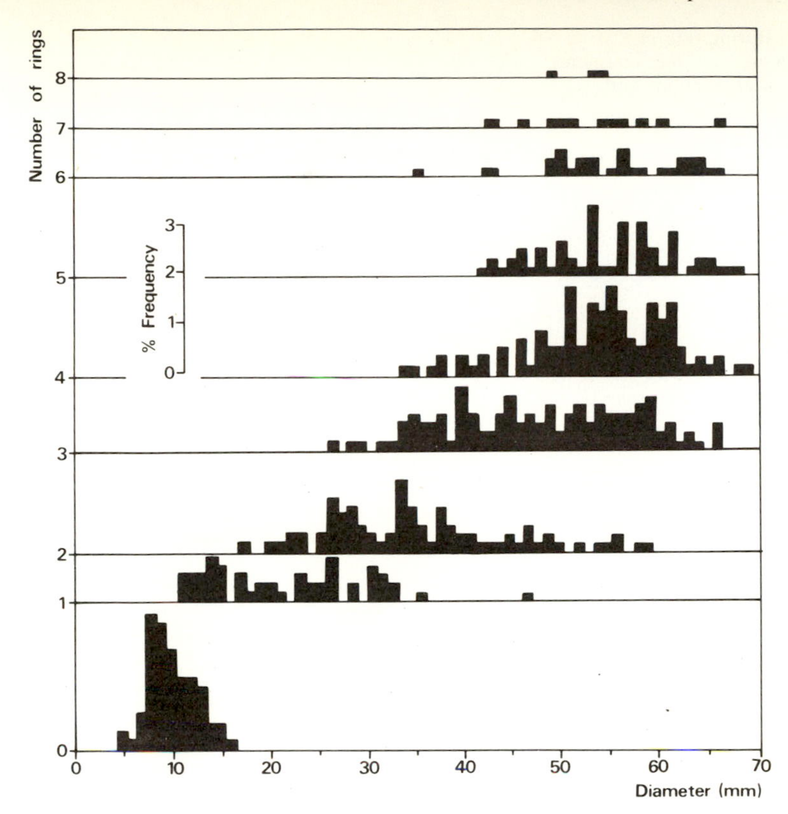

Fig. 9.37 Growth rate of *Paracentrotus* for specimens collected at Bantry Bay, Eire (modified from Crapp and Willis, 1975).

Byssate bivalves (e.g. mussels – *Mytilus*).
Encrusting barnacles (e.g. *Tetraclita, Balanus, Chthalamus*).
Calcareous algae (e.g. *Lithothamnion, Lithophyllum, Porolithon*).
Encrusting tube worms/vermetid gastropods (e.g. *Pomatoceros, Dendropoma*).
Mat-forming organisms (e.g. zooanthids, hydroids)
Sand binders (e.g. filamentous algae which may bind sand in a protective mat)

Patelliform (limpet-like) molluscs cannot be included in this list because although they can encrust the rock they excavate a home scar of up to 2–5 mm deep. The protective roles of barnacles and mussels are probably important in a mid-intertidal temperate environment, although the relationships between biological cause and geomorphological effects are difficult to establish. The organisms commonly occur on the tops of pinnacles, though it is difficult to ascertain whether the pinnacles form by other means and the organisms then preferentially encrust the top (where water circulation and filter feeding are encouraged) or whether they colonise patchily initially and then the rock where they have not colonised is preferentially eroded. It is probable that neither extreme view is wholly tenable and that the organisms help to reinforce differential erosion by the colonisation and protection of pinnacle tops while simultaneously receiving some benefit from this strategy. In addition, it is by no means an absolute rule that that the barnacles and mussels co-

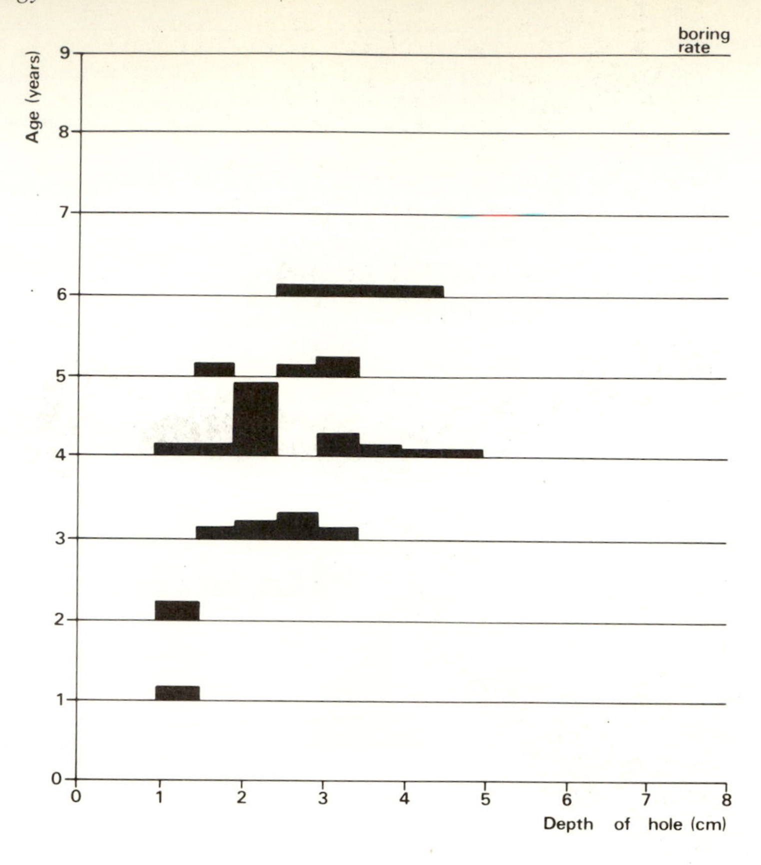

Fig. 9.38 Boring rate of *Paracentrotus*, from measurements of depths of excavations and growth ring counts.

lonise the pinnacle tops alone; it is only a tendency, as can be seen in Fig. 9.2. (p. 126).

9.3.9 Zonation

The study of biological zonation is obviously crucial to the understanding of morphological zonation if the morphology is strongly influenced by biological action; this is discussed more fully in section 10.3. At this stage of the discussion, the zonation of the organisms thus far discussed will be clarified and predictions can be made about the probable zonation of morphology, emphasising again, of course, that simple correlation of biological and morphological zonation in itself does not necessarily indicate that a biological cause–geomorphological effect relationship may exist; morphology produced by other means may provide preferential sites for colonisation by bioeroders. Having examined the biology of many organisms involved, it is clear where erosive mechanisms do exist and thus where cause–effect inferences about the biological influences upon morphology can be made.

The average intertidal distribution of the organisms discussed so far can be listed as follows:

Grazing molluscs: widespread intertidally, often concentrated in the mid-intertidal.
Algal borers: widespread, often noticeable in the upper intertidal but subtidal to several metres.
Sponges: mostly lower intertidal and subtidal.
Bivalves and other tube borers: mid- and lower intertidal and subtidal.
Echinoderms: mid- and lower intertidal and subtidal.

It should be stressed that this pattern can be altered by local conditions of topography and especially sand cover; for example, organisms having a wide range may be absent in the lower range if a sand cover is present in the lower intertidal and thus they may appear to be concentrated in the mid-intertidal. As a further example, the distribution of algae used by grazers as a food source may be influenced strongly by exposure to wave energy and desiccation, and local modifications of exposure by topography, and thus the distribution of grazers may also be modified, concentrating, for instance, lower down the intertidal zone where there is less desiccation.

In general, the patterns outlined above occur in relation to adaptations to exposure and dessication and they also relate to feeding strategies. The sedentary borers are more or less confined to subtidal positions or to pools in the intertidal (where subtidal zonation conditions are replicated), that is where desiccation is minimal or absent and where filter feeding in water is possible. Molluscan grazers can withstand desiccation by possessing behavioural strategies or physical adaptations such as occupying a home scar during low tide or isolating their body fluids from the atmosphere by the closing of the operculum at the shell entrance; thus they can occupy positions further up the intertidal than can the boring animals. Boring algae have a wide range provided they can photosynthesise at depth in the subtidal and withstand desiccation; though desiccation can, in part, be minimised as the boring habit is partly a hydrotrophic adaptation in order to escape desiccation at the rock surface.

The net effect of all the patterns described above is that bioerosion is often concentrated in the mid- or lower intertidal zones, the precise patterns varying with local conditions, and with the adaptations of the organisms involved and environmental factors such as exposure to the sun (through aspect of the shore), wave energy and variations in substrate hardness.

9.4 Summary

Intertidal limestone erosion cannot be understood without reference to biological processes. Biological action may remove rock surface material by grazing and scraping and also remove material from within the rock by boring and excavation. A wide variety of organisms are involved, including molluscs, algae, sponges and echinoderms and both feeding and protection strategies are involved. The intertidal and subtidal distribution of bioerosive organisms (and also of those organisms affording some protection to the rock) has a strong influence upon the intertidal distribution of intertidal morphology, as discussed in Chapter 10.

10 Process and form in coastal limestone landforms

10.1 Introduction

Solution processes and biological processes are often emphasised in discussions on limestone coasts because they are frequently more noticeable on this rock type than on others. However, this emphasis neglects two things: first, that dissolution and bioerosion occur on other rocks (solution processes being common to many rocks, especially, for example, on basalt where soluble minerals such as olivine are present, and bioerosion being common in several other rocks, especially, for example, boring in softer sandstones); second, the processes common to all coasts are often neglected in discussions on limestone coasts. These processes include abrasion, wave and spray impact, wetting and drying and salt weathering (Zenkovitch, 1967). The relative importance of individual processes has received little attention but existing information is discussed in section 10.2. The role of many marine processes involving energy (especially wave action and abrasion) are best discussed in terms of the role of exposure to dominant wind direction (section 10.5). In all cases the intertidal distribution of the processes involved has a marked effect on morphology (sections 10.3 and 10.4). In addition, the role of sea-level changes has to be considered (section 10.6.) in order to give a full explanation of current morphology. An eroded notch may, for example, be visible in the supra-tidal zone, but it may not be understandable in terms of present-day processes, which will be focused lower down the shore. In such a case the notch would have been formed in the mid-intertidal zone during a former high sea-level and is now stranded at a higher level due to a fall in sea-level or rise in land level.

Many factors combine to produce landforms – not only the distribution of processes and sea-level changes but also the distribution of lithological weaknesses and the stage of adjustment of form to process over time. These factors are discussed together in section 10.7.

As stressed in Chapter 8, process–form relationships form the focus of discussion in many geomorphological texts, but the relationships may be complex, with many different processes involved in various combinations; also the relationships may not be clear if recent changes in conditions mean that form and processes are not completely adjusted to new conditions.

10.2 The relative importance of individual processes

It is undoubted that the divisions between the actions of individual processes are not clear cut in nature. This is especially true of biological and chemical processes, as shown in sections 9.2. and 9.3, the effectiveness of physical actions is also undoubtedly influenced by preconditioning by chemical and biological processes. The divisions between the stresses set up in rock surfaces by wetting and drying (swelling and shrinking), hydration (water uptake), the growth of salt crystals (prising rock grains apart) and chemical dissolution are also not clear cut; one process may weaken a cement between grains and another may dislodge the grains. Thus, precise partitioning of overall intertidal erosion processes and rates into different distinct causal agents is difficult, even if it were legitimate with reference to the co-operative way in which they work in reality. However, some useful statements can be made as to how the relative importance of different groups of processes may vary and these are largely in relation to energy environment. These statements, although admittedly difficult to make, are important since they may assist with the explanations of morphology. Morphology can be seen to vary with exposure and it is thus likely that the relative importance of the various physical, chemical and biological erosion processes may vary. As a general statement, it is clear that as exposure to wave energy increases then biological and chemical processes tend to decrease in relative importance. This is not only because the relative importance of physical processes increases with wave energy, biological erosion also decreases with increasing wave energy as the populations of organisms with a low ability to withstand wave energy decrease. In addition, as physical action increases more the fretting associated with chemical action decreases and thus the numbers of protective crevices which afford shelter for organisms decreases. The relationship with physical action and boring is not a simple one, however, as boring depth increases with exposure in order to provide greater shelter for the boring organism. Thus net biological action tends to increase with a progression from a sheltered environment to a moderately exposed environment, decreasing with a progression from a moderately exposed environment to a highly exposed one.

Quantification of these relationships is not easy since estimates of rates for all the different processes operating at one site are necessary and research workers have tended to focus upon just one or two particular aspects of erosion. However, recent advances in measurement techniques has enabled some quantification to be undertaken.

The roles of direct and indirect biological erosion processes have been stressed by many research workers (for example, Wiens, 1959; Zenkovitch, 1969; Emery, 1962; Neumann, 1968; Schneider, 1976; Trudgill, 1976a). Newell and Imbrie (1955) suggested that as much as 50% of the rock in an intertidal notch in reef limestones is removed by rock-boring barnacles and worms; they also pointed out that the algal boring and molluscan grazing is concentrated in the mid-intertidal zone. However, the importance of mechanical and hydraulic erosion should be emphasised; Wiens (1959), while reviewing the erosion of reefs and reef limestones, stressed the role of biological erosion, but he also observed that blocks of limestone of up to 10 m × 3 m^2 could be broken off and moved up to 100 m by storms.

This observation highlights a classic geomorphological problem of ascertaining the relative importance of continued steady erosion of a moderate nature and discontinuous, infrequent erosion of very great impact. Little work has been done in this field in marine limestone erosion in order to be able to give a definite answer to this problem, but in tropical areas hurricanes and cyclones have catastrophic effects on reefs and reorganise much of the unconsolidated sediments; the effect on consolidated reef rock is much less clear but it is likely that the tops of undercut notches may be broken off by storms. The relative role of storms undoubtedly increases with exposure to dominant winds and storm directions, leading to the survival of the more delicately sculptured landforms only in relatively sheltered areas.

Many research workers tend to stress the importance of one particular aspect of erosion processes on which they have selected to work, rather than emphasising a process having taken an overall view and evaluating the importance of a particular process. Thus, Revelle and Emery (1957) emphasise chemical erosion in their paper entitled 'Chemical erosion of beachrock'; though this does not necessarily imply that they believe this is the only or most important process operating. Similarly, Schneider (1976) tends to emphasise biological and biochemical action and does not measure mechanical processes, which although they appear in his case to be subordinate are not evaluated. Various attempts at measurement of erosion rates of reefs and reef limestones are reviewed by Trudgill (1984), and are listed in Table 10.1. Some idea of the relative importance of individual biological processes can be gained by comparing rates for individual organisms with overall rates though many of the data are not comparable as they are derived from different geographical locations with various substrate hardness and exposures. However, it is apparent that the few estimates for overall surface lowering can often be matched, and therefore, arguably, accounted for, by rates of biological erosion.

Using micro-erosion meters (High and Hanna, 1970; Trudgill, High and Hanna, 1981; Trudgill, 1983) and also weight loss limestone tablets (Trudgill, 1975, 1982a), protected variously from biological action, abrasion and/or intertidal wetting and drying processes, Trudgill (1976a) estimated the overall rates of surface retreat in an intertidal notch to be around 1 mm per year. Of this, about two-thirds could be accounted for by grazing where sand was absent and, where sand was present, the overall rate rose to 1.25 mm per year, of which about one-third could be accounted for by grazing and one-third by abrasion (Table 10.2). These data are for surface retreat only and do not take boring into account. The data were obtained by comparing micrometer readings of rock surface lowering relative to steel reference studs inserted into the rock with erosion rates derived from experimental tablets. Rock tablet weight loss can be converted to a rate of surface retreat in millimetres per year using weight loss per year data if the density, volume and dimensions of the limestone tablets are known (Trudgill, 1976a, p. 187), distributing volume loss over the surface area. Tablets were emplaced away from grazing action, away from abrasive sand and subtidally away from intertidal wetting and drying in order to complete the experimental design of evaluating the role of individual process, the differences between the weight losses being used to indicate the relative effects of the processes controlled for. Figures 10.1 and 10.2 show the suggested trends of rates of erosion in notches and the allocation of processes in terms of relative importance. The 'other processes' category shown in the figures is the shortfall between the calculated tablet rates for abrasion and grazing and includes other intertidal processes such as dissolution, salt weathering, wetting drying, spray and wave action. Observations on weight loss of subtidally placed Aldabra Limestone tab-

Locality	*Substrate*	*Erosion type/agent*	*Rate*	*Author*
I. **Growing reefs**				
Bermuda	Reef	Fish	2–3 tonnes ha a^{-1}	Bardach, 1961
Bermuda	Reef	Bioerosion (fish, Clionids)	1.3 mm a^{-1}	Bromley, 1978
Mariana Islands	Atoll	Fish	1.1–1.6 tonnes m^2 a^{-1}	Cloud, 1959
Orpheus Island, GBR	Reef	*Tridacna crocea*	100 cm^3 m^2 a^{-1}	Hamner and Jones, 1976
Florida	Reef	*Cliona* boring	746–4303 mm^3 reworked	Hein and Risk, 1975
Florida	Reef	Bioerosion, esp. *Cliona*	1 m coral head in 150 years	Hudson, 1977
Barbados	Fringing reef	*Diadema*	97 tonnes sediment ha a^{-1}, 40% reworked	Hunter, 1977
Barbados	Fringing reef	*Cliona* boring	80–377 gm m^2 a^{-1}	Stearn and Scoffin, 1977
II. **Carbonate rocks**				
Puerto Rico	Reef limestone	Intertidal notch retreat	1.0 mm a^{-1}	Kaye, 1959
Red Sea	Coral reef limestone	Surface lowering	Surface lowering of *Tetraclita squamosa*, if 10–15 yrs old = 1 mm a^{-1}	MacFadyen, 1930
Barbados	Beachrock	*Echinometra* boring	4.9 cm a^{-1}; 9.96 cc a^{-1}; 24.0 g a^{-1}	McLean, 1967
Barbados	Beachrock	*Acmaea* grazing	1.5 mm a^{-1}; 0.99 cc a^{-1}; 2.4 g a^{-1}	McLean, 1967
Barbados	Beachrock	*Littorina ziczac*	0.4 cm^3 a^{-1}	McLean, 1967
Barbados	Beachrock	*L. meleaguis*	0.15 cm^3 a^{-1}	McLean, 1967
Barbados	Beachrock	*Nodollitorina tuberculata*	0.6 cm^3 a^{-1}	McLean, 1967
Barbados	Beachrock	*Nerita tesselata*	0.4 cm^3 a^{-1}	McLean, 1967
Barbados	Beachrock	*Nerita versicolor*	0.8 cm^3 a^{-1}	McLean, 1967
Barbados	Beachrock	*Cittarium pica*	1.3 cm^3 a^{-1}	McLean, 1967
Barbados	Beachrock	*Acmaea*	2.0 cm^3 a^{-1}	McLean, 1967
Barbados	Beachrock	*Fissurella*	5.0 cm^3 a^{-1}	McLean, 1967
Barbados	Beachrock	*Anacthopleura*	13.0 cm^3 a^{-1}	McLean, 1967
Barbados	Beachrock	*Chiton*	8.0 cm^3 a^{-1}	McLean, 1967
Barbados	Beachrock	*Echinometra lucunter*	14.0 cm^3 a^{-1}	McLean, 1967
Barbados	Beachrock	Surface grazers	1–2 mm a^{-1}	McLean, 1967
Heron Island, GBR	Beachrock	*Acanthozostera*	18.0 cm^3 a^{-1}	McLean, 1967
Virgin Islands	Reef limestone	Sponge boring	Up to 7 kg m^2 a^{-1}	Moore and Shedd, 1977
Bermuda	Eolianite	*Cliona* boring	1.0–1.4 cm a^{-1}	Neumann, 1966
GBR	Beachrock	*Lithophaga*	1.5 cm a^{-1}	Otter, 1937, in McLean, 1974
Bikini Atoll	Beachrock	Surface lowering	0.3 mm a^{-1}	Revelle and Emery, 1957
SW Australia	Reef limestone	Surface lowering	270–670 $cm^3$100 cm^{-2} a^{-1}	Revelle and Fairbridge, 1957
Bermuda	Iceland spar Calcite	Sponge boring	7 kg m^2a^{-1}	Rutzler, 1975
Heron Island, GBR	Beachrock	Surface lowering	0.5 mm a^{-1}	Stephenson, 1961
Aldabra	Reef limestone	Intertidal surface retreat	0.5–4.00 mm a^{-1} (micro-erosion meter)	Trudgill, 1976a
Aldabra	Reef limestone	*Lithophaga* boring	0.9 cm a^{-1}; 0.87 cc a^{-1}	Trudgill, 1976a
Aldabra	Reef limestone	*Lithotrya* boring	0.8 cm a^{-1}; 0.78 cc a^{-1}	Trudgill, 1976a
Aldabra	Reef limestone	Subaerial surface lowering	0.26 mm a^{-1}	Trudgill, 1976a
Oman	Reef limestone	*Lithophaga*	0.0025 m a^{-1}	Vita-Finzi and Cornelius, 1973

* Listed alphabetically

Table 10.1 Erosion rates measured on reefs and reef limestones* (From Trudgill, 1984)

Table 10.2 Erosion processes responsible for surface removal, (mm a^{-1})

Results: sand present			*Inferences*		
1. All intertidal processes					
	(MEM)	1.25			
2. Minus grazing					
	(Tablet)	0.80	Effect of grazing:		0.45
3. Minus grazing and abrasion					
	(Tablet)	0.39	Effect of grazing and abrasion:		0.81
			Effect of abrasion:	0.41	
Conclusions:		mm a^{-}			% Erosion
Effect of abrasion		0.41			32.8
Effect of grazing		0.45			36.0
Effect of other processes		0.39			31.2
	Total	1.25			100.0

Results: sand absent			*Inference*		
1. All processes					
	(MEM)	1.01			
2. Minus grazing					
	(Tablet)	0.40	Effect of grazing:		0.61
Conclusions:			mm a^{-1}		% Erosion
Effect of grazing			0.61		64
Effect of other processes			0.40		36
			Total 1.01		100

(Modified from Trudgill, 1976a)

lets showed noticeable weight loss, suggesting that dissolution processes and possibly algal boring could be playing an important role here. Also, the mean subtidal weight loss rate was 0.012 g per year, that is about 30 times less than the mean intertidal rate of 0.360 g per year. This emphasises the potency and significance of non-biological intertidal processes. It should be emphasised that these data were gained over a time period of a few months and in no way can they be taken as an indicator of the long-term relative importance of processes when storm events may alter the picture markedly. However, specific as the data are to the measurement period and the place involved, they confirm the relative importance of biological processes in notch formation. In addition, it is felt that the relative importance of biological processes must decrease with exposure: Fig. 10.3 shows how erosion rate increases with exposure to southeast winds on Aldabra and, based on drops in numbers of bio-eroders with increasing exposure, how the relative importance of bio-erosion is thought to decrease.

10.3 Biological zonation and form

While the existence of biological zonation on rocky shores is well established (e.g. Lewis, 1964) there is some doubt as to whether the zonation is a simple response to critical tide levels, that is to say, whether different species are zoned according to their ability to withstand differing periods of emersion and desiccation. In common with many branches of ecology, theories of intertidal distribution have become less environ-

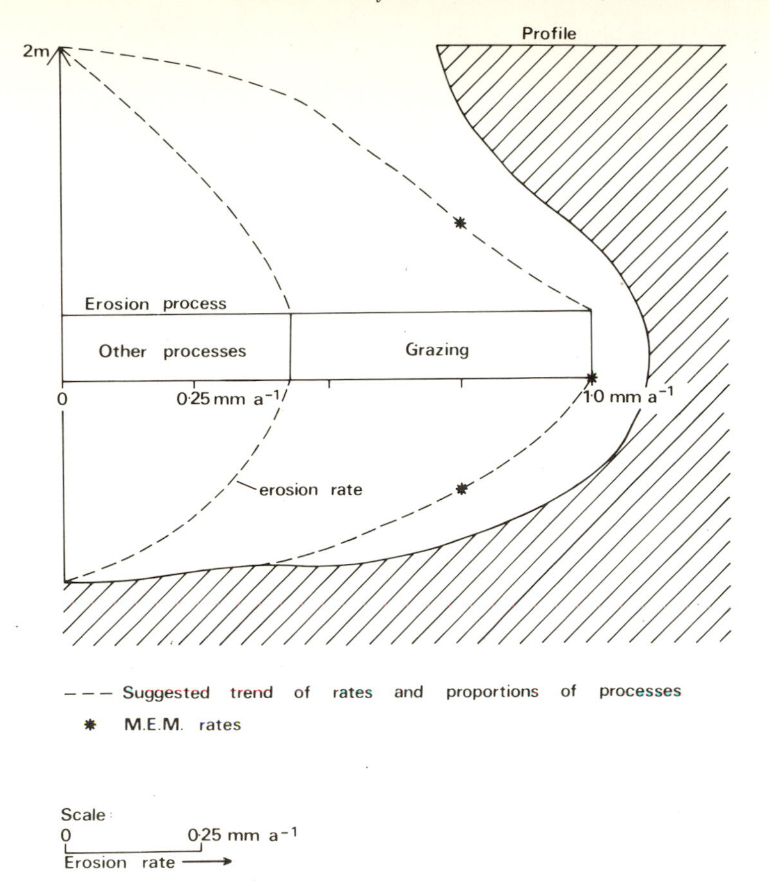

Fig. 10.1 Erosion processes in an intertidal notch, Aldabra Atoll; the rate of 1 mm a^{-1} has a substantial contribution of over 0.5 mm a^{-1} grazing. MEM = micro erosion meter (from Trudgill, 1976a).

mentally deterministic and involve more concepts of inter-specific competition and predation. The role of critical tide levels is questioned by Underwood (1978), who cites the way in which species distribution can be made to vary markedly up and down the intertidal zones by the removal or modification of predator pressure. This is of interest to the geomorphologist, who proposes that there is a cause and effect relationship between biological zonation and morphological zonation: the situation is likely to be more complex and less deterministic than the coincidence of biologial and morphological zones would at first sight suggest. Underwood shows that there is a continuous smooth emersion curve in the intertidal zone, with no discontinuities which could explain biological zonation patterns (Fig. 10.4). This can mean two things in a geomorphological context: (a) that intertidal processes and zonations dependent on water cover are likely to vary smoothly up the intertidal, and neither do they have any marked discontinuities; (b) that the distribution of individual bioeroders is liable to vary in different locations in relation to competition factors rather than to tide levels. Thus, zonation of such eroders as echinoids, boring bivalves, sponges and grazing molluscs and that of possible protective mussels and barnacles need not be the same from place to place and should be interpreted in terms of competitive advantages afforded not only by water level but also site characteristics such as shelter provided by morphological variations and the adaptations of each organism. It is especially necessary to look beyond environmental factors to wider predator–prey relationships to fully

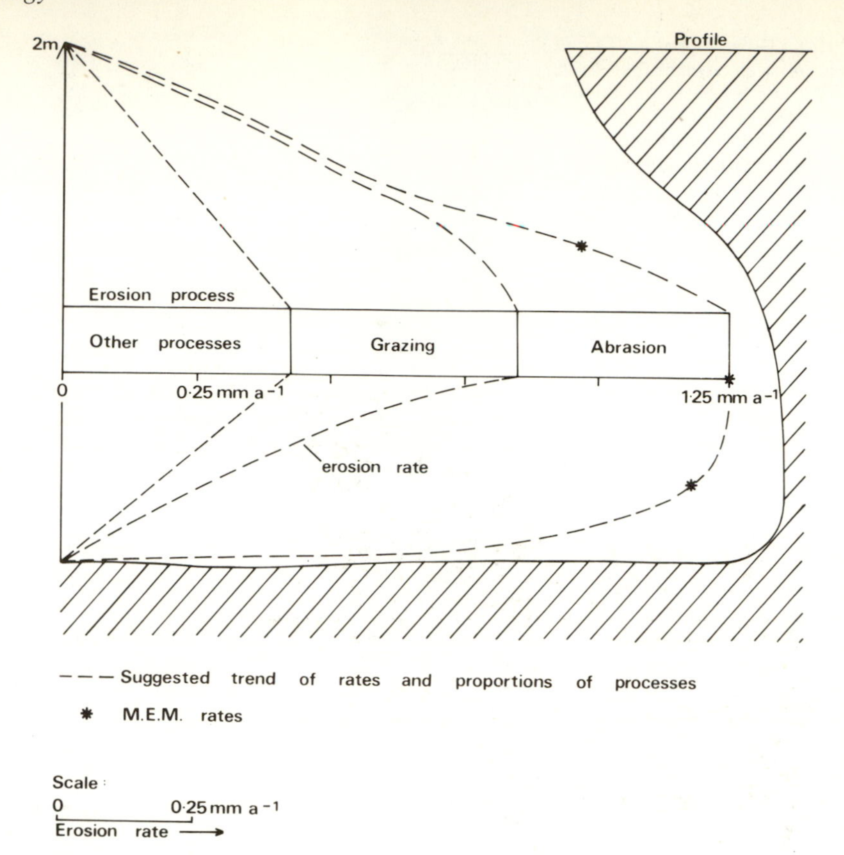

Fig. 10.2 As for Fig. 10.1 but with sand present, causing abrasion (from Trudgill, 1976a).

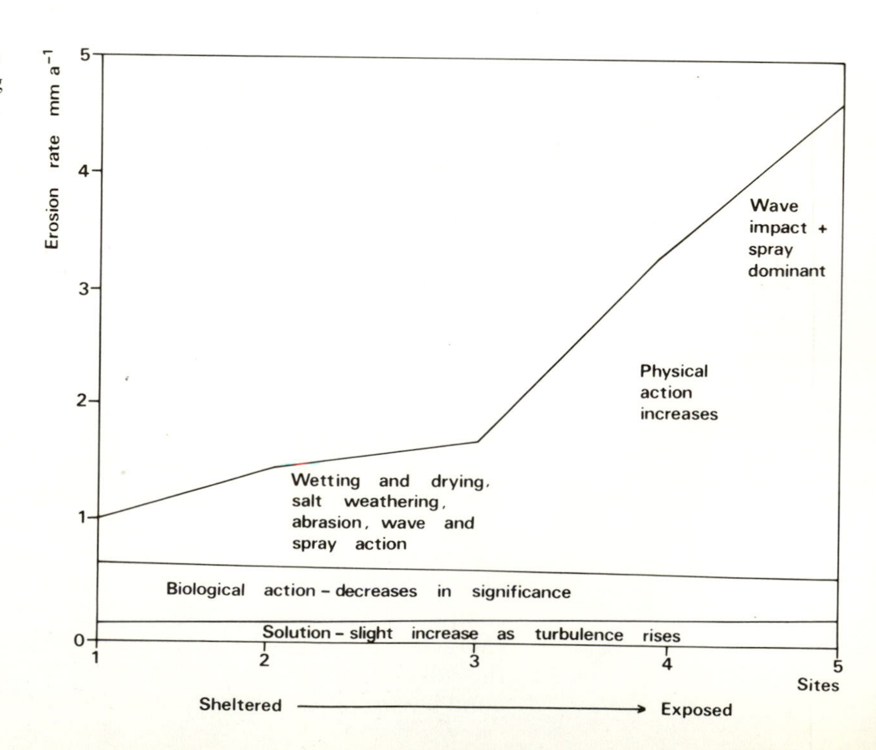

Fig. 10.3 Increase of erosion rate from 1 to 4 mm a^{-1} with increasing exposure, with suggested trends in importance of processes, Aldabra Atoll, Indian Ocean: for sites, see figure 10.13, p.168 (from Trudgill, 1976a).

understand the zonation of bioeroders. There is clearly scope for further work in this field, especially to ascertain whether morphological zones do correspond strictly with tide levels while biological zonation varies according to other factors, or whether morphological zonation varies strictly with biological zonation.

It has been shown by many workers that morphological zones vary broadly with biological zones, but the precise variations and boundaries have not always been specified. For example, Lundberg (1977; Fig. 10.5) and Trudgill (1977b), working in County Clare, Eire (Fig. 10.6.) plotted intertidal morphological and biological changes in temperate regions while Emery (1962) (Fig. 10.7), on Guam, and Trudgill (1976a), on Aldabra (Fig. 10.8) have undertaken similar work in tropical regions. Schneider (1976), Fig. 10.9, suggests that the distribution of bioeroders varied with microclimate, that is, populations increase with increasing moisture down the intertidal or locally in pools. Neumann (1966) illustrates general coastal morphology and zonation at Bermuda (Fig. 10.10), showing that the presence of an undercut notch coincides with the zonation of infauna (borers). Similar subtidal notches occur with *Hiatella arctica* in County Clare (Fig. 10.6).

In general, in tropical regions the presence of an undercut notch coincides with the presence of bio-eroders; in temperate regions dissection is most marked where barnacles and mussels (probable protectors) are present together with boring algae and the excavated pools of erosive echinoids.

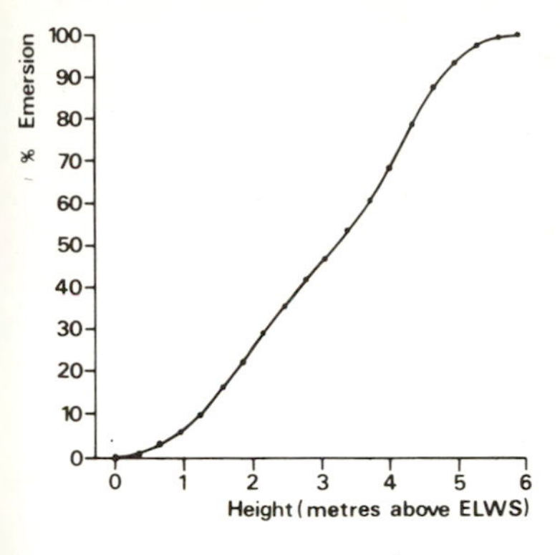

Fig. 10.4 Emersion curve – mean annual percentage time of emersion at different heights on the shore calculated by Underwood for British shores (modified from Underwood, 1978).

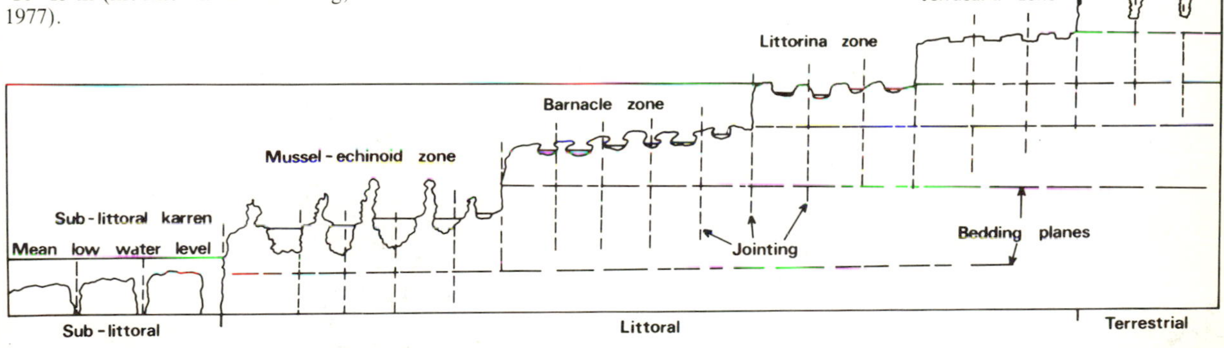

Fig. 10.5 Intertidal biological and morphological zonation, County Clare, Eire. The deepest dissection is in the echinoid (sea urchin) zone. Profile relief 2–3 m high, length, °10–15 m (modified from Lundberg, 1977).

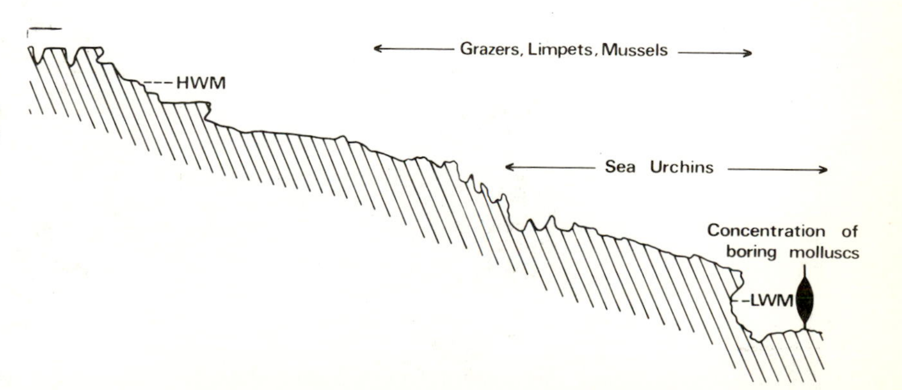

Fig. 10.6 Intertidal biological and morphological zonation, County Clare, Eire. A small subtidal notch is seen, produced by *Hiatella arctica* borings. HWM = High Water Mark and LWM = Low Water Mark. Profile height °2 m, length °10 m (modified from Trudgill, 1977b).

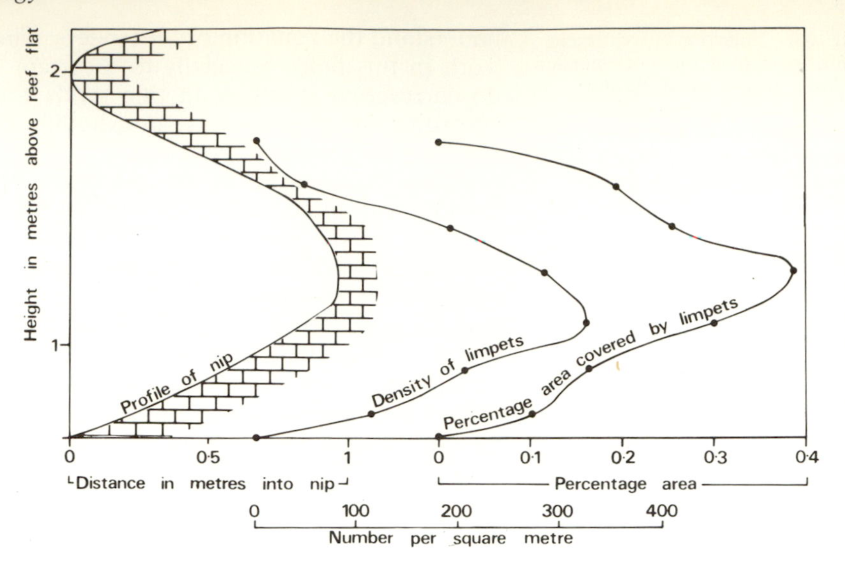

Fig. 10.7 Limpet zonation and morphology on Guam. Maximum density coincides with maximum erosion (after Emery, 1962).

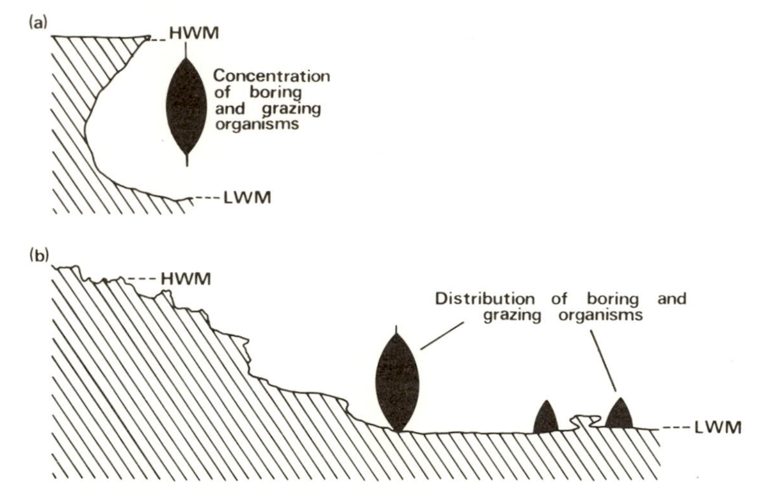

Fig. 10.8 Distribution of bioeroders and morphology, Aldabra Atoll, Indian Ocean. (a) sheltered coasts with borers and grazers coincident with 2 m high notch. (b) exposed coasts where physical forces are more important and the relationship between bioeroders and morphology is not well developed, profile °20 m long (from Trudgill, 1977b).

10.4 Differential erosion rate and form

Overall erosion rates are often quoted in the literature, i.e. for surface retreat, when the understanding of the actual evolution of a dissected morphology depends upon *differential* erosion rates. In the case of undercut notches the question is: how fast, and why, is the centre of the notch eroding when compared with the notch top and base? In the case of pinnacles and potholes, the question is: what are the relative rates of erosion of each type of form? Very few differential measurements of this type are reported in the literature. Robinson (1976) used a micro-erosion meter (High and Hanna, 1970; Trudgill, High and Hanna, 1981) in a littoral environment in order to understand differential erosion on shales; Kirk (1977) quotes Robinson's rates as 1.5–15.0 mm per year. However, the relationship with this spread of rates and the evolution of microtopography is not clear, largely because the data are regarded as a sample of a population and sample means are usually calculated and compared with other groups of data. However, Trudgill, High and Hanna (1981) give data for differential erosion at one site, suggesting

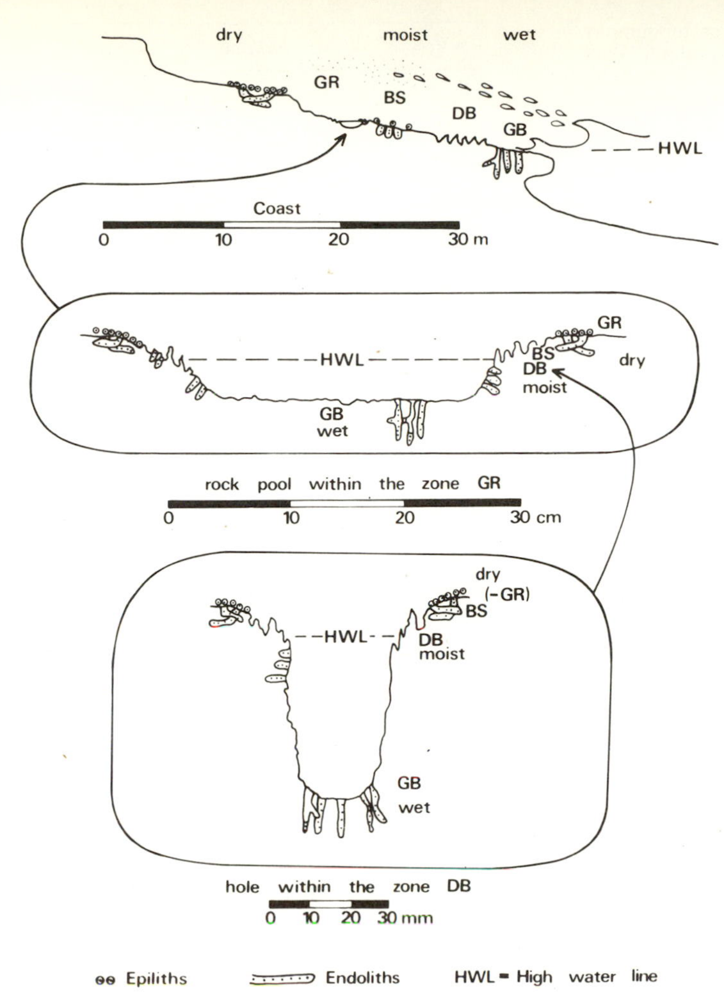

Fig. 10.9 Schematic distribution of bioeroders and microclimate. Upper profile – coastal section with two successive enlargements (see insets below, with scales). Key to algal and lichen colour zones recognised by Schneider: GR = grey zone (dry): BS = blue – black (moist): DB = dark brown (moister): GB = yellow-brown (wet). Moister areas are preferentially colonised and therefore preferentially eroded. Epiliths: algae on rock surface: endoliths: rock boring algae (modified from Schneider, 1976).

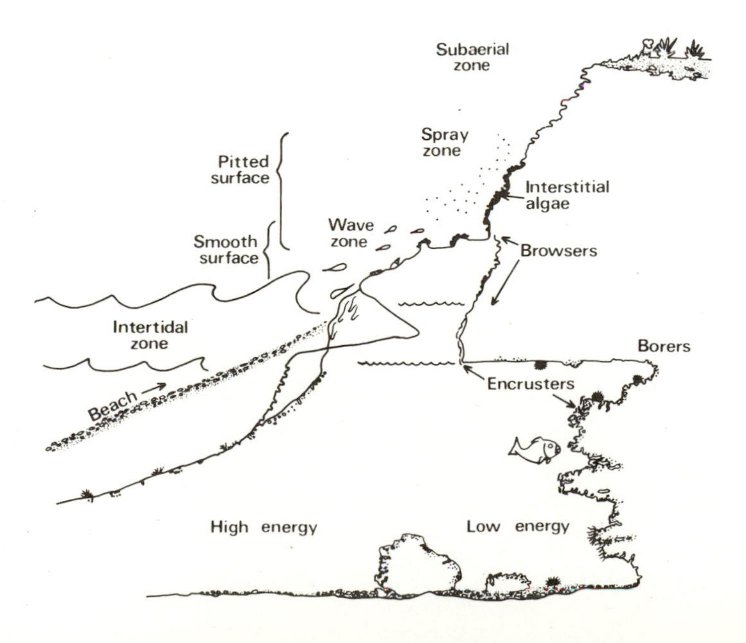

Fig. 10.10 Coastal sections, Bermuda. Several sections are shown superimposed (land is on the right-hand side), showing a deeply incised subtidal profile where borers exist under low energy conditions. In exposed, high energy situations a transition is seen towards the beach profile (modified from Neumann, 1966).

Table 10.3 Microtopography and erosion rate (mm a^{-1})

Hollows	Flats	Raised areas
0.70	0.18	0.37
0.24	0.17	0.16
0.21	0.15	0.08
	0.13	0.08
	0.12	0.07
	0.12	
mean 0.383	mean 0.145	mean 0.152

(modified from Trudgill, High and Hanna, 1981)

how the surface may evolve with time (Table 10.3). Certainly, one problem with micro-erosion meters is that the reference studs are more difficult to insert on dissected topography where irregular fragile surfaces occur, with greater differential erosion than on flat smooth surfaces, and this is probably one of the reasons why differential erosion has had relatively little study. Kirk (1977) has made extensive measurements of 31 micro-erosion meter sites on the Kaikoura peninsula on the northeast coast of South Island, New Zealand. A wide variety of rates were recorded on Tertiary limestones with a mean of 1.53 mm per year, but here the concern was with mean rate and planation, rather than with dissection. On Aldabra, Trudgill (1976a) showed how the location of the fastest rate of erosion varied with exposure (Fig. 10.11), with differential erosion varying from accentuating the notch to the removal of the notch top as exposure increased. Extrapolation of current rates suggests that notches will continue developing (with occasional fracture of the top to renew the cycle) and that the ramp form will probably degrade to a form similar to adjacent beaches (Fig. 10.12). In Puerto Rico, Kaye (1959) also suggests that undermining of the notch leads to the loss of the notch top in a cyclic fashion. Dissection and differential erosion is then perhaps

Fig. 10.11 Changes in the location of the most rapid erosion rate, Aldabra Atoll, Indian Ocean. The focus of erosion moves from the mid to the upper intertidal (from Trudgill, 1976a).

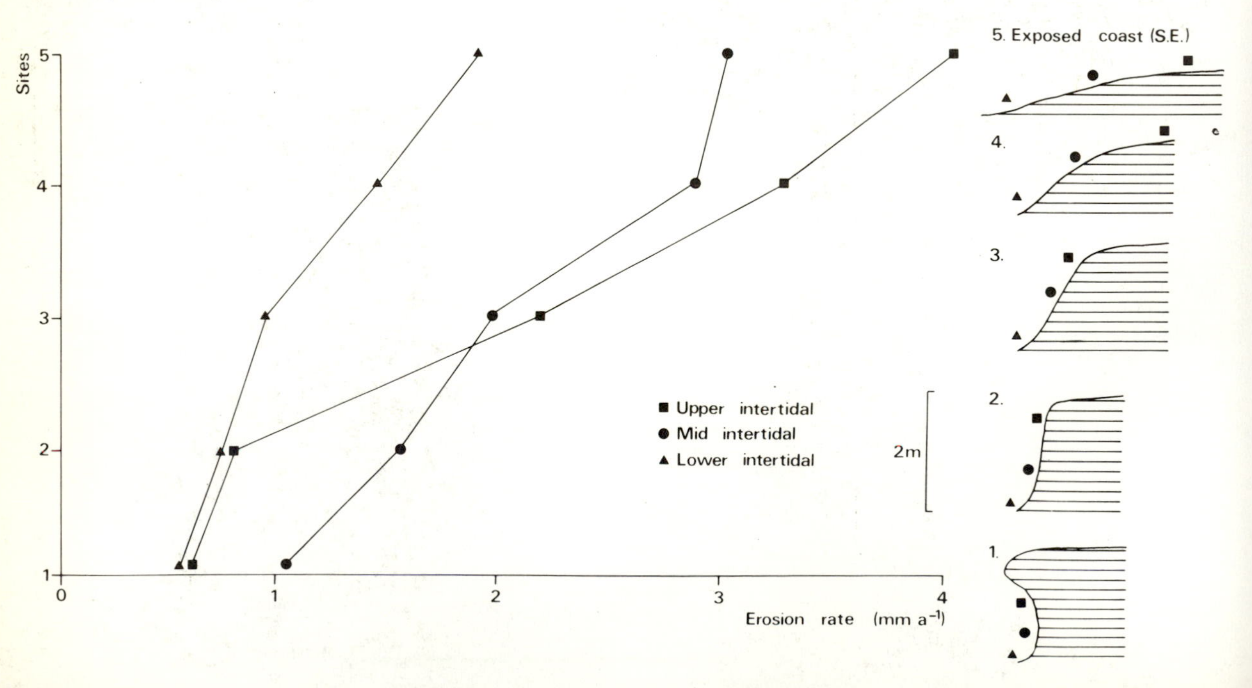

Fig. 10.12 Extrapolation of current erosion rates and suggested resultant forms (from Trudgill, 1976a).

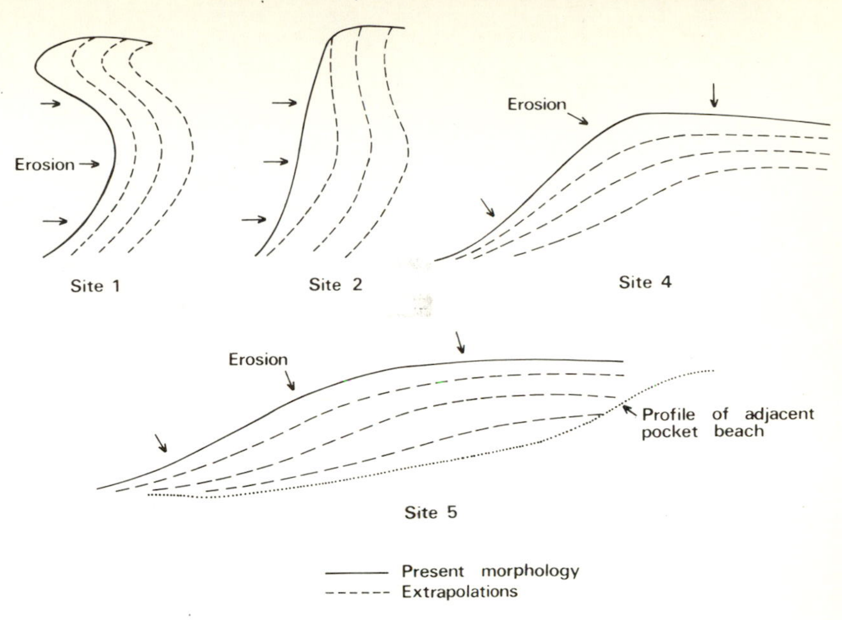

a neglected feature, since much marine erosion work is concerned with planation and the formation of platforms; however, if the considerable dissection evident on limestone shores is to be fully understood, then more work will have to be undertaken on this topic.

10.5 The role of exposure

In this context the word exposure is used to indicate exposure to dominant wind direction and therefore to wave action and storm action. Trudgill (1976a) records how a notch is replaced by a ramp upon increasing exposure to southeast trades on Aldabra (Fig. 10.13). Guilcher (1958) records how a notch is replaced by a tidal platform as exposure increases (Fig. 10.14) as do Focke (1977) and Loenhoud *et al.* (1976). Davies (1980) has classified coastal form according to energy environment. Based on the frequency of onshore winds of Beaufort force 4 and above (Fig. 10.15), coasts classified as exposed occur in Atlantic-facing western Europe, northeastern South America, southwestern South America, southwest Iceland, eastern Madagascar and those parts of central East Africa not sheltered to the east by Madagascar, parts of south Australia and southern New Zealand and, seasonally, west- and east-facing coasts of India and South East Asia. Tidal range also varies considerably in the world and areas with low tidal range, such as the Mediterranean, can be expected to have vertically limited undercut notches. Increased tidal range (Fig. 10.16), such as in parts of East Africa, northwest Australia and the British Isles, can be expected to lead to more extensive intertidal erosion features. Deep notches can thus be expected in locations where bioerosive organisms are concentrated intertidally and those of limited exposure where biological activity is not discouraged; the height of the notch will be related to tidal range.

10.6 Sea-level changes

It is doubtful whether there are any limestone coasts which do not show some evidence of the influence of former sea-levels on morphology; the evidence may, however, be far from clear and may thus give rise to

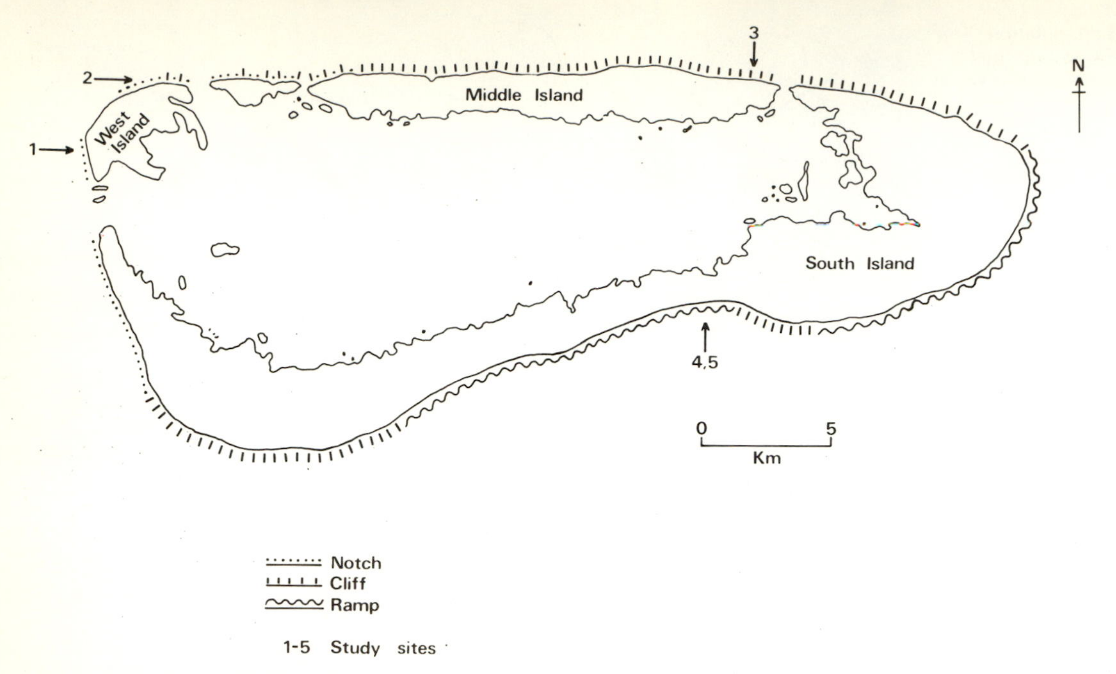

Fig. 10.13 Exposure and coastal form, Aldabra Atoll, Indian Ocean. The sites 1 to 5 are the micro-erosion meter sites shown in Fig. 10.11 (from Trudgill, 1976a).

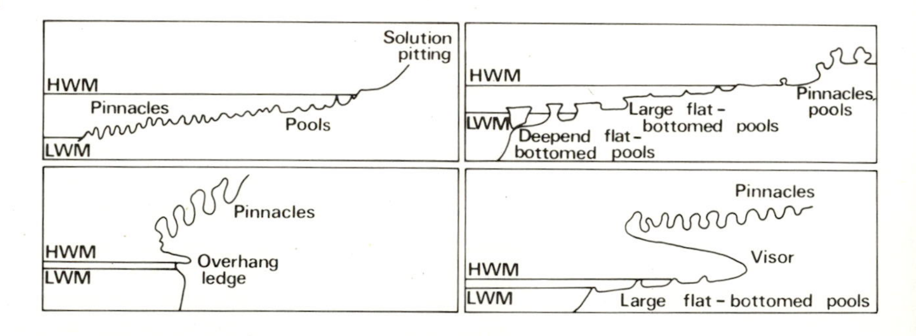

Fig. 10.14 Exposure and morphology. The notch, or visor, occurs on sheltered sites with low tidal range (lower two diagrams); upper diagrams: platforms on exposed shores with higher tidal range (from Guilcher, 1958).

misleading conclusions about current process–form relationships. A clear example of this is shown in Fig. 10.17 from the west coast of South Island, New Zealand. An originally subtidal notch is now seen just above high tide level due to land uplift. Schwartz (1967) interprets much contemporary erosion as an adjustment to recent sea-level rise as seen in Fig. 10.18 which, although concerned primarily with beaches, is also relevant to rock erosion as Wright (1967) shows by discussion how rock platforms may form in relation to sea-level changes. In many cases rock platforms exist at high or low tide levels away from a probable mid-intertidal focusing of processes; in these cases the high or low bench is very probably a relict of former sea-levels, though without precise dating the evidence is circumstantial. It is however clear that if a morphological feature does not coincide with observed current processes then it is not necessarily warranted to invoke the action of alternative current processes. Indeed, in reef morphology, Backshall, *et al.* (1979) show how drowned subaerial forms can appear in marine situations. Drowned dolines appear as 'blue holes' or deep circular subtidal holes inherited from

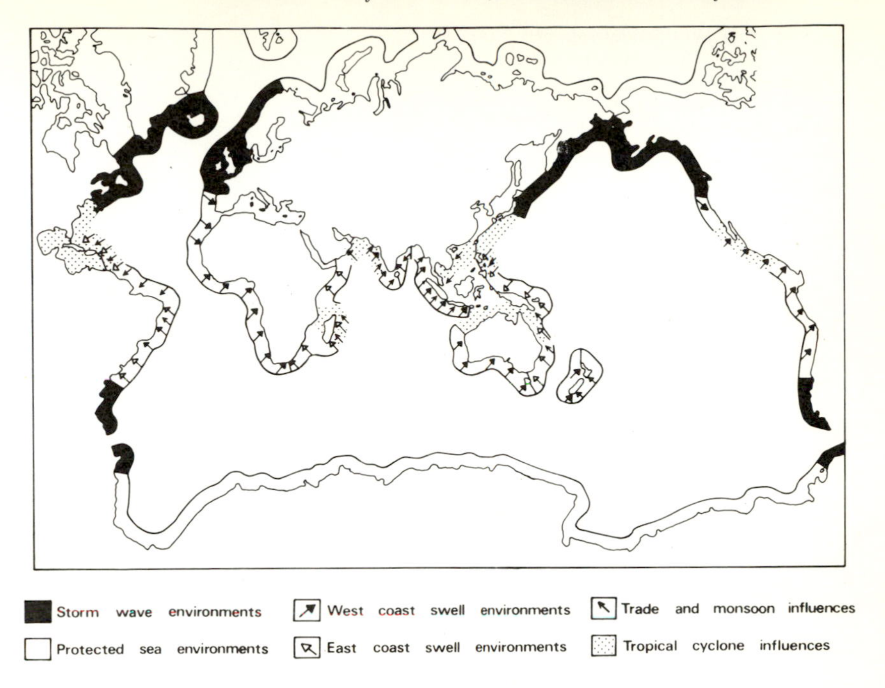

Fig. 10.15 Coastal classification according to exposure (from Davies, 1980).

former low sea levels. In an area of rapid uplift, on the Huon peninsula in New Guinea, Chappell (1974) has used dating of carbonate material to demonstrate the relationships between rapid uplift and periods of erosion.

10.7 Tropical and temperate coasts – variations in climate, structure, lithology and stage

Geomorphologists may readily fall into the trap of attempting to assign universal explanations to diverse and unlike phenomena in different parts of the world on an unsound basis; the alternative trap is to provide unique explanations for each individual phenomena. Further work should therefore attempt to achieve a level of generalisation appropriate to the current level of knowledge without falling into either trap. The basic difficulty lies in the fact that different research workers have worked on different aspects of erosion in different parts of the world and thus it is virtually impossible to fit their data together in any valid or comparable way. Thus, comparison of tropical and temperate environments is often undertaken, but not only does climate but also lithology and the nature of biological populations (species, actions and numbers) and the importance of each process vary markedly as well. In addition, local variations may be greater than global variations in some cases.

Tropical limestone coasts are commonly comprised of younger limestones than temperate coasts, and are frequently of Tertiary limestones and Quaternary and Recent reef limestones, aeolianite and beachrock. These younger rocks are less well cemented than the older Mesozoic limestone more common in temperate regions and are less extensively recrystallised and metamorphosed. Thus they are relatively heterogeneous and often preserve the original relationships between clasts and cements. This heterogeneity encourages differential erosion, especially the preferential removal of cements, leaving the clasts proud of the surface.

Both Guilcher (1958) and Zenkovitch (1969) reviewed work on the

Fig. 10.16 (a) World tidal ranges (from Davies, 1980). (b) Plot of examples of tidal ranges; in sheltered sites similar notch profiles might be expected (from Wright, 1967).

(a)

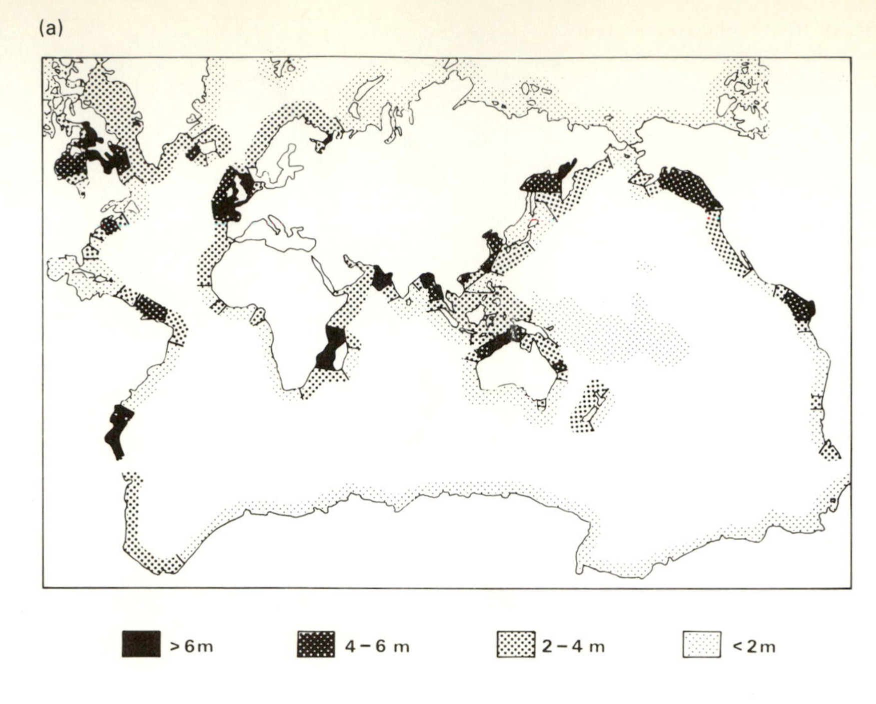

(b)

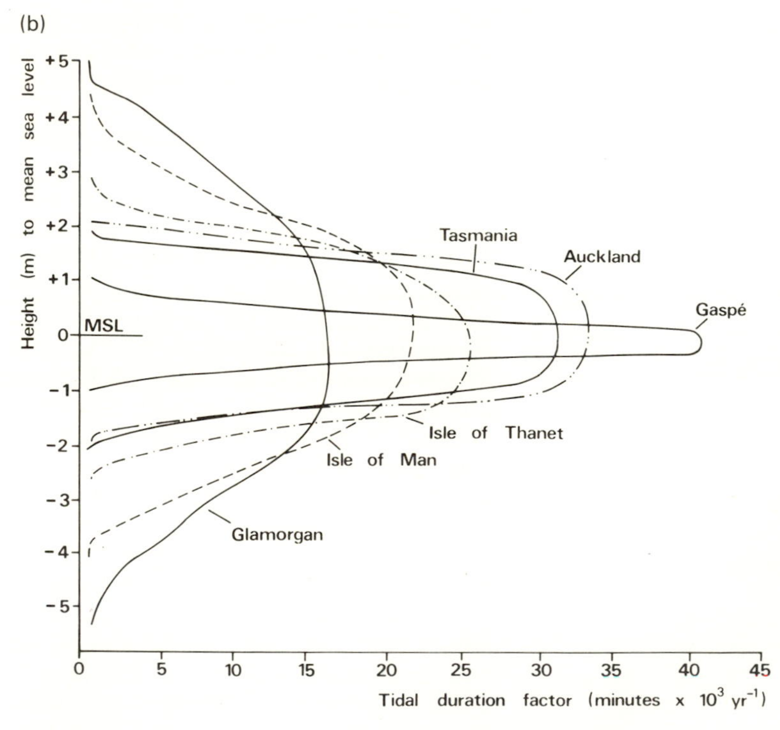

erosion of limestone in temperate regions and attempt to classify coasts in terms of morphology without detailed reference to some of the process and stage complexities involved; Guilcher refers to zonation of corrosion forms in different regions (British Isles, Morocco, Mediterranean, tropical seas) without reference to differences in lithology or biological

Fig. 10.17 Former intertidal notch seen in profile on a marine stack and on right-hand side cliff; notch is now abandoned because of uplift. Its origin as a subtidal notch is indicated by the occurrence of numerous old borings by sponges; west coast of South Island, New Zealand. (Photo: S. T. Trudgill.)

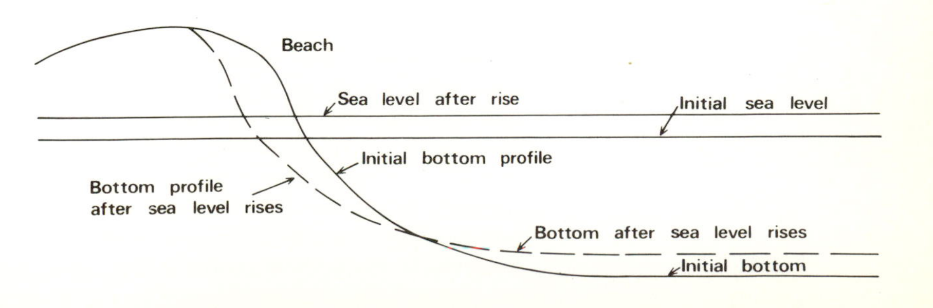

Fig. 10.18 Sea-level rises and equilibrium profile responses. Erosion is seen in the context of adjustment to new sea levels (modified from Schwartz, 1967).

action. One important feature is emphasised, however, and that is tidal range. For example, tidal range is very limited in the Mediterranean and any intertidal notch formed here has a correspondingly limited vertical development. Other factors are mentioned by Dalongeville (1977a, b; 1978) who emphasises the importance of inclination of slope of the shore, tidal range and constructional features made by coralline algae in Yugoslavia and Lebanon. Schneider (1967) suggests that in the Adriatic moisture conditions provide the limiting factor for erosion since most of the erosion is biological. He proposes the term 'biokarst' for intertidal erosional phenomena. As moisture retention increases a positive feedback exists so that deeper areas retain more moisture, encourage plant and animal colonisation and growth and therefore bioerosion, and become deeper. Each pool thus enlarges, often breaking into the next. His generalised scheme is summarised below:

Stage 1. The primary depressions in the rock surface are first colonised since they retain moisture longer than their higher surroundings. Conditions for boring and grazing organisms prevail longest here. As a result each depression becomes the site of more intense biological corrosion and abrasion.
Stage 2. The pools enlarge laterally and the small depressions coalesce. Any wet area is preferentially bioeroded and relief is thus intensified.
Stage 3. This is the stage of maximal relief. It shows maximal contrast in ecological conditions and thus in destructive processes. The water in the depressions is changed at most high tides thus bringing fresh seawater to organisms.

This sequence provides an important conceptual framework which can be used elsewhere, based as it is on a simple feedback process. The limiting factor is usually provided by bedding planes. The pools become deeper and deeper until they break through to the next plane and frequently drain down these; alternatively a water-holding edge of the pool may be breached. Indeed, empty sea urchin-excavated pockets may be seen stranded above current water-levels. It can thus be suggested that pinnacles simply form as a residual feature involving excavation of the surrounding rock. Protection by mussels and barnacles need not be invoked – and not all pinnacles have such organisms on them – and they are most probably simply taking advantage of the site for colonisation where there is good water circulation for filter feeding. The fact that their strong means of attachment to the rock contributes to their survival on pinnacle sites when other organisms would be dislodged will also be a contributary factor.

10.8 Summary

Many morphological descriptions have been made of limestone coasts and many process studies have been undertaken; dissolution of limestones appears to be possible in many situations, especially in relation to biological activity; direct biological erosion appears to be of paramount importance in intertidal limestone erosion. The precise relationship between process and form may be obscured by sea-level changes and modified by lithology and exposure but the evidence suggests that bioerosion is concentrated in the mid-intertidal zone. Where there is an initial near-vertical surface in a sheltered position a notch is then formed, its precise position being dependent upon the intertidal zonation and competitive advantages and adaptations of the bioeroders. On more horizontal surfaces maximal relief appears to be correlated with a balance of moisture retention and the adaptation of the bioeroding organisms; pools, once established, become deeper as more water is retained and bioerosion encouraged, leaving upstanding pinnacles between. Thus, the two types of limestone coast described at the beginning of this chapter (Figs. 10.1 and 10.2) can be interpreted largely in terms of zonations and adaptations of bioerosive organisms; the notch being a simple zonation concentration effect, the pinnacles being simply a residual topography with preferential biological erosion in the surrounding pools. Such basic relationships may well be modified in highly exposed conditions and/or in storm conditions and obscured where abrasive sand covers are present. In this latter case smooth wave worn and potholed features will be seen and delicate fretting will be absent. It is to be hoped that further quantitative work will be able to test more rigorously the suggestions and interpretations of marine limestone landforms discussed in this chapter.

11 Limestone geomorphology – perspectives and applications

11.1 Introduction

While the primary task of the geomorphologist is to provide explanations of landforms, the results of the studies undertaken often have applications in terms of management. This chapter therefore aims to give some perspectives on the current status of geomorphological work (as, for example, covered by Trudgill and Brack, 1977), indicating some of the ways in which knowledge gained in geomorphological study may be of use in the understanding and management of limestone environments.

Limestone landforms provide important mineral resources, the limestone itself being quarried for road building material, agricultural lime and for building stone as well as for industrial uses. There may also be useful minerals deposited in veins running through limestone. In addition, limestone aquifers may provide important water supplies. Limestone landforms also produce some spectacular landscapes and recreation opportunities. Moreover, their vegetation, soils and landforms provide unique assemblages which are worth conserving for their intrinsic interest and for scientific study. A knowledge of geomorphological processes and landform distributions can often be useful in the context of some of the management problems which may be posed by the use of limestone landscapes, as indicated by some of the examples outlined below.

11.2 Hydrogeology and quarrying

Limestones vary widely in their purity and porosity, but the purest, that is those limestones dominantly made up of $CaCO_3$, are often used extensively for quarrying. This is especially true of the Carboniferous Limestone in Britain. Quarrying activity may, however, interact with the use of a limestone area for water abstraction. Here, a knowledge of the nature of the diversions of water courses to an underground location may be of considerable applied use. An important factor is the way the nature of the rock influences the routes through which water flows, and especially whether the water moves in relatively discrete conduits or whether it percolates through the rock mass as a whole. The extent of these various flow routes can be assessed by a study of primary and secondary permeability (Fig. 11.1). Primary permeability covers the flow through the rock mass as a whole, involving intergranular porous flow,

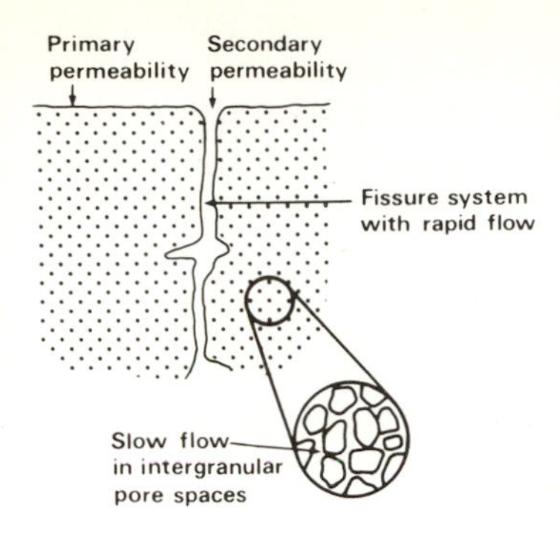

Fig. 11.1 Primary and secondary permeability.

as shown in the inset in Fig. 11.1. Secondary permeability involves water flow through fissures and joints and it is usually much more rapid than flow through the bulk rock. In Britain, porous rocks, such as the Cretaceous Chalk and the Jurassic Cotswolds oolites, have higher primary permeabilities than the crystalline Carboniferous Limestones but all these rocks have secondary fissures systems developed to a greater or lesser extent. A range of permeability values is shown in Fig. 11.2. Permeability values, often given the notation K, increase from a range of zero to 0.01 m water flow per day for primary permeability and from 0.008 to 60 m day^{-1} for secondary permeability. Where caverns exist, values may increase from 10^5–10^6 m day^{-1}. Secondary permeabilities are of most significance in crystalline limestones where primary values are close to zero. These considerations closely influence the value and nature of a limestone mass as an aquifer.

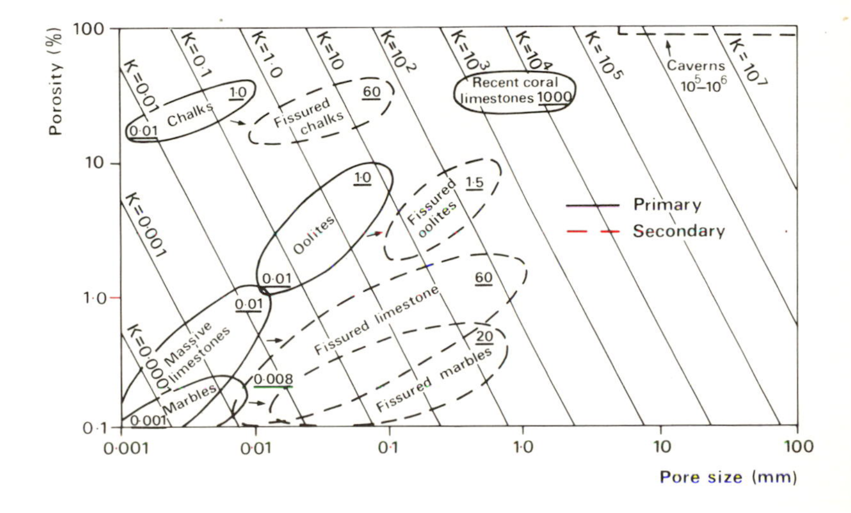

Fig. 11.2 Hydraulic conductivity (K in m day^{-1}), for limestones, showing the increase of permeability with fissuring (from Smith *et al.*, 1976).

In limestones of low primary permeability and with a developed cave network system, water yield may be erratic, with the timing of output peak water flows following closely behind the timing of input rainfall events. There is little storage in the system and the supply may thus diminish considerably during dry weather, while there may be flooding in wet weather. At the output point there will be very peaked hydrographs, with rapid travel times of water and rapid recession of flow. By contrast, where primary permeability is higher, there will be more storage and a more damped response to rainfall. This will also be true if primary permeability is low but there is a poorly connected fissure system or a large number of small fissures. In these situations, there will be a less rapid response to rainfall and a slower recession after rainfall. In Britain, the Carboniferous Limestone is often of the type where there is a well integrated flow net and low primary permeability, though the extensive development of small fissures may increase water storage capacity considerably and moderate the pattern of flow (Smith *et al.* 1976). Here, water output may often be divisible into a rapidly travelling swallet component, flowing through an integrated underground fluvial network from a point of surface stream engulfment, and a more slowly travelling percolation component, flowing directly from the surface through small fissures. The Chalk is usually of the type where primary porosity is high, but larger fissures may be present, giving rise to more rapid travel times in some localities. Because of the high primary porosity of the Chalk, it is a much more important aquifer in Britain than is the Carboniferous Limestone, as the amount of water storage is considerably higher. Most of the outflow is derived from percolation water and an integrated underground fluvial network is often largely absent as water can flow through the bulk of the rock and is not confined to joints and fissures.

The presence of solutionally enlarged fissure systems and rapidly flowing integrated networks of cave systems means that groundwater in many limestones travels faster than that in other aquifers of higher primary permeability, such as the Triassic Bunter Sandstone in Britain. Thus, on limestone there may be dangers of pollution from point sources because mixing, decomposition and sorption processes will all be minimised during rapid flow. This is illustrated by a well-developed fissure system which exists in the Hampshire Chalk, where rapid flows of approximately 2–3 km day^{-1} were recorded by Atkinson and Smith (1974) using Rhodamine WT as a tracer dye (Smart and Laidlaw, 1977), as shown in Fig. 11.3. The dye was inserted at the point shown and travelled around 6 km in around 2 days, or 3×10^3 m day^{-1}, a value greatly in excess of those shown for unfissured chalk in Fig. 8.2 (0.01–1.0 m day^{-1}). The peaked shape of the trace (Fig. 8.3 b) also indicates that there was a rapid movement of the dye, with very little mixing or dispersion. Great care should therefore be used in the management of limestone aquifers, especially if upstream pollution sources may exist. Great caution should be applied to water use in such situations and attention should be paid to pollution sources, with dumping closely controlled or eliminated on limestone bedrocks.

The development of a fissure system in crystalline limestone also means that water supply from wells is often unreliable, as illustrated in Fig. 11.4. Water supply depends upon chance intersection of water-filled fissures and, unless fissuring is extensive or well integrated, supplies will be limited.

Hydrogeology and engineering geology become closely connected subjects during quarrying procedures in massive, fissured limestones. Groundwater levels may be unpredictable and those in one area may be unconnected with those in another. If a conduit is intercepted during quarrying, the quarry may be flooded and the water-levels lowered, curtailing water supplies, and in a manner not necessarily predictable by the considerations normally applied to rocks with a higher primary porosity. If a fissure system is intercepted, water supply may be intercepted from a much wider area than the immediate draw-down zone, as illustrated in Fig. 11.5.

The production of limestone by the UK quarrying industry doubled from 1961 to 1969 (Fig. 11.6), mainly from Derbyshire, Somerset and Gloucester. Quarry extension is limited by watertable depth and also slope failure propensity (Fig. 11.7). Here, the dip of the rock has a strong influence on how high the quarry face may be worked, steeply dipping rock being prone to failure. Water-level and geological dip thus may limit the depth to which a quarry may be worked, tending to encourage the lateral extension of quarries. Here, however, amenity considerations become important, together with conservation and agricultural considerations since more land is lost by lateral extension of quarries than with vertical extension. Amenity considerations are

Fig. 11.3 Dye tracing in the Chalk, Hampshire. (a) map of dye tracing location (in Southern Britain); (b) graph of tracer dye recovery at two springs (modified from Atkinson and Smith, 1974).

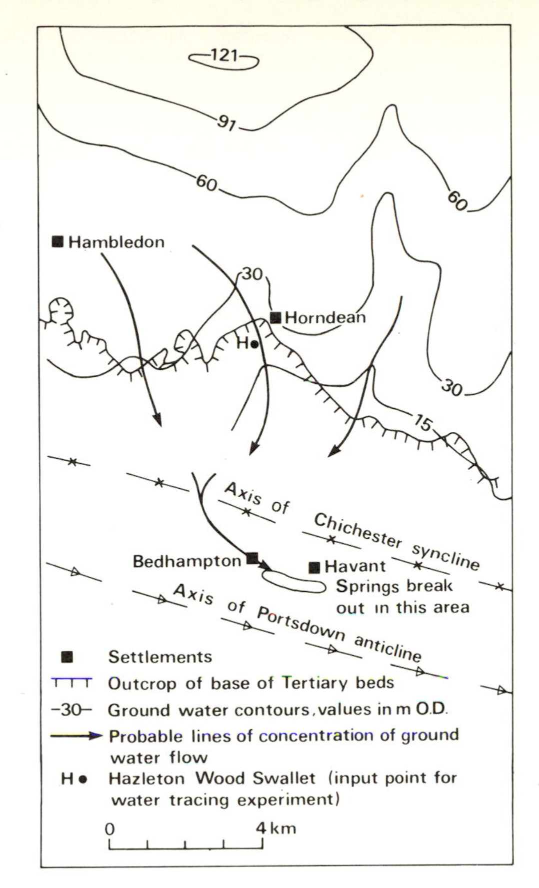

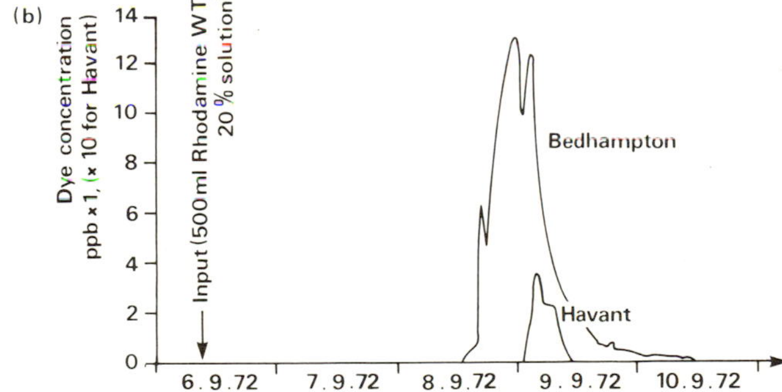

Fig. 11.4 Intersection of water-filled fractures by wells (modified from Atkinson *et al.*, 1973).

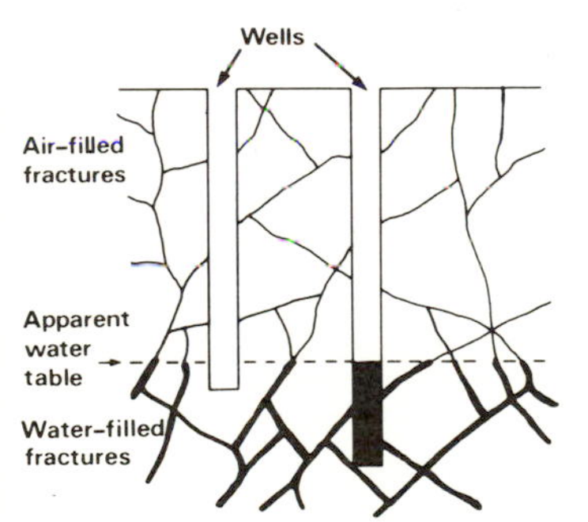

often hotly discussed since many limestone deposits are in areas of outstanding beauty and the operations can be extensive (Fig. 11.8).

11.3 Wider considerations

Agriculturally, the value of limestone areas is often limited to pasture for sheep and cattle by the thin, stony nature of the soil if the soil has developed *in situ*; by contrast, soils developed on glacial drift derived from limestone may be deep and productive, if they are not too stony to cultivate. Harder, massive crystalline limestones tend to produce thin-

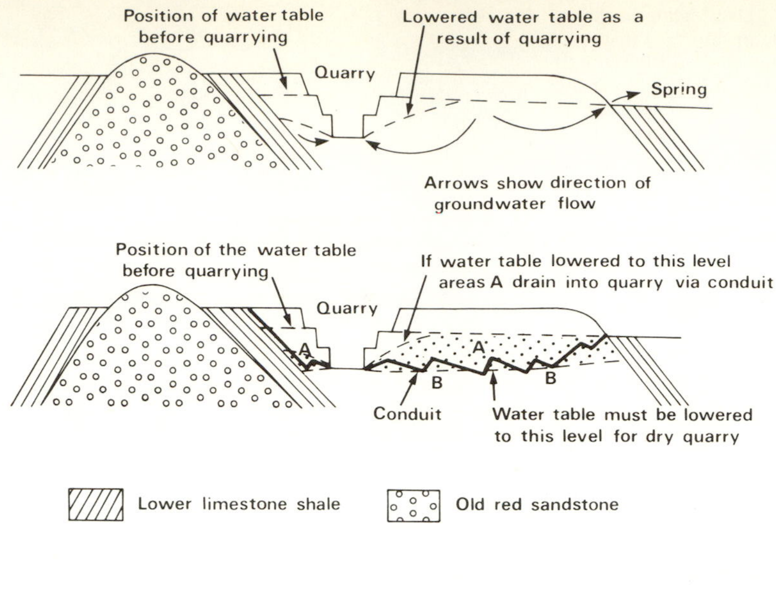

Fig. 11.5 Schematic diagram of interaction of water levels and quarrying (modified from Atkinson *et al.*, 1973).

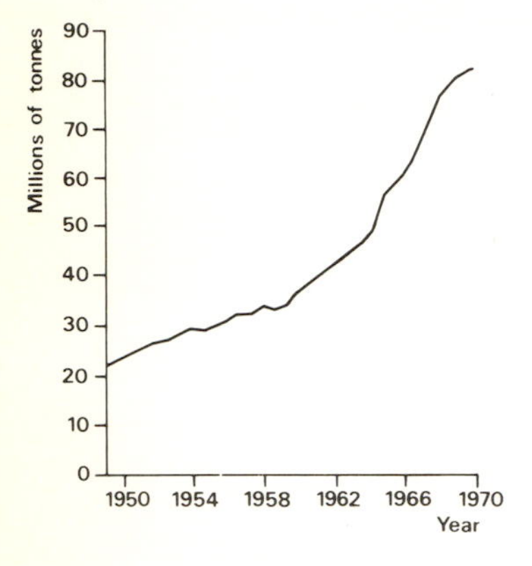

Fig. 11.6 Output from UK quarries, 1950–1970 (modified from Somerset County Council, 1971).

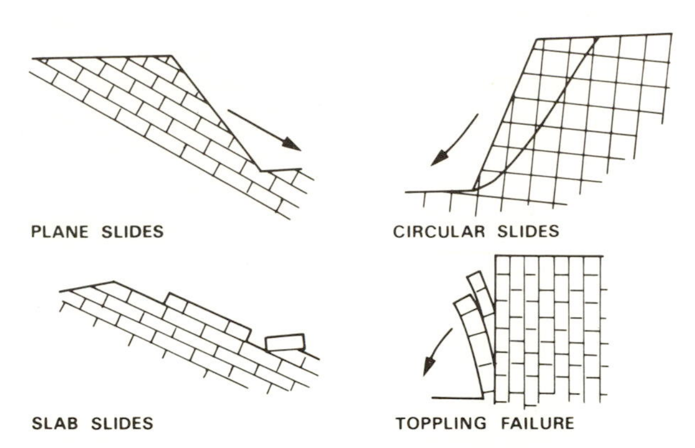

Fig. 11.7 Limitations on vertical quarry extension (modified from Atkinson *et al.*, 1973).

Fig. 11.8 Quarry operation in a national park, Peak District, UK. (Photo: S. T. Trudgill.)

ner soils than more closely bedded limestones which break up more readily to produce deeper soils. Rock type and geomorphological history may thus have a strong influence on land capability for agriculture.

Geomorphological history has also left considerable conservation interest in terms of limestone pavements and caves. Limestone pavements often have many rare plant species present, with a range of micro-habitats in the joints and crevices giving shelter to a distinctive flora. The pavements themselves are under threat from ornamental stone merchants who extract and sell solutionally sculptured stone for garden use, leaving behind denuded areas of little geomorphological or botanical interest. In Britain, the Nature Conservancy Council has identified those sites of greatest floristic and geomorphological interest, with a view to their long-term protection from this form of extraction and other threats.

Caves provide well-used routeways for exploration and adventure as well as for scientific research. Caving also poses conservation problems, particularly in caves which do not now have active streamways in them but in which unique assemblages of sediments are preserved, acting as records of deposition during past environments (Sweeting, 1980). It is these sediments which can provide such useful information on the evolution of caves and limestone landforms as a whole as well as indicating how past environments have changed (Bull, 1980). In addition, many important fossils are to be found in cave sediments, including significant remains of fossil man. Proper training of cavers and the use of active streamways away from unique and scientifically valuable sediment sequences helps to minimise the problems involved.

There is thus a relationship between management and the understanding of environmental processes and history. Lack of application of geomorphological knowledge of how limestone systems work can lead to management which is well short of the optimum returns, as perhaps typified by the illustration of the Connemara canal (Fig. 11.9), built on limestone and irretrievably dry because of the extensive fissure network in the limestone, acting as a drainage outlet for the canal. Careful management of limestone systems is also necessary for the conservation of the landscape and the preservation of evidence which will assist the geomorphologist in the understanding of the origin and age of landforms.

Theories of development of limestone landform development (Pow-

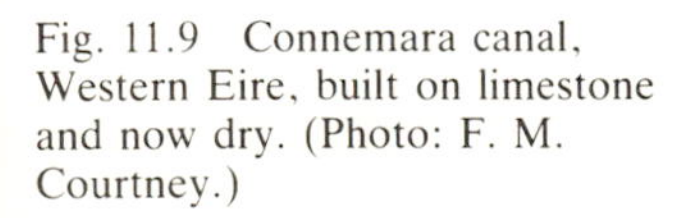

Fig. 11.9 Connemara canal, Western Eire, built on limestone and now dry. (Photo: F. M. Courtney.)

ell, 1980) have involved the identification of sequences of downcutting through rock masses. Overall models of landform evolution, as discussed by Ford (1980), Douglas (1980) and Brunsden (1980), for example, may be hampered by the lack of precise historical evidence on sequences of evolution. In limestone geomorphology, the future is interesting in that not only will geomorphologists be able to continue to provide information which is of use in environmental management, there exists the possibility, through the study of dated cave sequences, that whole dated sequences of limestone landscape evolution may be studied, indicating more closely the nature of the relationships between erosion processes, rates of erosion, rates of adjustment to climatic changes, geological structure, stages of evolution and the production of limestone landforms visible at the present day.

Appendix
Solute data and water sample analysis

The reporting of data on solution concentrations in limestone solution studies can vary. Some workers use mg l^{-1} Ca^{2+}, others mg l^{-1} $CaCO_3$. The latter does not imply that $CaCO_3$ is the solute species present in solution, the practice of reporting in $CaCO_3$ units in limestone geomorphology stems from the use of such data in the calculation of denudation estimates (see p. 46). Here, annual run-off loads are converted to losses of solid calcium carbonate and so it is convenient to present the data in the $CaCO_3$ form. The dominant ion species in solution are Ca^{2+} and HCO_3^- (see p. 17). Analyses of limestone run-off water are for calcium ions in solution, either by EDTA titration (see below) or by atomic absorption spectrophotometry. Calcium data may be converted to calcium carbonate data using an appropriate conversion factor. The conversion factors are given below (from Hem, 1970, pp. 81–83).

mg l^{-1} Ca^{2+} to $CaCO_3$ = $Ca^{2+} \times 2.497$
mg l^{-1} $CaCO_3$ to Ca^{2+} = $CaCO_3/2.497$

Data may also be presented in millimoles per litre (mM l^{-1}, or m mol l^{-1}) where a mole is the molecular weight in grams:

Atomic weights are: (rounded to nearest whole number).
calcium 40
carbon 12
oxygen 16

Molecular weights:

carbonate	$CO_3 = 12 + (16 \times 3) = 60$
hydrogen carbonate	$HCO_3 = 1 + CO_3 = 61$
calcium carbonate	$40 + CO_3 = 100$

Thus 1 mole of calcium is 40 g.

Ca^{2+} mg $l^{-1} \times 0.02495$ = mM l^{-1}

Milli equivalents (meq) are also sometimes reported where an equivalent is the atomic weight/valency (for calcium = 40/2) which has relevance in calculations of balances of ion species and how they combine,

Ca^{2+} mg $l^{-1} \times 0.04990$ = meq l^{-1}

Data may be reported in parts per million (ppm) or mg l^{-1}, the former being mg solute per kg solution; assuming 1 litre solution to weigh 1 kg, for all practical purposes the data in either units are regarded as interchangeable (see Hem, 1970, pp. 79–80).

Calcium is regarded as a metal as it is an element which, in solution, gives a positive cation. A base is defined as a substance which can accept a proton (H^+) from another substance. An acid can donate a proton. An alkali reacts to yield (OH^-) in solution (see Stumm and Morgan, 1970, pp. 70–3 for a full explanation).

Quantitative analysis of calcium and magnesium in solution by titration with EDTA (see also Douglas, 1968)

Titration with EDTA is a standard technique for the quantitative analysis of Ca^{2+} and Mg^{2+} ions in solution. The method can also be used for the analysis of rock and soil samples if the samples are first ground and then dissolved in hydrochloric acid to bring the calcium and magnesium into solution (see Bisque, R. E., 1961). The technique has become widely used in the study of erosion processes in limestone areas, especially with the analysis of karst waters. It is also used in the study of the geochemistry of carbonate rocks and soils, and has thus become useful to geomorphologists, geologists and pedologists.

EDTA (Ethylene-diamine-tetra-acetic acid, in the disodium salt form) forms stable complexes with a large number of metallic cations (calcium, magnesium, iron, cadmium, zinc, mercury, copper, cobalt, silver, nickel, aluminium, platinum, titanium, beryllium, manganese and lead). In karst waters it is normal to assume that calcium and magnesium are the dominant cations present but in heavily contaminated waters and in the analysis of soils and rocks it is necessary to use masking substances in order to inactivate the interfering elements. Triethanolamine masks iron, manganese and copper; sodium (D+) tartrate masks aluminium; hydroxylammonium hydrochloride masks iron, manganese and copper and while potassium cyanide masks copper, cobalt, nickel and zinc it is not normally used because of its poisonous nature (details of masks are found in Bisque).

The standard procedure is to titrate two aliquots from the same sample, one for calcium and one for calcium + magnesium. This is achieved by adding a strong alkali to one aliquot which has the effect of precipitating the magnesium so that it does not take part in the titration. The titration for calcium + magnesium is often referred to as the total hardness and the figure for magnesium is gained by subtracting that for the calcium from the total.

The calcium titration proceeds by adding potassium hydroxide to alkalise the solution and then adding an indicator which forms a complex with the free calcium. Upon adding the EDTA to the solution the calcium is complexed by the EDTA (which forms a stronger complex than the indicator–calcium complex) and upon losing calcium the indicator changes colour. When the colour change is complete the end point has been reached and the amount of EDTA used is noted and the amount of calcium present in the solution is computed from a knowledge of the ratio in which calcium and EDTA combine.

The calcium + magnesium titration is undertaken on the same principles but a mildly alkaline solution is used instead of a strongly alkaline one.

Apparatus and chemicals needed

CAUTION: potassium hydroxide is caustic to the skin and ammonia causes irritation to eyes and lungs.

100 ml measuring cylinder or pipette (or use a 50 ml one twice).
250 ml conical flask.

Retort stand, bosshead and clamp.
White tile or white piece of paper.
100 ml or 50 ml burette.
0.025 M EDTA (9.306 g l^{-1}).
Ammonium purpurate indicator for calcium titration (0.02 g in 10 ml distilled water).
Potassium hydroxide buffer for calcium titration (80 g l^{-1}).
Erio-T indicator for calcium + magnesium titration (0.02 g in 10 ml alcohol).
Buffer for calcium + magnesium titration (70 g ammonium chloride in 570 ml of concentrated ammonia and made up to 1 litre).
Two small beakers.
pH paper.
Small funnel.

Procedure

Set up retort stand with clamp and bosshead, clamp in burette so that tip is just above level of 250 ml flask, placing a white tile or clean paper under the flask to facilitate estimation of colour change.
Remove flask from under burette. Fill burette with EDTA using the small funnel, taking care not to overfill funnel. Check that tap flows freely. Rinse out the measuring cylinder, or pipettes, flask and beakers with distilled water.

Calcium

Prepare two colour standards

Take approximately 25 ml of sample, add potassium hydroxide buffer to pH 14 (use pH papers to check) and ammonium purpurate indicator (a few drops to give a good colour). Repeat this in the other beaker. To one beaker add EDTA till a strong colour change occurs. The beaker without EDTA should be a reddish purple colour. The one with an excess of EDTA should be pure mauve without any trace of red. Place these beakers one either side of the flask and use these as colour reference points to ascertain the end point during the titration.

Titrate the sample

Take 100 ml of the water sample, measured in the measuring cylinder or pipette. Pour or pipette into the conical flask. Add potassium hydroxide to pH 14. Add a few drops of indicator to colour. Perform a rough titration first and then an accurate one. Note the burette reading and compute the EDTA used. Prepare a fresh 100 ml of sample and add buffer and indicator as appropriate. Note the burette reading (R1). Run in most of the EDTA to be used but approach the end point carefully adding the last amount of EDTA accurately, drop by drop (if you are unsure whether the end point has been reached, read the burette before adding the next drop). R2 is the final burette reading at the end point. Compute:

$R_2 - R_1 = V_1$ (V_1 = cc EDTA used in titrating calcium)
$V_1 \times 25 = CaCO_3$ content of water sample in parts per million (ppm)

Magnesium

Prepare two colour standards as above, but using ammonia buffer to pH 10 and a few drops Erio-T indicator. Before addition of EDTA the solution should be purple and after it should be a clear blue.

Perform a rough titration as before and then an accurate one, approaching the end point dropwise.

Read burette before titration (R3) and after (R4), then calculate:

R4 – R3 = V2 (cc EDTA used on calcium + magnesium).
V2 – V1 = V3 (cc EDTA used on magnesium alone).
V3 × 21.1 = $MgCO_3$ content of water sample in ppm.

Write up environmental interpretation of results.

Notes: Burette reading can be facilitated by placing a spliced piece of paper on the burette column.

With very low concentrations of magnesium it is best to adopt the following procedure: Take R3, add most of the EDTA amount used in calcium titration, add buffer and indicator, then complete titration drop-wise (Smith and Mead, 1962).

References

Abou-el-Enin, Hassan, S. 1973. *Essays on the geomorphology of the Lebanon*. Beirut Arab University.

Adams, C. S. and **Swinnerton, A. C.** 1937. The solubility of limestone. *Transactions of the American Geophysical Union*, **11**, 504–8.

Ahr, W. M. and **Stanton, R. J.** 1973. The sedimentologic and paleoecologic significance of *Lithotrya*, a rock boring barnacle. *Journal of Sedimentary Petrology*, **43**, 20–23.

Alexander, M. 1961. *Introduction to soil microbiology*. Wiley.

Alexandersson, E. T. 1975. Marks of unknown carbonate-decomposing organelles in cyanophyte borings. *Nature*, **254**, 237–8.

Alexandersson, E. T. 1976. Actual and anticipated petrographic effects of carbonate undersaturation in shallow seawater. *Nature*, **262** (5570), 653–7.

Allain, J. Y. 1972. Sur les populations de *Paracentrotus lividus* (Lamarck) et de *Psammechinus miliaris* (Gmelin) de Bretagne nord (Échinodermes). *Bulletin du Muséum National D'Histoire Naturelle*, 3e série, No. 32, Mars-Avril, Zoologie **26**, 305–15.

Allen, J. R. L. 1972. On the origin of cave flutes and scallops by the enlargement of inhomogeneities, *Rassegna Speleologica Italiana*, XXIV, 1, 1–19.

Ansell, A. D. and **Nair, N. B.** 1969. A comparative study of bivalves which bore mainly by mechanical means. *American Zoologist*, **9**, 857–68.

Atkinson, T. C. 1971. 'Hydrology and erosion in a limestone terrain'. Unpublished Ph.D. thesis, University of Bristol.

Atkinson, T. C. 1977. Carbon dioxide in the atmosphere of the unsaturated zone: an important control of groundwater hardness in limestones. *Journal of Hydrology*, **35**, 111–23.

Atkinson, T. C., **Bradshaw, R.** and **Smith, D. I.** 1973. *Quarrying in Somerset. Supplement 1. Hydrology and rock stability ¬ Mendip Hills. A Review of Existing Knowledge*. Somerset County Council, Taunton, U.K.

Atkinson, T. C., **Harmon, R. S.**, **Smart, P. L.** and **Waltham, A. C.** 1978. Palaeo-climatic and geomorphic implications of $^{230}Th/^{234}U$ dates on speleothems from Britain. *Nature*, **5648**, 24–28.

Atkinson, T. C., and **Smith, D. I.** 1974. Rapid groundwater flow in fissures in chalk: an example from South Hampshire, *Quarterly Journal of Engineering Geology*, **7**, (2), 197–205.

Atkinson, T. C. and **Smith, D. I.** 1976. The erosion of limestones. Ch. 5. In: Ford, T. D. and Cullingford, C. H. (eds.), *The Science of Speleology*, Academic Press.

Aubert, D. 1969. Phénomènes et formes du karst jurassien. *Eclogae geologicae Helvetiae*, **62**, 325–99.

Backshall, D. G., **Barnett, J.**, **Davies, P. J.**, **Duncan, D. C.**, **Harvey, N.**, **Hopley, D.**, **Isdale, P. J.**, **Jennings, J. N.** and **Moss, R.** 1979. Drowned dolines – the blue holes of Pompey Reefs, Great Barrier Reef. *BMR Journal of Australian Geology and Geophysics*, **4**, 99–109.

Balazs, D., 1971. Relief types of tropical karst areas. *Geographical Union, European Regional Conference, Hungary 1971, Symposium on karst morphogenesis*, 1–22.

Bathurst, R. G. C. 1966. Boring algae, micrite envelopes and lithification of molluscan biosparites. *Geological Journal*, **5**, 15–32.

Bardach, J. E. 1961. Transport of calcareous fragments by reef fishes. *Science*, **133**, 98–99.

Baver, L. D. 1927. Factors affecting the hydrogen-ion concentration of soils. *Soil Science*, **23**, 399–414.

Beckinsale, R. P. 1972. The limestone bugaboo: surface lowering or denudation or amount of solution? *Transactions of the Cave Research Group*

of Great Britain, **14**, 55–58.
Bell, M. and **Limbrey, S.**, 1982. Archaeological aspects of woodland ecology. *BAR International Series*, **146**, 115–27.
Ben-Yaakov, S. and **Kaplan, I. R.** 1969. Determination of carbonate saturation of seawater with a carbonate saturometer. *Limnology and Oceanography*, **14**, 874–82.
Berner, R. A. 1965. Activity coefficients of bicarbonate, carbonate and calcium ions in sea water. *Geochimica et Cosmochimica Acta*, **29**, 947–65.
Berner, R. A. 1967. Comparative dissolution characteristics of carbonate minerals in the presence and absence of aqueous magnesium ion. *American Journal of Science*, **265**, 45–70.
Berner, R. A. 1971. *Principles of chemical sedimentology*. McGraw-Hill.
Berner, R. A. 1978. Rate control of mineral dissolution under earth surface conditions. *American Journal of Science*, **278**, 1235–52.
Bisque, R. E. 1961. Analysis of carbonate rocks for calcium, magnesium, iron and aluminium with EDTA. *Journal of Sedimentary Petrology*, **31**, 113–22.
Bissel, H. J. and **Chilingar, G. V.**, 1967. Classification of carbonate rocks. An 4: In Chilingar, G. V., Bissel, H. J., and Fairbridge, R. W. (eds.) *Carbonate Rocks*, Developments in sedimentology, 9A, Elsevier, 150–68.
Blake, J. A. 1969. Systematics and ecology of shell-boring polychaetes from New England. *American Zoologist*, **9**, 813–20.
Blumberg, P. N. and **Curl, R. L.** 1974. Experimental and theoretical studies of dissolution roughness. *Journal of Fluid Mechanics*, **65**, 735–51.
Bögli, A. F. 1960. Kalklösung und Karren Bildung. *Zeitschrift für Geomorphologie, Suppl.* **2**, 4–21.
Bögli, A. 1964. Mischungs Korrosion: ein Beitrag zum Ver Kastungsproblem. *Erkunde*, **18**, 83–92.
Bolin, B. and **Keeling, C. D.** 1963. Large scale atmospheric mixing as deduced from the seasonal and meridional variations of carbon dioxide. *Journal of Geophysical Research*, **68**, 3899–3920.
Bradfield, R. 1941. Calcium in the soil. I. Physico-chemical relations. *Soil Science Society of America Proceedings*, **6**, 8–15.
Brady, N. C. 1974. *The nature and properties of soil* (esp. pp. 384–7). Macmillan (8th edn).
Bray, L. G. 1969. Some notes on the chemical investigation of cave waters. *Transactions of the Cave Research Group of Great Britain*, **11**, 165–174.
Bray, L. G. 1971. Some problems encountered in a study of the chemistry of solution. *Transactions of the Cave Research Group of Great Britain*, **13**, 115–22.
Bray, L. G. 1972. Preliminary oxidation studies on some cave waters from South Wales. *Transactions Cave Research Group Great Britain*, **14**, 59–66.
Briggs, D. J. 1977. *Sediments*. Butterworths.
Bromley, R. G. 1978. Bioerosion of Bermuda Reefs. *Palaeogeography, palaeoclimatology and palaeoecology*, **23**, 169–97.
Brook, D. 1977. Caves and karst of the Hindenburg Ranges. *Geographical Journal*, **143**, 27–41.
Brook, D. A. and **Waltham, A. C.**, 1978. *Caves of Mulu*. Royal Geographical Society, London, U.K.
Brook, G. A. 1977. Preliminary thoughts on a structural-lithological model of karst landform development. *Proceedings 7th International Speleological Congress, Sheffield 1977. British Cave Research Association*, ISU, 81–83b.
Brook, G. A. and **Ford, D. C.**, 1976. The Nahanni north karst: a question mark on the validity of the morphoclimatic concept of karst development. *Proceedings of the 6th International Congress of Speleology, Olomouc*. Academia, Prague. II, 43–57.
Brook, G. A. and **Ford, D. C**, 1977. The sequential development of karst landforms in the Nahanni Region of Northern Canada and a remarkable size hierarchy. *Proceedings 7th International Speleological Congress, Sheffield 1977. British Cave Research Association*, ISU, 77–81.
Brunsden, D. 1980. Applicable models of long term landform evolution. *Zeitschrift für Geomorphologie, Suppl.* **36**, 16–26.
Bryan, R. 1967. Climosequences of soil development in the Peak District of Derbyshire. *East Midlands Geographer*, **4**, 251–61.
Bull, P. A. 1975. Birdseye structures in caves. *Transactions British Cave Research Association*, **2**, 35–40.
Bull, P. A. 1976. An electron microscope study of cave sediments from Agen Allwedd, Wales. *Transactions of the British Cave Research Association*, **3**, 7–14.
Bull, P. A. 1977. Boulder chokes and doline relationships. *Proceedings 7th International Speleological Congress, Sheffield 1977. British Cave Research Association*, 93–6.
Bull, P. A. 1980. The antiquity of caves and dolines in the British Isles. *Zeitschrift für Geomorphologie, Suppl.* **36**, 217–32.
Bull, P. A. 1981. Some fine grained sedimentation phenomena in caves. *Earth Surface Processes and Landforms*, **6**, 11–22.
Bullock, P. 1964. A study of the origin and development of soils over Carboniferous Limestone in the Malham District of Yorkshire. Unpublished M.Sc. thesis, University of Leeds.
Bullock, P. 1971. The soils of the Malham Tarn area. *Field Studies*, **3**, 381–408.
Burke, A. R. 1970. Deposition of stalactite and related forms of peat; genesis and bacterial oxidation. *Transactions Cave Research Group Great Britain*, **12**, 247–58.
Busenberg, E. and **Plummer, L. N.** 1982. The kinetics of dissolution of dolomite in CO_2-H_2O systems at 1.5 to 65 °C and 0 to 1 Atm. P_{CO_2} *American Journal of Science*, **282**, 45–78.

Campbell, A. C., Dart, J. K. G., Head, S. M. and **Ormond, R. F. G.** 1973. The feeding activity of *Echinostrephus molaris* (de Blainville) in the central Red Sea. *Marine Behavioural Physiology*, **2**, 155–69.

Carriker, H. R. and **Smith, E. H.** 1969. Comparative calcibiocavitology: summary and conclusions. *American Zoologist*, **9**, 1011–20.

Carson, M. A. and **Kirkby, M. J.**, 1972. *Hillslope form and process*. Cambridge.

Chaberek, S. and **Martel, A. E.** 1959. *Organic sequestering agents*. Wiley.

Chappell, J. 1974. Geology of Coral terraces, Huon Peninsula, New Guinea; a study of Quaternary tectonic movements and sea-level changes. *Bulletin of the Geological Society of America*, **85**, 555–70.

Charity, R. A. P. and **Christopher, N. S. J.** 1977. The stratigraphy and structure of the Ogof Ffynuon Ddu area. *Transactions of the British Cave Research Association*, **4**, 403–16.

Chave, K. E. and **Suess, E.** 1970. Calcium carbonate saturation in sea water: effects of dissolved organic matter. *Limnology and Oceanography*, **15**, 633–7.

Chétail, M. and **Fournié, J.** 1969. Shell-boring mechanism of the gastropod, *Purpura* (*Thais*) *Lapillus*: a physiological demonstration of the role of carbonic anhydrate in the dissolution of $CaCO_3$. *American Zoologist*, **9**, 983–90.

Chilingar, G. V., Bissel, H. J. and **Fairbridge, R. W.**, 1967a. *Carbonate Rocks*, Developments in sedimentology, 9A. Elsevier, Amsterdam.

Chilingar, G. V., Bissel, H. J. and **Fairbridge, R. W.**, 1967b. *Carbonate Rocks*, Developments in sedimentology 9B. Elsevier, Amsterdam.

Ciric, M. 1967. Characteristics of soil formation on limestones and principles of limestone classification. *Soviet Soil Science*, **1**, 57–78.

Clark, M. J. and **Small, R.J.**, 1982. *Slopes and Weathering*. Cambridge.

Cloud, P. E. 1959. Geology of Saipan, Mariana Islands. Part 4. Submarine topography and shoal water ecology. *US Geology Survey Professional Papers*, 280-K, 361–445.

Cloud, P. E. 1965. Carbonate precipitation and dissolution in the marine environment. In: Riley, J.P., and Skirrow, G. (Eds.), *Chemical Oceanography*, Volume 2. Ch. 17, 127–58. Academic Press, NY.

Coase, A. and **Judson, D.** 1977. Dan yr Ogof and its associated caves. *Transactions of the British Cave Research Association*, **4**, 244–344.

Cobb, W. R. 1969. Penetration of calcium carbonate substrates by the boring sponge, *Cliona. American Zoologist*, **9**, 783–90.

Cole, C. V. 1957. Hydrogen and calcium relationships of calcareous soils. *Soil Science*, **83**, 141–50.

Comfort, A. 1957. The duration of life in molluscs. *Proceedings of the Malacological Society*, **32**, 219–41.

Corbel, J., 1957. Les karsts du nord-ouest de l'Europe. *Memoir Institute Etudes Rhodianiennes*, **12**, 544 pp.

Corbel, J. 1959. Erosion en terrain calcaire, *Annales Geographie*, **68**, 97–120.

Crabtree, R. W. 1981. Hillslope solute sources and solutional denudation on Magnesian Limestone, Unpublished Ph.D. Thesis, University of Sheffield.

Crapp, G. B. and **Willis, M. E.** 1975. Age determination in the Sea Urchin. *Paracentrotus lividus* (Lamarck) with notes on the reproductive cycle. *Journal Experimental Marine Biology and Ecology*, **20**, 157–58.

Craig, G. Y. and **Hallam, A.** 1963. Size frequency and growth ring analyses of *Mytilus edulis* and *Cardium edule* and their palaecological significance. *Palaeontology*, **6**, 731–750.

Craig, A. K., Dobkin, S., Grimm, R. B. and **Davidson, J. B.** 1969. The gastropod *Siphonaria pectinata*: a factor in destruction of beach rock. *American Zoologist*, **9**, 895–901.

Crisp, D. J. (Ed.) 1964. *Grazing in marine and terrestrial environments*. Blackwell.

Culkin, F. and **Cox, R. A.** 1966. Sodium, potassium magnesium, calcium and strontium in sea water. *Deep-Sea Research*, **13**, 789–804.

Curl, R. L. 1966. Scallops and flutes. *Transactions Cave Research Group Great Britain*, **7**, 121–60.

Curl, R. L. 1977. The dissolution kinetics of calcite in carbonic acid. In: Tolson, J. S. and Doyles, F. L. (Eds.). *Karst hydrogeology*. Proceedings of the 12th Congress of the International Association of Hydrogeologists. University of Alabama.

Curtis, L. F., Courtney, F. M. and **Trudgill, S. T.** 1976. *Soils in the British Isles*, Longman.

Dakyns, J. R. 1890. In: Dakyns, J. R., Tiddeman, R. H., Gunn, W. and Strahan, A. *The geology of the country round Ingleborough*. Memoir Geological Survey of Great Britain, **25.**

Dalongeville, R. 1977a. Formes littorales de corrosion dans les roches carbonates au Liban. Étude morphologique. *Méditerranée*, **3**, 21–33.

Dalongeville, R. 1977b. Trottoir-encorbellement-encoche: aspects du littoral actual Libanais. *Bulletin du Laboratoire Rhodanien de Géomorphologie, Lyon*, **1**, 53–58.

Dalongeville, R. 1978. Le littoral actuel Yougoslave. *Bulletin du Laboratoire Rhodamien de Géomorphologie, Lyon*, **3**, 5–11.

Davies, J. L. 1980. *Geographical variation in coastal development* (2nd ed.) Longman.

Day, M. J. 1978. Morphology and distribution of residual limestone hills (mogotes) in the karst of northern Puerto Rico. *Geological Society of America Bulletin*, **89**, 426–432.

Day, M. J. 1981. Rock hardness and landform development in the Gunong Mulu National Park, Sarawak, E. Malaysia. *Earth Surface Processes and Landforms*, **6**, 165–72.

de Burgh, M. E., West, A. B. and **Jeal, F.** 1977.

Absorption of L-ALANINE and other dissolved nutrients by the spines of *Paracentrotus lividus* (Echinoidea). *Journal of the Marine Biological Association*, U.K., **57**, 1031–45.

Douglas, I. 1968. Field methods of water hardness determination. *British Geomorphological Research Group, Technical Bulletin*, 1.

Douglas, I. 1980. Climatic geomorphology. Present-day processes and landform evolution. Problems of interpretation. *Zeitschrift für Geomorphologie, Suppl.* **36**, 27–47.

Drouet, F. and **Dayly, W. A.** 1956. Revision of the coccoid Myxophyceae. *Butler Chemistry Botanical Studies*, **10**, 1–218.

Drew, D. P., 1983. Accelerated soil erosion in a karst area: The Burren, western Ireland. *Journal of Hydrology*, **61**, 113–24.

Duff, R. B., Webley, D. M. and **Scott, R. O.** 1963. Solubilisation of minerals and related minerals by 2-ketogluconic acid-producing bacteria. *Soil Science*, **95**, 105–14.

Ede, D. P. 1975. Limestone drainage systems. *Journal of Hydrology*, **27**, 247–18.

Edwards, N. T. and **Sollins, P.** 1973. Continuous measurement of carbon dioxide evolution from partitioned forest floor components. *Ecology*, **54**, 406–12.

Emery, K. O. 1946. Marine solution basins. *Journal of Geology*, **54**, 209–28.

Emery, K. O. 1962. Marine geology of Guam. *U.S. Geological Survey Professional Paper*, 403–13.

Emmett, W. W. 1970. The hydraulics of overland flow on hillslopes. *U.S. Geological Survey Professional Paper*, 662–7, 1–68.

Engh, L. 1980. Can we determine solutional erosion by a simple formula? *Transactions of the British Cave Research Association*, **7**, 31–2.

Farrow, G. E. 1971. Periodicity structures in the bivalve shell: experiments to establish growth controls in *Cerastoderma edule* from the Thames Estuary. *Palaeontology*, **14**, 571–88.

Findlay, D. C. 1965. *The soils of the Mendip District of Somerset.* Memoir Soil Survey of Great Britain, Harpenden.

Flugel, E. (Ed.) 1977. *Fossil algae*. Springer-Verlag, Berlin.

Focke, J. W. 1977. The effect of a potentially reef-building Vermitid – Coralline Algal community on an eroding limestone coast, Curaçao, Netherlands Antille. *Proceedings, 3rd International Coral Reef Symposium, University of Miami, Florida, U.S.A.*, 239–45.

Focke, J. W. 1978. Limestone cliff morphology on Curaçao (Netherlands Antilles), with special attention to the orgin of notches and vermetid/coralline algal surf benches ('cornices', 'trottoirs'). *Zeitschrift für Geomorphologie*, **22**, 329–49.

Fogg, G. E. 1973. In: Carr, N. G. and Whitton, B. A., *The Biology of blue-green algae*, University of California press, 368n78.

Folk, R. L. 1959. Practical petrographic classification of limestones. *Bulletin of the American Association of Petroleum Geologists*, **43**, 1–38.

Folk, R. L. 1962. Spectral subdivision of limestone types. In: Ham, W. E. (Ed.). Classification of carbonate rocks. A symposium. *American Association of Petroleum Geologists, Memoir 1*.

Folk, R. and **Land, L. S.** 1975. Mg/Ca ratio and salinity: two controls over crystallisation of dolomites. *American Association of Petroleum Geologists Bulletin*, **59**, 60–8.

Folk, R. L., Roberts, H. H. and **Moore, C. H.** 1973. Black phytokarst from Hell, Cayman Islands, British West Indies. *Geological Society of America, Bulletin*, **84**, 2351–60.

Ford, D. C. 1971a. Characteristics of limestone solution in the southern Rocky Mountains and Selkirk Mountains, Alberta and British Columbia. *Canadian Journal of Earth Science*, **8**, 585–609.

Ford, D. C., 1971b. Alpine karst in the Mt. Castleguard–Columbia Icefield area, Canadian Rocky Mountains. *Arctic & Alpine Research*, **3**, 239–52.

Ford, D. C., 1979. A review of alpine karst in the southern Rocky Mountains of Canada. *The National Speleological Society*, Bulletin, **41**, 53–65.

Ford, D. C. 1980. Threshold and limit effects in karst geomorphology. In: Coates, D. R. and Vitek, J. D. (Eds.) *Thresholds in Geomorphology*. George Allen & Unwin, 345–62.

Ford, D. C., 1983. Effects of glaciations upon karst aquifers in Canada. *Journal of Hydrology*, **61**, 149–58.

Ford, D. C. and **Drake, J. J.** 1982. Spatial and temporal variations in karst solution rates: the structure of variability. In: Thorn, C. E. (Ed.). *Space & time in Geomorphology*. Allen & Unwin.

Ford, T. D., Gascoyne, M. and **Beck, J. S.** 1983. Speleothem dates and Pleistocene chronology in the Peak District of Derbyshire. *Cave Science: Transactions of the British Cave Research Association*, **10**, 103–15.

Ford, T. D. 1975. Sediments in caves. *Transactions of the British Cave Research Association*, **2**, 41–6.

Ford, T. D. 1976. The geology of caves. In: Ford, T. D. and Cullingford, C. H. (Eds.) *The Science of Speleology*, Academic Press, Ch. 2, 11–60.

Ford, T. D. 1978. Chillagoe – a tower karst in decay. *Transactions of the British Cave Research Association*, **6**, 61–84.

Froment, A. 1972. Soil respiration in a mixed oak forest. *Oikos*, **23**, 273–77.

Gagarina, E. I. 1968. Study of weathering of carbonate rocks in soil. *Soviet Soil Science*, **9**, 1300–7.

Gardiner, M. J. and **Ryan, P.** 1962. Relic soil on limestone in Ireland. *Irish Journal of Agricultural Research*, **1**, 181–8.

Garrels, R. M. and **Christ, C. L.** 1965. *Solutions, minerals and equilibria*. Harper and Row.

Garrett, H. E. and **Cox, G. S.** 1973. Carbon dioxide evolution from the floor of an oak-hickory forest. *Proceedings of the Soil Science Society of America*, **37**, 641–4.

Gascoyne, M., Schwarcz, H. P. and **Ford, D. C.** 1978. Uranium series dating and stable isotope studies of speleothems: Part I Theory and Techniques. *Transactions of the British Cave Research Association*, **5**, 91–111.

Glew, J. R. 1977. Simulation of rillenkarren. *Proceedings 7th International Speleological Congress, Sheffield 1977. British Cave Research Association*, 218–19.

Glew, J. R. and **Ford, D. C.** 1980. A simulation study of the development of rillenkarren. *Earth Surface Processes*, **5**, 25–36.

Golubic, S. 1969. Distribution, taxonomy and boring patterns of marine endolithic algae. *American Zoologist*, **9**, 747–51.

Golubic, S. 1973. The relationship between blue-green algae and carbonate deposits. In: Carr, N. G. and Whitton, B. A. *The Biology of blue-green algae*, Botanical monographs, **9**, 434–472, Blackwell, Oxford.

Gorham, E. 1955. On the acidity and salinity of rain. *Geochimica et Cosmochimica Acta*, **7**, 231–9.

Gorham, E. 1957. The chemical composition of rain from Rosscahill in Co. Galway. *Irish Naturalists Journal*, **12**, 122–6.

Gorham, E. 1961. Factors influencing supply of major ions to inland waters with special reference to the atmosphere. *Bulletin of the Geological Society of America*, **72**, 795–840.

Gosden, M. S. 1968. Peat deposits of Scar Close, Ingleborough, Yorkshire: *Journal of Ecology*, **56**, 345–53.

Grime, J. P. 1963. Factors determining the occurrence of calcifuge species on shallow soils over calcareous substrates. *Journal of Ecology*, **51**, 375–90.

Grime, J. P. and **Hodgson, J.** 1969. An investigation of the ecological significance of lime-chlorosis by means of large-scale comparative experiments. In: Rorison, I. (Ed.) *Ecological aspects of the mineral nutrition of plants*. Symposium British Ecological Society, 9. Blackwell.

Groom, G. E. and **Ede, D. P.** 1972. Laboratory simulation of limestone solution. *Transactions of the Cave Research Group of Great Britain*, **14**, 89–95.

Grubb, P. J., Green, H. E. and **Merrifield, R. J. C.** 1969. The ecology of chalk heath: its relevance to the calcicole-calcifuge and soil acidification problems. *Journal of Ecology*, **57**, 175–212.

Guilcher, A. 1958. *Coastal and submarine morphology*. Methuen.

Gunn, J. 1980. Comment on Leif Engh's question 'Can we determine solutional erosion by a simple formula?' *Transactions of the British Cave Research Association*, **7**, 205.

Gunn, J. 1981a. Limestone solution rates and processes in the Waitomo district, New Zealand. *Earth Surface Processes and Landforms*, **6**, 427–45.

Gunn, J. 1981b. Hydrological processes in Karst depressions. *Zeitschrift für Geomorphologie*, **25**, 3, 313–31.

Gunn, J., 1981c. Prediction of limestone solution rates from rainfall and runoff data: some comments. *Earth Surface Processes and Landforms*, **6**, 595–7.

Gunn, J., 1982. Magnitude and frequency properties of dissolved solids transport. *Zeitschrift für Geomorphologie*, **26**, 505–11.

Hale, M. E. 1967. *The biology of lichens*. Arnold.

Hamner, W. M. and **Jones, M. S.** 1976. Distribution, borrowing and growth rates of the clam *Tridacna crocea* on interior reef flats. *Oceologia* (Berl.), **24**, 207–27.

Handley, W. R. C. 1954. Mull and mor in relation to forest soils. *Forestry Commission Bulletin*, **23**, HMSO.

Hanwell, J. D. and **Newson, M. D.** 1970. The great storm and floods of July 1968 on Mendip. *Wessex Cave Club, Occasional Publication*, 1 (2), 1–72.

Harmon, R. S., Ford, D. C. and **Schwarcz, H. P.** 1977. Interglacial chronology of the Rocky and Mackenzie Mountains based on ^{230}Th-^{234}U dating of calcite speleothems. *Canadian Journal of Earth Sciences*, **14**, 2543–52.

Harmon, R. S., Hess, J. W., Jacobson, R. W., Shuster, E. T., Haygood, C. and **White, W. B.** 1972. Chemistry of carbonate denudation in North America. *Transactions of the Cave Research Group of Great Britain*, **14**, 96–103.

Harmon, R. S., Schwarcz, H. P. and **Ford, D. C.** 1978. Late pleistocene sea level history of Bermuda. *Quaternary Research*, **9**, 205–18.

Healy, T. R. 1968. Bioerosion on shore platforms developed in the Waitemata formation, Auckland. *Earth Science Journal*, **2**, 26–37.

Hein, F. J. and **Risk, M. J.** 1975. Bioerosion of coral heads: inner patch reefs, Florida reef tract. *Bulletin of Marine Science*, **25**, 133–8.

Hem, J. D. 1970. *Study and interpretation of the chemical characteristics of natural water*. United States Geology Survey, Water Supply Paper, 1473. U.S. Government Printing Office, Washington.

Henderson, M. E. K. and **Duff, R. B.** 1963. The release of metallic and silicate ions from minerals, rocks and soils by fungal activity. *Journal of Soil Science*, **14**, 236–46.

High, C. J. and **Hanna, K. K.** 1970. *A method for the direct measurement of erosion on rock surfaces*. British Geomorphological Research Group, Technical Bulletin, 5.

Hodgkin, N. M. 1962. Limestone boring by the mytilid *Lithophaga*. *Veliger*, **4**, 123–9.

Holmes, A., 1965. *Principles of physical geology*. Nelson.

Hope Simpson, J. F. and **Willis, A. J.** 1955. Vegetation. Ch. 6 in *Bristol and its adjoining counties*. British Association, Bristol (esp. p. 103).

Horton, R. E. 1945. Erosional development of streams and their drainage basins: hydrophysical approach to drainage basins and their morphology. *Bulletin, Geological Society of America*, **56**, 275–370.

Hudson, J. H. 1977. Long-term bioerosion rates on a Florida reef: a new method. *Proceedings, Third International Coral Reef Symposium*, University of Miami, Florida, U.S.A., 491–7.

Hughes, T. McK. 1901. Physical geography of Ingleborough. *Proceedings Yorkshire Geological Society*, **14**, 125.

Hunter, I. G. 1977. Sediment production by *Diadema antillarum* on a Barbados fringing reef. *Proceedings, Third International Coral Reef Symposium*, University of Miami, Florida, U.S.A., 105–9.

Hunter, W. R. 1949. The structure and behaviour of *Hiatella gallicana* (Lamarck) and *H. arctica* (L.), with special reference to the boring habit. *Proceedings of the Royal Society of Edinburgh*, B63, III, (19), 271–89.

Hutchins, P., 1974. A preliminary report on the density and distribution of invertebrates living on coral reefs. *Proceedings: Second International Coral Reef Symposium*, **1**, 285–96.

Ireland, P. 1979. Geomorphological variations of 'case-hardening' in Puerto Rico. *Zeitschrift für Geomorphologie, Suppl.* **32,** 9–20.

Jacobson, R. C. and **Langmuir, D.** 1972. An accurate method for calculating saturation levels of groundwaters with respect to calcite and dolomite. *Transactions of the Cave Research Group of Great Britain*, **14**, 104–8.

Jakucs, L. 1977. *Morphogenetics of Karst regions*. Adam Hilger, Bristol.

Jennings, J. N. 1971. *Karst*, M.I.T. Press, London.

Jennings, J. N. 1972. The character of tropical humid Karst. *Zeitschrift für Geomorphologie*, **16**, 336–41.

Jennings, J. N. 1975. Doline morphometry as a morphogenetic tool: New Zealand examples. *New Zealand Geographer*, **31**, 6–28.

Jennings, J. N., 1976. A visit to China. *Journal of the Sydney Speleological Society*. **20** (5), 119–39.

Jennings, J. N. 1977. Limestone tablet experiments at Cooleman Plain, New South Wales, Australia and their implications. *Abhandlungen zur Karst und Höhlenkunde; Reihe A-Speläologie Heft 15; Festchrift für Alfred Bögli*, 526–38.

Jennings, J. N., 1982. Principles and problems in reconstructing karst history. *Helictite*, **20**, 37–52.

Jensen, M. 1969. Age determination of echinoids. *Sarsia*, **37**, 41–4.

Jones, R. I. 1965. Aspects of the biological weathering of limestone pavements. *Proceedings of the Geologists' Association*, **76**, 421–34.

Kaye, C. A. 1957. The effect of solvent motion on limestone solution. *Journal of Geology*, LXV, 35–46.

Kaye, C. A. 1959. Shoreline features and Quaternary shoreline changes, Puerto Rico. *U.S. Geological Survey, Professional Paper*, 317-B.

Kerpen, W. and **Scharpenseel, W. H.** 1967. Movements of ions and colloids in undisturbed soil and parent material columns. In: *Isotope and radiation techniques in soil physics and irrigation studies*. Proceedings Symposium FAO/IAEA, Istanbul, 213–26.

Kester, D. R. and **Pytkowicz, R. M.** 1969. Sodium magnesium and calcium sulfate ion-pairs in sea water at 25 °C. *Limnology and Oceanography*, **14**, 686–92.

Kirk, R. M. 1977. Rates and forms of erosion on intertidal platforms at Kaikoura peninsula, South Island, New Zealand. *New Zealand Journal of Geology and Geophysics*, **20**, 571–613.

Kleeman, K. 1973. *Lithophaga lithophaga* (L.) (Bivalvia) in different limestones. *Malacologia*, **14**, 345–7.

Kobluk, D. R. and **Risk, M. J.** 1977a. Rate and nature of infestation of a carbonate substratum by a boring alga. *Journal of Experimental Marine Biology and Ecology*, **27**, 107–15.

Kobluk, D. R. and **Risk, M. J.** 1977b. Calcification of exposed filaments of endolithic algae, micrite envelope formation and sediment production. *Journal of Sedimentary Petrology*, **47**, 517–28.

Kohlmeyer, J. 1969. The role of marine fungi in the penetration of calcareous substances. *American Zoologist*, **9**, 741–6.

Kononova, M. M. 1966. *Soil organic matter*. Pergamon.

Kowsmann, R. O. 1972. The burrowing mechanism of sea urchins and its ecological significance: a review. *Atas da Sociedade de Biologia do Rio de Janeiro*, **16**, 39–41.

Kubiëna, W. L. 1953. *The Soils of Europe*. Thomas Murby and Co., London.

Kuznetzov, S. I. 1962. *The geological activity of micro-organisms*. USSR Academy of Science Press, Translation, Consultants Bureau, New York.

Lang, S. 1977. Relationship between world-wide karstic denudation (corrosion) and precipitation. *Proceedings 7th International Speleogical Congress, Sheffield*, 282–283. British Cave Research Association.

La Valle, P. 1968. Karst depression morphology in south central Kentucky. *Geografiska Annaler*, **50A**, 94–108.

Lawrence, J. M. 1975. On the relationships between marine plants and sea urchins. Oceanography and Marine Biology, Annual Review, **13**, 213–86.

Lewis, J. R. 1964. *The ecology of rocky shores*. E.U.P.

Li, Y.-H., Takahashi, T. and **Broecker, W. S.** 1969. Degree of saturation of $CaCO_3$ in the oceans. *Journal of Geophysical Research*, **74**, 5507–25.

Loenhoud, N. P. J., d. San De, J. C. P. N. V. and **Focke, J. W.** 1976. Rocky shore zonation on Curaçao, Netherlands Antilles. *Abstracts, 12th Meeting Association Island Marine Laboratories Caribbean*. Caribbean Marine Biological Institute, Curaçao.

Lowenstam, H. A. 1962. Magnetite in denticle capping in Recent chitons (Polyplacophora). *Bulletin Geological Society of America*, **73**, 435–8.

Lundberg, J. 1977. Karren of the littoral zone, Burren District, Co. Clare, Ireland. *Proceedings 7th International Speleological Congress*, Sheffield, 1977, British Cave Research Association, 291–3.

MacFadyen, W. A. 1930. The undercutting of coral reef limestone on the coasts of some islands in the Red Sea. *Geographical Journal*, **75**, 27–34.

MacGeachy, J. K. 1977. Factors controlling sponge boring in Barbados reef corals. *Proceedings, Third International Coral Reef Symposium*, University of Miami, Florida, U.S.A., 477–83.

Manskaya, S. M. and **Drozdova, T. V.** 1969. Organic matter as a factor in rock weathering. In: Khitarov, N. I. (Ed.) *Problems of Geochemistry, Israel Programme of Scientific Translation.*

McDonald, R. C., 1976. Hillslope base depressions in tower karst topography of Belize. *Zeitschrift für Geomorphologie Suppl.* **26**, 98–103.

McLean, R. F. 1967. Measurements of beachrock erosion by some tropical marine gastropods. *Bulletin of Marine Science*, **17**, 551–61.

McLean, R. F. 1974. Geologic significance of bioerosion of beach rock. *Proceedings of the Second International Coral Reef Symposium, 2.* Great Barrier Reef Committee, Brisbane, 401–8.

Mercado, A. 1977. The kinetics of mineral dissolution in carbonate aquifers as a tool for hydrological investigations, II: Hydrogeochemical models. *Journal of Hydrology*, **35**, 365–84.

Mercado, A. and **Billings, G. U.** 1975. The kinetics of mineral dissolution in carbonate aquifers as a tool for hydrological investigations, I: Concentration-time relationships. *Journal of Hydrology*, **24**, 303–31.

Merrill, A. S., Posgat, T. A. and **Nichi, F. F.** 1962. Annual marks on shell and ligament of sea scallop (*Placopecten magellanicus*). *US Fish & Wildlife Service, Fisheries Bulletin*, **62**, 299–311.

Miller, L., 1952. A portion of the system calcium carbonate–carbon dioxide–water, with geological implications. *American Journal of Science*, **250**, 161–203.

Miotke, F. D. 1973. The subsidence of the surface between mogotes in Puerto Rico east Arecibo. *Caves and Karst*, **15**, 1–12.

Miotke, F. D. 1974. Carbon dioxide and the soil atmosphere. *Abhlandung für Karst und Hohlen Kunde, Reihe A., Heft* 9, Munich, 1–49.

Miserez, J. J. 1970. Utilisation d'une électrode spéciale pour la mesure de p CO_2 dans les eaux et l'atmosphère – Application à l'étude des phénomènes karstiques du Jura Suisse. *Laboratoire de Mineralogie, Institute de Géologie, Neuchâtel, Suisse*, 555/1.

Monroe, W. H., 1966. Formation of tropical karst topography by limestone solution and reprecipitation. *Caribbean Journal of Science*, **6** 1–7.

Moore, C. H. and **Shedd, W. W.** 1977. Effective rates of sponge bioerosion as a function of carbonate production. *Proceedings, Third International Coral Reef Symposium*, University of Miami. Florida, U.S.A., 499–505.

Morgan, J. J. 1967. Applications and limitations of chemical thermo-dynamics in natural water systems. In: Gould, R. F. (Ed.) *Equilibrium concepts in natural water systems.* American Chemical Society, Advances in Chemistry Series, **67**, 1–29.

Morse, J. W., Mucci, A. and **Millero, F. J.** 1980. The solubility of calcite and aragonite in sea water of 35% salinity at 25 °C and atmospheric pressure. *Geochimica et Cosmochimica Acta*, **44**, 85–94.

Moseley, F. 1973. Orientations and origins of joints, faults and folds in the Carboniferous Limestones of N.W. England. *Transactions of the Cave Research Group of Great Britain*, **15**, 99–106.

Murray, A. N. and **Love, W. W.** 1929. Action of organic acids upon limestone. *Bulletin of the American Association of Petroleum Geologists*, **13**, 1467–75.

Nernst, W. 1904. Theorie der Reaktionsgeschwindigkeit in heterogenen Systemen. II. *Zeitschrift für Physikalische Chemie*, **47**, 52–55.

Neumann, A. C. 1966. Observations on coastal erosion in Bermuda and measurements of the boring rate of the sponge *Cliona lampa. Limnology and Oceanography*, II, 92–108.

Neumann, A. C. 1968. Biological erosion of limestone coasts. In: Fairbridge, R. W. (Ed.) *Encyclopedia of Geomorphology*, 75–81. Reinhald Book Corporation.

Newell, N. D. and **Imbrie, J.** 1955. Biogeological reconnaisance in the Bimini area, Great Bahama Bank. *New York Academy of Science Transactions*, **18**, 3–14.

Newson, M. D. 1970. Studies in chemical and mechanical erosion by streams in limestone terrains. Unpublished Ph.D. thesis, University of Bristol.

Newson, M. D. 1971. The role of abrasion in cavern development. *Transactions Cave Research Group of Great Britain*, **13**, 101–7.

Nicholson, F. H. and **Nicholson, H. M.** 1969. A new method of measuring soil carbon dioxide for limestone solution studies, with results for Jamaica and the United Kingdom. *Journal of the British Speleological Association*, **6**, 136–148.

Noel, M., Homonko, P. and **Bull, P.** 1979. The palaeomagnetism of sediments from Agen Allwedd, Powys. *Transactions of the British Cave Research Association*, **6**, 85–92.

Orton, J. H. 1928. On rhythmic periods in shell growth in *Ostrea edulis* with a note on fattening. *Journal of Marine Biological Association U.K.*, **15**, 365–427.

Otter, G. W. 1932. Rock-burrowing echinoids. *Biological Review*, **7**, 89–107.

Otter, G. W. 1937. Rock-destroying organisms in relation to coral reefs. *Great Barrier Reef*

Expedition, 1928–29, Scientific Reports, Volume 1, British Museum (Natural History), 323–352.

Paine, R. T. and **Vadas, R. L.** 1969a. The effects of grazing by sea urchins, *Strongylocentrotus* spp., on benthic algal populations. *Limnology and Oceanography*, **14**, 710–19.

Paine, R. T. and **Vadas, R. L.** 1969b. Calorific values of benthic marine algae and their postulated relation to invertebrate food preference. *Marine Biology*, **4**, 79–86.

Palmer, A. N. 1975. The origin of maze caves. *National Speleological Society Bulletin*, **37**, 56–76.

Pang, R. K. 1973. The ecology of some Jamaican excavating sponges. *Bulletin of Marine Science*, **23**, 227–43.

Pannella, G. P. and **MacClintock, C.** 1966. Biological and environmental rhythms reflected in molluscan shell growth. *The Palaeontological Society, Memoir*, **2**, 64–81.

Parry, J. T. 1960. Limestone pavements of North West England. *Canadian Geographer*, **15**, 14–21.

Peterson, M. N. A. 1966. Calcite: rates of dissolution in a vertical profile in the central Pacific. *Science*, **154**, 1542–4.

Pfeffer, K-H., 1981. Relikte tropischer Karstformen auf der Frankischen Alb im Pegnitzgebiet. *Sonderveröff Geologische Institute Universität Koln*, **41**, 155–72.

Picknett, R. G. 1964. A study of calcite solutions at 10 °C. *Transactions of the Cave Research Group of Great Britain*, **7**, 41–62.

Picknett, R. G. 1972. The pH of calcite solutions with and without magnesium carbonate present, and the implications concerning rejuvenated aggressiveness. *Transactions of the Cave Research Group of Great Britain*, **14**, 141–150.

Picknett, R. G. 1973. Saturated calcite solutions from 10 °C to 40 °C: a theoretical study evaluating the solubility product and other constants. *Transactions of the Cave Research Group of Great Britain*, **15**, 67–80.

Pigott, C. D. 1962. Soil formation and development on the Carboniferous Limestone of Derbyshire. Part I. Parent materials. *Journal of Ecology*, **50**, 145–56.

Pigott, M. E. and **Pigott, C. D.** 1959. Stratigraphy and pollen analysis of Malham Tarn and Tarn Moss. *Field Studies*, **1**, 84–101.

Pitty, A. F. 1966. *An approach to the study of karst water*. Occasional paper 5, Geography Department, University of Hull.

Pitty, A. F. 1966. *An approach to the study of karst* solutional loss from a limestone tract of the southern Pennines. *Proceedings of the Geologists' Association*, **79**, 153–77.

Plummer, L. N. and **MacKenzie, F. T.** 1974. Predicting mineral solubility from rate data: application to the dissolution of magnesian calcites. *American Journal of Science*, **274**, 61–83.

Pomponi, S. A. 1976. An ultrastructural study of boring sponge cells and excavated substrata. *Scanning Electron Microscopy*, VIII. Proceedings, Workshop and Zoological Application of SEM, IIT Research Institute, Chicago, Illinois 60616, USA, 569–75.

Pomponi, S. A. 1977. Etching cells of boring sponges: an ultrastructural analysis. *Proceedings, Third International Coral Reef Symposium*, University of Miami, Florida, USA, 485–90.

Powell, R. L. 1980. Theories of development of karst topography. In: Melhorn, W. N. and Flemal, R. C. (Eds.) *Theories of Landform Development*, 217–47. George Allen and Unwin.

Price, N. J. 1966. *Fault and joint development in brittle and semi-brittle rock*. Pergamon Press, Oxford.

Purdy, E. G. and **Kornicker, L. S.** 1968. Algal disintegration of Bahamian Limestone coasts. *Journal of Geology*, **66**, 97–9.

Pytkowicz, R. M. 1969. Chemical solution of calcium carbonate in sea water. *American Zoologist*, **9**, 673–9.

Pytkowicz, R. M. and **Kester, D. R.** 1971. The physical chemistry of sea water. *Oceanography and Marine Biology Annual Review*, **9**, 11–60.

Randazzo, A. F. and **Hickey, E. W.** 1978. Dolomitization in the Floridan aquifer. *American Journal of Science*, **278**, 1177–84.

Rashid, M. A. 1971. Role of humic acids of marine origin and their different molecular weight fractions in complexing di- and tri-valent metals. *Soil Science*, **111**, 298–306.

Rauch, H. and **White, W. B.** 1977. Dissolution kinetics of carbonate rocks. 1. Effects of lithology on dissolution rate. *Water Resources Research*, **13**, 381–94.

Raymont, J. E. G. 1963. *Plankton and productivity in the oceans*. Pergamon.

Read, H. H. 1962. *Rutley's Elements of Mineralogy*. Allen & Unwin, London.

Regis, M. B. 1972. Croissance de *Paracentrotus lividus* Lmk. II – Le Système Apical. *Tethys*, **4**, 481–92.

Revelle, R. and **Emery, K. O.** 1957. Chemical erosion of beach rock and exposed reef rock. *U.S. Geological Survey Professional Paper*, 260-T.

Revelle, R. and **Fairbridge, R. W.** 1957. Carbonates and carbon dioxide. In: Hedgpeth, W. (Ed.) *Treatise on Marine Ecology and Paleoecology*. Geological Society of America, Memoir 67, Volume 1, 239–96.

Rice, M. E. 1969. Possible boring structures of sipunculids. *American Zoologist*, **9**, 803–12.

Richards, B. N. 1974. *Introduction to the soil ecosystem*. Longman.

Robinson, L. A. 1976. The micro-erosion meter technique in a littoral environment. *Marine Geology*, **22**, 51–58.

Roques, H. 1969. A review of present-day problems in the physical chemistry of carbonates in solution. *Transactions of the Cave Research Group of Great Britain*, **11**, 139–63.

Rowland, R. W. and **Hopkins, D. M.** 1971. Comments on the use of *Hiatella arctica* for determining

Cenozoic sea temperatures. *Palaeogeography, palaeoclimatology, Palaeoecology*, **19**, 59–64.

Russell, E. W. 1961. *Soil conditions and plant growth* (9th ed.) Longman.

Rutzler, K. 1975. The role of burrowing sponges in bioerosion. *Oecologia* (Berl.) **19**, 203–16.

Ryder, P. F. 1975. Phreatic network caves in the Swaledale area, Yorkshire. *Transactions British Cave Research Association*, **2**, 177–92.

Schatz, A. 1963. Soil micro-organisms and soil chelation. The pedogenic action of lichens and lichen acids. *Agriculture, Food, Chemistry*, **11**, 112.

Schmalz, R. F. and **Swanson, F. J.** 1969. Diurnal variation in the carbonate saturation of sea water. *Journal of Sedimentary Petrology*, **39**, 255–67.

Schneider, J. 1976. Biological and inorganic factors in the destruction of limestone coasts. *Contributions to sedimentology*, **6**, 1–112.

Schnitzer, M. and **Skinner, S. I. M.** 1963. Organo-metallic interactions in soils. 1. Reactions between a number of metal ions and the organic matter of a podzol Bh horizon. *Soil Science*, **96**, 86–93.

Schwartz, M. L. 1967. The scale of shore erosion. *Journal of Geology*, **76**, 508–17.

Seatz, L. F and **Peterson, H. B.** 1964. Acid, alkaline, saline and sodic soils. In: Bear, F. E. (Ed.) *Chemistry of the Soil*. American Chemical Society Monograph, 160, Reinhald, New York, Ch. 7.

Shaw, W. M. 1960. Rate of reaction of limestone with soils. *University of Tennessee Agricultural Experimental Station, Bulletin* 319.

Shimwell, D. W. 1971. Festuco-Brometea Br. – Bl. and R. Tx. 1943 in the British Isles: The phytogeography and phytosociology of limestone grasslands. *Vegetatio*, **23**, 1–60.

Sillen, L. G. 1961. Physical chemistry of sea water. *Oceanography, American Association Advancement of Science*, 549–80.

Smart, C. C. and **Ford, D. C.** 1983. The Castleguard Karst, Main Ranges, Canadian Rocky Mountains. *Journal of Hydrology*, **61**, 193–97.

Smart, P. L. and **Laidlaw, I. M. S.** 1977. An evaluation of some fluorescent dyes for water tracing. *Water Resources Research*, **13**, 15–33.

Smith, B. J., 1978. The origin and geomorphic implications of cliff foot recesses and tafoni on limestone hamadas in the northwest Sahara. *Zeitschrift für Geomorphologie*, **22**, 21–43.

Smith, D. I. 1969. The solutional erosion of limestone in an arctic morphogenetic region. In: Stelcl, O. (Ed.) *Problems of karst denudation*, Proceedings 5th International Speleological Congress, Brno, Czechoslovakia, Studia Geographica, 5.

Smith, D. I. and **Atkinson, T. C.** 1976. Process, landform and climate in limestone regions. Ch. 13. in Derbyshire, E. (Ed.) *Geomorphology and climate*, Wiley, 367–409.

Smith, D. I., Atkinson, T. C. and **Drew, D. P.** 1976. The hydrology of limestone terrains. In: Ford, T. D. and Cullingford, C. H. *The Science of speleology*, Academic Press, Ch. 6, 179–212.

Smith, D. I., Drew, D. P. and **Atkinson, T. C.** 1972. Hypothesis of karst landform development in Jamaica. *Transactions of the Cave Research Group of Great Britain*, **14** (2), 159–73.

Smith, D. I. and **Mead, D. G.** 1962. The solution of limestone with special reference to Mendip. *Proceedings of the University of Bristol Speleological Society*, **9**, 188–211.

Somerset County Council 1971. *Quarrying in Somerset*. Somerset County Council, Taunton, U.K.

Southward, A. J. 1964. Limpet grazing and the control of vegetation on rocky shores. In: Crisp, D. J. (Ed.) *Grazing in terrestrial and marine environments*, Blackwell, 265–273.

Spencer, T. 1981. Micro-topographic change on calcarenites, Grand Cayman Island, West Indies. *Earth Surface Processes and Landforms*, **6**, 85–94.

Standing, P. A. 1969. Poll Ballynahown, Co. Clare, Ireland. *Proceedings University of Bristol Speleological Society*, **12**, 117–22.

Stearn, C. W. and **Scoffin, T. P.** 1977. Carbonate budget of a fringing reef, Barbados. *Proceedings, Third International Coral Reef Symposium*, University of Miami, Florida, U.S.A., 471–6.

Stenner, R. D. 1969. The measurement of aggressiveness of water towards calcium carbonate. *Transactions of the Cave Research Group of Great Britain*, **11**, 175–200.

Stephenson, W. 1961. Experimental studies on the ecology of intertidal environments at Heron Island. *Australian Journal Marine Freshwater Research*, **11**, 241–67.

Stevenson, C. M. 1968. An analysis of the chemical composition of rain water and air over the British Isles and Eire for the years 1959–1964. *Quarterly Journal of the Royal Meteorological Society*, **94**, 56–70.

Stringfield, V. T. and **Legrand, H. E.** 1969. Hydrology of carbonate rock terranes – a review. *Journal of Hydrology*, **8**, 349–376.

Stumm, W. and **Morgan, J. J.** 1970. *Aquatic Chemistry*, Wiley-Interscience, New York.

Sweeting, M. M. 1966. The weathering of limestones. In: Dury, G. (Ed.), *Essays in Geomorphology*. Heinneman, 177–210.

Sweeting, M. M. 1970. Recent developments and techniques in the study of karst landforms in the British Isles. *Geographica Polonica*, **18**, 227–41.

Sweeting, M. M. 1972. *Karst landforms*. Macmillan.

Sweeting, M. M. 1974. Karst geomorphology in North-West England. In: Waltham, A. C. *Limestones and Caves of North-West England*. David and Charles.

Sweeting, M. M. 1980. Karst and climate – a review. *Zeitschrift für Geomorphologie, Suppl.* **36**, 203–16.

Sweeting, M. M. and **Sweeting, G. S.** 1969. Some aspects of the Carboniferous Limestone in relation to its Landforms (with particular reference to N.W. Yorkshire and Co. Clare). *Étudee et Travaux de 'Mediteranée'*, 7, Revue Géographique des payes Mediteranéens, 201–9.

Swinchatt, J. P. 1969. Algal boring: a possible depth indicator in carbonate rocks and sediments. *Geological Society of America Bulletin*, **80**, 1391–6.

Syers, J. V. 1964. A study of soil formation on Carboniferous Limestone, with particular reference to lichens as pedogenic agents. Unpublished Ph.D. thesis, University of Durham.

Thomas, T. M. 1970. The limestone pavements of the North Crop of the South Wales coalfield with special reference to solution rates and processes. *Transactions, Institute of British Geographers*, **50**, 87–105.

Thomas, T. M. 1974. The South Wales interstratal karst. *Transactions British Cave Research Association*, **1**, 131–152.

Thorne, M. J. 1967. Homing in the chiton *Acanthozostera gemmata* (Blainville). *Proceedings of the Royal Society of Queensland*, **79**, 99–108.

Thrailkill, J. 1977. Relative solubilities of limestone and dolomite. In: Tolson, J. S. and Doyle, F. L. (Eds.) *Karst Hydrogeology*, Proceedings 12th Congress International Association of Hydrogeologists. University of Alabama, 491–500.

Tratman, E. K. 1969. *The Caves of N.W. Co. Clare*. David and Charles, Newton Abbott, U.K.

Travis, D. F. and **Gonsalves, M.** 1969. Comparative ultrastructure and organisation of the prismatic region of two bivalves and its possible relation to the chemical mechanism of boring. *American Zoologist*, **9**, 635–61.

Trudgill, S. T. 1971. Poulcraveen, Co. Clare, Ireland. *Proceedings of the University of Bristol Speleological Society*, **12**, 293–5.

Trudgill, S. T. 1972. The influence of drifts and soils on limestone weathering in N.W. Clare, Ireland. *Proceedings University of Bristol Speleological Society*, **13**, 113–118.

Trudgill, S. T. 1975. Measurement of erosional weight-loss of rock tables. *British Geomorphological Research Group, Technical Bulletin*, **17**, 13–19.

Trudgill, S. T. 1976a. The marine erosion of limestones on Aldabra Atoll, Indian Ocean. *Zeitschrift für Geomorphologie, Suppl.* **26**, 164–200.

Trudgill, S. T. 1976b. The subaerial and subsoil erosion of limestones on Aldabra Atoll, Indian Ocean. *Zeitschrift für Geomorphologie Suppl.* **26**, 201–10.

Trudgill, S. T. 1976c. Rock weathering and climate: quantitative and experimental aspects. Ch. 3 in Derbyshire, E. *Geomorphology and Climate*, Wiley.

Trudgill, S. T. 1976d. The erosion of limestones under soil and the long term stability of soil–vegetation systems on limestone. *Earth Surface Processes*, **1**, 31–41.

Trudgill, S. T. 1976e. Limestone erosion under soil. In: Panos, V. (Ed.) *Proceedings of the 6th International Congress of Speleology*. II. Ba, 409–22. Academia/Prague.

Trudgill, S. T. 1977a. The role of a soil cover in limestone weathering, Cockpit Country, Jamaica. *Proceedings 7th International Speleological Congress, Sheffield, 1977, British Cave Research Association*, 4a–404.

Trudgill, S. T. 1977b. A comparison of tropical and temperate marine Karst erosion. *Proceedings 7th International Speleological Congress, Sheffield, 1977, British Cave Research Association*, 403a, 404–5.

Trudgill, S. T., 1977c. *Soil and Vegetarian Systems*, Oxford.

Trudgill, S. T. 1979a. Spitzkarren on calcarenites, Aldabra Atoll, Indian Ocean. *Zeitschrift für Geomorphologie, Suppl.* **32**, 67–74.

Trudgill, S. T. 1979b. Chemical polish of limestone and interactions between calcium and organic matter in peat drainage waters. *Transactions British Cave Research Association*, **6**, 30–35.

Trudgill, 1979c. Surface lowering and landform evolution on Aldabra. *Philsophical Transactions of the Royal Society, B*, **286**, 35–45.

Trudgill, S. T. 1983. *Weathering and erosion* (Sources and Methods in Geography). Butterworths, London.

Trudgill, S. T. 1984. Erosion rates on reefs. *Proceedings Coral Reef Workshop*, Australian Institute of Marine Science.

Trudgill, S. T. and **Brack, E. V.** 1977. *A bibliography of British karst 1960–1977*. Geo Abstracts, Norwich, U.K.

Trudgill, S. T., Crabtree, R. W. and **Walker, P. J. C.** 1979. The age of exposure of limestone pavements – a pilot lichenometric study in Co. Clare, Eire. *Transactions of the British Cave Research Association*, **6**, 10–14.

Trudgill, S. T., High, C. J. and **Hanna, F. K.** 1981. Improvements to the micro-erosion meter. *British Geomorphological Research Group, Technical Bulletin*, **29**, 3–17.

Turner, R. C., Nichol, W. E. and **Miles, K. E.** 1958. Leachates from calcareous soils and related materials. *Canadian Journal of Soil Science*, **38**, 94–99.

Underwood, A. J. 1978. A refutation of critical tidal levels as determinants of the structure of intertidal communities on British shore. *Journal of Experimental Marine Biology and Ecology*, **33**, 261–76.

Vita-finzi, C. and **Cornelius, P. F. S.** 1973. Cliff sapping by molluscs in Oman. *Journal of Sedimentary Petrology*, **43**, 31–32.

Wadge, G. and **Draper, G.** 1977a. Tectonic control of speleogenesis in Jamaica. *Proceedings of the 7th International Speleological Congress, Sheffield, 1977*. British Cave Research Association, 416–19.

Wadge, G. and **Draper, G.** 1977b. The influence of lithology on Jamaican cave morphology. *Proceedings of the 7th International Speleological Congress, Sheffield, 1977*. British Cave Research Association, 414–16.

Waksman, S. A. 1931. *Principles of Soil Microbiology*. Ballière, Tindall and Cox.

Waltham, A. C. 1971. *British karst research expedition to the Himalaya, 1970*. Published: Waltham, A. C. Trent Polytechnic, Nottingham.

Waltham, A. C. 1981. Origin and development of limestone caves. *Progress in Physical Geography*, **5** (2), 242–56.

Waltham, A. C. and **Ede, D. P.** 1973. The karst of Kuh-E-Parau, Iran. *Transactions of the Cave Research Group of Great Britain*, **15**, 27–40.

Warme, J E. and **Marshall, N. F.** 1969. Marine borers in calcareous terrigenous rocks of the Pacific coast. *American Zoologist*, **9**, 765–774.

Watson, R. A. 1972. Limitations on substituting chemical reactions in model experiments. *Zeitschrift fur Geomorphologie*, **16**, 103–8.

Weaver, J. D. 1973. The relationship between jointing and cave passage frequently at the Head of the Tawe Valley, South Wales. *Transactions Cave Research Group of Great Britain*, **15**, 169–73.

West, A. B. and **Jeal, F.** 1973. Observations on the nutrition of the echinoid *Paracentrotus lividus*. *Proceedings of the Challenger Society*, **4**, 122–3.

West, B., de Burgh, M. and **Jeal, F.** 1977. Dissolved organics in the nutrition of benthic invertebrates. In: Keegen, B. F., Ceidiga, P. O. and Boaden, P. J. J. (Eds.) *Biology of benthic organisms*. Pergamon.

Weyl, P. K. 1961. The carbonate saturometer. *Journal of Geology*, **69**, 32–44.

White, W. B. 1976. Cave minerals and speleothems. In: Ford, T. D. and Cullingford, C. H. (Eds.), *The science of speleology*, Academic Press, Ch. 8, 267–327.

White, W. B. 1977. Role of solution kinetics in the development of karst aquifers. In: Tolson, J. S. and Doyle, F. L. (Ed.) *Karst hydrogeology*. Proceedings 12th Congress International Association of Hydrogeologists. University of Alabama, 503–17.

Whitney, R. S. and **Gardner, R.** 1943. The effect of carbon dioxide on soil reaction. *Soil Science*, **55**, 127–41.

Wiens, H. J. 1959. Atoll development and morphology; *Annals of the Association of American geographers* **49,** 31–54.

Wigley, T. M. L. and **Plummer, L. N.** 1976. Mixing of carbonate waters. *Geochimica et Cosmochimica Acta*, **40**, 989–95.

Wilkinson, M. 1974. Investigations on the autecology of *Eugomontia sacculata* Kornm., a shell-boring alga. *Journal of Experimental Marine Biology and Ecology*, **16**, 19–27.

Williams, J. E. 1949. Chemical weathering at low temperatures. *Geographical Review*, **34**, 129–135.

Williams, P. W. 1966. Limestone pavements with special reference to Western Ireland. *Transactions, Institute of British Geographers*, **40**, 155–172.

Williams, P. W. 1969. The geomorphic effects of ground water. In: Chorley, R. J. (Ed.) *Introduction to fluvial processes*, Methuen, 6. II, 108–23.

Williams, P. W., 1971. Illustrating morphometric analysis of karst with examples from New Guinea. *Zeitschrift für Geomorphologie*, **15**, 40–61.

Williams, P. W. 1978a. Interpretations of Australian Karsts. Ch. 13 in: Davies, J. L. and Williams, M. A. J. *Landform Evolution in Australasia*, ANU Press, Canberra, 259–286.

Williams, P. W. 1978b. Karst research in China. *Transactions, British Cave Research Association*, **5**, 29–46.

Williams, P. W., 1982. Speleothem dates, Quaternary terraces and uplift rates in New Zealand. *Nature*, **298** (5871), 257–60.

Williams, P. W., 1983. The role of the subcutaneous zone in karst hydrology. *Journal of Hydrology*, **61**, 45–67.

Williams, P. W. and **Dowling, R. K.** 1979. Solution of marble in the karst of the Pikikiruna Range, North west Nelson, New Zealand. *Earth Surface Processes*, **4**, 15–36.

Wolf, K. H., Chilingar, B. V. and **Beales, F. W.** 1967. Elemental composition of carbonate skeletons, minerals and sediments. Ch. 2: In: Chilingar, C. V., Bissell, H. J. and Fairbridge, R. W., *Carbonate rocks*. Developments in Sedimentology, **98**. Elsevier, 23–140.

Woo, R. C. and **Brater, E. F.** 1962. Spatially varied flow from controlled rainfall. *American Society of Civil Engineers Proceedings*, **88**, 476, 31–56.

Wright, L. W. 1967. Some characteristics of the shore platforms of the English Channel coast and the northern part of New Zealand. *Zeitschrift für Geomorphologie*, **11**, 36–46.

Yonge, C. M. 1955. Adaptation to rock boring in *Botula* and *Lithophoga* (Lamellibranchia: mytilidae) with a discussion on the evolution of this habit. *Quarterly Journal Microscope Science*, **96**, 383–410.

Yonge, C. M. 1961. *The Sea Shore*. Collins, New Naturalist Series, Ch. 12.

Zenkovitch, V. P. 1967. *Processes of coastal development*. Oliver and Boyd.

Index

Index page numbers printed in **bold** indicate a major reference in a group of references